'One of the great tragedies of the post-war era is how psychedelic research died in the heat of the cold war – and how psychedelics were abused by US authorities. But that is only a fraction of the history of psychedelics. To get the full picture you have to do just one thing: drop LSD – or read this book!'

Norman Ohler, author of *Blitzed: Drugs in Nazi Germany*

'A fascinating and detailed history of how psychiatry, psychotherapy and neuropharmacology have long been inextricably linked with psychedelic experiments. Thoroughly researched and written with style, the book makes for an eye-opening and engrossing read.'

**Łukasz Kamieński, author of *Shooting Up:
A History of Drugs in Warfare***

'Barber shines light on a fascinating period of scientific history which inspired a cultural revolution. This book delves into the lives of the Saskatchewan scientists, and the major influence their early LSD research had on psychiatry, biochemistry, and scientific ethics.'

Amanda Feilding, Director of the Beckley Foundation

'Barber beautifully transcends the oft relied upon tropes of psychedelic culture in order to contextualize and honor a foundational, often overlooked chapter in psychedelic history.'

Sean Dunne, documentary filmmaker and director of *Oxyana*

'An intimate look into the early pioneering LSD research of Abram Hoffer, Humphrey Osmond, and Duncan Blewett, showing how they laid the groundwork upon which today's modern science of psychedelics rests. A must read for anyone wishing to understand the history of LSD.'

**David E. Nichols, founder and Chairman of the Board,
Heffter Research Institute**

About the author

P.W. Barber has spent the better part of a decade
researching, pondering, and writing on the history of
hallucinogenic science in Saskatchewan, the birthplace
of "psychedelic." He lives in Buena Vista, Saskatchewan.

PSYCHEDELIC REVOLUTIONARIES

Three Medical Pioneers, the Fall of
Hallucinogenic Research and the Rise of Big Pharma

P. W. BARBER

ZED

Originally published by University of Regina Press as *Psychedelic Revolutionaries: LSD and the Birth of Hallucinogenic Research* by P.W. Barber.

This edition has been published by arrangement with the University of Regina Press, Regina, Saskatchewan S4S 0A2 Canada.

www.uofrpress.ca

This edition published in 2018 by Zed Books Ltd, The Foundry, 17 Oval Way, London SE11 5RR, UK.

www.zedbooks.net

Text design by Duncan Campbell, University of Regina Press & John van der Woude, JVDW Designs
Index by Patricia Furdek
Cover design by Andy Allen
Cover photo © Hein Nouwens/Shutterstock

A catalogue record for this book is available from the British Library

ISBN 978-1-78699-435-6 hb
ISBN 978-1-78699-436-3 pb
ISBN 978-1-78699-437-0 pdf
ISBN 978-1-78699-438-7 epub
ISBN 978-1-78699-439-4 mobi

For Jody, Samuel, Avrie, and Sasha

CONTENTS

LIST OF FIGURES

ACKNOWLEDGEMENTS

The seed for this book was planted as I was completing my master's thesis in History, and I am grateful to University of Regina historians James Pitsula and William Brennan and psychologists William Smythe and Angelina Baydala for sparking the initial idea to pursue publishing the work. A debt of gratitude is also owed to acquisitions editor extraordinaire Jean Wilson, for her unwavering support and assistance in navigating those early steps on the academic publishing path.

As the journey unfolded, there were many people who helped to make this book a reality, be it through their ongoing encouragement and support, their openness to discussing the subject, their assistance in locating and connecting with sources and archival materials, their review and valuable input on early iterations of the manuscript, or their innate ability to keep me sane in the process and provide that extra nudge to carry the task through to completion. They include, in no particular order: Arthur Allen, Erika Dyck, John Mills, Barbara Kahan, Abram Hoffer, Miriam Hoffer, L. John Hoffer, Fee Blackburn, Duncan and June Blewett, Mary Lowe, Lea Bill, Karen Ryder, anonymous former patients, Ben Sessa, Connie Littlefield, Elizabeth Seitz, Nadine Charabin, Christine Charmbury, Tim Novak, and the anonymous peer reviewers.

A special thanks to Giselle Marcotte for pointing me in the direction of Bruce Walsh and the University of Regina Press and recommending them as a potential publisher. It was truly a good move, and I am indebted to acquisitions editor Karen Clark and the rest of the team at URP for all of their help and support in the final stages of this project. Another nod of appreciation also goes to Dallas Harrison for his assistance and skill in slicing and dicing the manuscript down to a more manageable size.

Thanks to my family as well for their support and to my parents in particular for the use of a cabin for a writer's residence, along with all of the required amenities (which sometimes included the use of a vehicle to get to the destination).

Lastly, I would like to thank my partner, my "rock," Jody Greenman-Barber, and our three children, Samuel, Avrie, and Sasha, for their patience, support, and understanding in what has indeed been a long, and occasionally strange, trip.

INTRODUCTION

It is only to the uninitiated and the novice that the development of science appears to be wholly smooth and even. In the perspective of great distance, as with any large picture observed, the finer lines are not perceptible.... As we gain knowledge of it from the actual experience of research or from reading its history, science shows a rougher picture.
—Bernard Barber, "Sociology of Science" (99)

Saskatchewan's experimental research in the 1950s and 1960s with psychedelic drugs, and the roles played by psychiatrists Humphry Osmond and Abram Hoffer and psychologist Duncan Blewett in that research, present an interesting and formidable challenge for the historian. A whole mythology has grown up around these experiments and individuals, yet little about them is understood. To those who lived in the province at the time and are most familiar with the subject, their names evoke a number of contradictory responses, anything from "charlatan" and "eccentric" to "genius" and "saviour." Some experts have credited their research with stunning achievements in mental health, in fields such as psychotherapy and neurochemistry, whereas others have claimed that their findings were exaggerated and even glaring examples of "pseudoscience." Some have said that their work with psychedelic drugs, particularly d-lysergic acid diethylamide (LSD), produced lasting results; others have argued that their experiments were flawed from the outset because of their poor methodology and the subjectivity of those involved. I aim in this book to shed light on this debate.

When I started down this research path as a graduate student in 2002, critical historical discourse on the subject was minimal—some informative and entertaining documentary films and media articles, yes, but few in-depth analyses of the Saskatchewan psychedelic research or the men and their roles in it. Assessments were simplistic at best. In general histories of the province, reference to the research was either absent or a minor footnote. Saskatchewan's foremost historian, John Archer, for example, noted that the research of Hoffer and Osmond "gained wide publicity" but said little else.[1] For a psychedelic research program that put the province on the world stage in the 1950s, the relative lack of attention is puzzling. What exactly did this research accomplish, where did it fail, and why did it engender the vociferous criticism that it did? For the most part, historians seem to have neglected to deal at length with these questions.

Periodically, over the years, the topic has generated curiosity (and will continue to do so) based upon the fact that Saskatchewan is the birthplace of the term "psychedelic."[2] Yet only in the past fifteen years—with the release of documentaries, journal and newspaper articles, and recent obituaries of Osmond, Blewett, and Hoffer—has there been more than occasional interest in this episode in Saskatchewan history.[3] Many interpretations have raised general, though somewhat limited, awareness of some of the main tenets of Saskatchewan's psychedelic drug research, from use of the drugs as model psychoses and the biochemically based hypothesis regarding schizophrenia to incorporation of the drugs as psychotherapeutic adjuncts in the treatment of chronic alcoholism and their use in architectural design (e.g., mental hospitals). These reports have also illustrated how the theories of Hoffer, Osmond, and Blewett fit into the larger psychedelic movement.

One stereotypical approach to these experiments, and others like them, has been to identify them with the reputation of psychedelic drugs as part of the 1960s counterculture. For the average layperson, the word *psychedelic* is associated with Timothy Leary, Ken Kesey and his Merry Pranksters, Haight Ashbury, and *Sgt. Pepper's Lonely Heart's Club Band* rather than groundbreaking scientific research. Reports on the Saskatchewan research have consistently relied on countercultural references and connotations to attract public attention, whether they are radio clips opening with "Lucy in the Sky with Diamonds" or articles introducing the topic with tales of personal acid trips and Jim Morrison lyrics. Some have used the countercultural connection to champion the subject, whereas others have used it for the opposite goal. One negative and sensationalist presentation of the psychedelic experiments and work of Hoffer, Osmond, Blewett, and their associates is Kenneth Bell's 1993 Canadian Broadcasting Corporation documentary *Acid Tests*, which begins by emphasizing how throngs of people gravitated to the Saskatchewan experiments to "turn on, tune in, and drop out."[4]

Admittedly, the experiments had curious links to many aspects of 1960s drug culture. These links are perhaps most aptly demonstrated by Osmond's coining the term "psychedelic" in 1957 and administering eminent author Aldous Huxley his first dose of mescaline, an experience that became the basis for the latter's countercultural classic *The Doors of Perception*. Loose associations can be drawn between Hoffer, Osmond, and Blewett and major '60s figureheads and intellectual elites such as beat poet Allen Ginsberg, radical Yippie Abbie Hoffman, and the psychedelic high priest himself, Timothy Leary. Many parallels have been made between Leary and Blewett because of the latter's reputation as the "Leary of the North." Hoffer, Osmond, and Blewett also maintained many of their views on the merits of psychedelic research when it was professional suicide to do so. As their views on the '60s drug scene played out in both the press and their frequent lectures on university campuses, their apparent anti-establishment stance resonated with many youth who rebelled against cultural and societal norms of the time.[5]

As intriguing as these "flashbacks" might be, they are only a fraction of what made up Saskatchewan's psychedelic drug tests. Overconcentration on the countercultural manifestations of psychedelic drugs has actually obfuscated the story and prevented serious examination of the science and people behind the research. The most detailed documentaries to date on Hoffer, Osmond, and Blewett and their part in the history of psychedelic drugs, Connie Littlefield's *Hofmann's Potion* (and to a lesser extent *Feed Your Head*), Gordon McLennan's *Psychedelic Pioneers*, and Kenneth Bell's *Acid Tests*, provide good backgrounds on the Saskatchewan experiments and highlight the research on schizophrenia and the use of psychedelics in the treatment of alcoholism.[6] These documentaries also exhibit opposite views on the subject: Littlefield and McLennan favour the experiments and their outcomes, whereas Bell judges the results of the research as harmful and grossly unethical. The most visible weakness of these studies is their lack of "scientific" explanations of the research. Littlefield and McLennan come the closest to presenting the accomplishments of Hoffer, Osmond, and Blewett in a scientific light and show how they compared with and contributed to the research outside the province; but by focusing mostly on interviews with the proponents the two filmmakers neglect to show why the opponents were so against the research and considered it unscientific. Bell, by pinning most of his argument on questionable conspiracy theories (e.g., the notion that the Saskatchewan research was part of a secretly funded CIA plot to employ psychedelics as brainwashing agents) and the testimony of a few people who suffered setbacks from the experiments, fails to consider the individuals who benefited from them. His documentary spends so much time demonizing the research that it is more sensationalist hype than meticulously researched work. His concluding statement that the experiments for many patients were a "nightmare from which they never recovered" is a gross overgeneralization of what happened.[7]

Understanding Saskatchewan's psychedelic drug research between 1951 and 1961 thus demands a fuller appreciation of the science behind it. Nearly fifty-five years after the research, few academics have attempted to show systematically and comprehensively why and how this experimentation unfolded, what its findings were, and how these findings were received both within and outside the scientific community. The same can be said about why Hoffer, Osmond, and Blewett and their theories were so controversial and either revered or rebuked. Fortunately, the more recent efforts of University of Saskatchewan's medical historian Erika Dyck,[8] Professor Emeritus of Psychology John Mills,[9] and others[10] have begun to remedy this gap in historical knowledge.

SASKATCHEWAN'S PSYCHEDELIC RESEARCH:
AN EPISODE IN SCIENTIFIC HISTORY

Although some might wonder whether the recent attention is deserved or not from a scientific standpoint, others, on becoming more familiar with the research, have cast it in a revolutionary light, with Hoffer, Osmond, and Blewett as the founding fathers of psychedelic science. Such views hinge on scientific expertise. The conventional narrative on the research has highlighted a meteoric rise to scientific fame followed by an equally rapid fall from scientific grace. Both the counterculture's psychedelic excesses and psychiatry's adoption of stricter, more empirically based, scientific standards have factored into this portrayal. Professional opinion on the research has tended to be negative: that is, most, if not all, of the discoveries produced by the research remain unfounded, and the experimentation represents a primitive stage in clinical and laboratory research science, especially compared with contemporary evidence-based standards.[11]

Whether there was, or might still be, scientific value in Saskatchewan's psychedelic research findings is at the heart of this book and is particularly relevant today given a resurgence of similar research.[12] Writing of the renaissance in psychedelic research in Europe and the United States, medical anthropologist Nicolas Langlitz insists that "it has contributed to an understanding of the cortical metabolism and neurochemical substrates of psychotic processes, elucidated the role of serotonin and glutamate receptors in cognitive and perceptual processing, and opened up new experimental venues for the treatment of post-traumatic stress disorder, cluster headaches and end-of-life anxiety in patients suffering from terminal illness."[13] These developments suggest why the historian should dig beneath the customary portrayals of Saskatchewan's psychedelic research. With closer examination of this research, a more complicated picture emerges.

It is not my intention here to confirm or deny the scientific validity of the research and related theories of Hoffer, Osmond, and Blewett. Their work remains highly controversial within medical and psychiatric communities, yet the historian interested in getting to its core cannot avoid engaging with questions of scientific veracity. Talcott Parsons, an American pioneer in the sociology of science, once wrote that "the knowledge [a scientist] possesses is only with difficulty if at all accessible to the untrained layman. The ultimate judgement of it must lie with his professionally qualified peers."[14] Philosopher of science Paul Feyerabend put the matter another way:

> To evaluate, the citizens need intellectual guides, they need standards. Now if standards for the evaluation of scientific research are research-immanent, if they change as research proceeds, and if their change can be controlled and understood only by those immersed in research, then a citizen who wants to judge science must either become a scientist himself, or he must defer to the advice of experts. A democratic control of science (and of other institutions) is then impossible.[15]

It would seem, then, the historian might have difficulty providing reliable assessments of scientific research. Many scientists are uneasy when historians try to do so; "scientists," as Bruno Latour remarked, "often have an aversion to what non-scientists say about science."[16] So it might be impossible to arrive at a final judgment on the scientific status of the research discussed here (even experts in psychology and psychiatry have not arrived at a consensus, and perhaps they never will). French historian of science Alexandre Koyré noted that "historical reconstructions" are "not only necessarily incomplete...[but also] necessarily partial."[17] Indeed, "historians do not state all the facts, not even all the facts that they know or might be able to know. How could they?"[18] Admitting that the historian cannot resolve the issue to everyone's satisfaction should not lead to a defeatist attitude, however, for the historian can still add much to the debate. Perhaps, when trying to address any scientific controversy, it is best to heed the advice of sociologists of science Donald Mackenzie and Barry Barnes: "The study of any particular episode in the history of science involves consideration of a wide range of activities...[that] must all be investigated in detail in the course of building up an overall reconstruction of what went on."[19]

One goal of this book, then, is to show how the Saskatchewan psychedelic research came about, how the theories of Hoffer, Osmond, and Blewett developed, and what some of their findings were. The research encompassed a variety of issues, many of which overlapped. As it evolved, it also raised questions that cut across disciplines (e.g., biochemistry, psychology, psychiatry, sociology, and religion). I recognize the complicated nature of the subject matter but strive to achieve a well-rounded understanding of the research conducted. The experiments with psychedelic drugs and their findings led to a number of major disagreements in various scientific and social circles in the 1950s and 1960s, and I attempt to provide insight into the debates, the positions taken on key issues, and why the participants were at odds.

By focusing on questions of scientific legitimacy, I also argue for the "revolutionary" nature of Saskatchewan's psychedelic experiments and the related work of Hoffer, Osmond, and Blewett. This is a difficult task because professional critics have denounced their work as scientifically and methodologically unsound, and it becomes problematic for a non-expert to assess whether the findings were scientifically valid or invalid. I thus place the experiments within the larger framework of scientific research as a whole and examine more closely the "scientific community" of which Hoffer, Osmond, and Blewett were a part that "controls inquiry."[20] More often than not, scientific discoveries provoke crises in any given field of study—all the more so if the particular discovery goes against established knowledge. Saskatchewan's psychedelic drug experiments were no exception. Contextualizing the research in this way can thus lead to more informed opinions on it.

"SCIENCE IN THE MAKING": THE NATURE OF SCIENTIFIC RESEARCH

What distinguishes good science from bad science? What is the guiding framework for scientists in their research? How are scientific discoveries to be validated and disseminated, and at what point are these discoveries accepted as established truths? What makes a scientific discovery revolutionary? Most scientists will concur that scientific research follows a certain course. One begins the quest for knowledge with an idea and examination of the available facts and then formulates an hypothesis, tested and retested through experimentation, data collection, and analysis. If the findings pass the preliminary tests, then, if the researcher is fortunate, they are published, disseminated, and subjected to the independent scrutiny of his or her scientific peers. The result is either to verify or to refute the claim put forward. If the claim is declared to possess scientific value, then it becomes part of the body of scientific knowledge; if attempts at verification fail, then the originator is forced to re-examine the claim and prove the critics mistaken. Is this process so straightforward, though? Does a definitive scientific method exist? Are there well-established methodological boundaries that no one must step across if her or his research is to be deemed scientifically relevant?

These very questions fuelled intense debate from the 1950s on, particularly among an increasing number of scientific historians, philosophers, and sociologists. As this book unfolded, I often found myself consulting the perspectives of key theorists in the debates over science such as Kuhn, Polanyi, Feyerabend, Barber, and many others for their assessments of the inner workings of scientific activity. This was never my intention in setting out to study Saskatchewan's psychedelic research, but I ended up delving deeper into the area in the hope that I might better grasp how scientists function as a collective unit and how the scientific establishment exercises control over individual members. While this approach led to moments of enlightenment, it created an equal measure of personal confusion over the issue when confronted with the varied and often conflicting interpretations, be it Sir Karl Popper's gold standard empirical criterion for scientific status (falsification) or Feyerabend's epistemological "anything goes" anarchism.

In spite of this ambiguity, this approach reaffirmed for me how important it was to ground Saskatchewan's psychedelic story (and that of Hoffer, Osmond, and Blewett) in the larger framework of scientific research and its philosophical underpinnings. It also reinforced in me the desire for a clearer sense of what historian of science Steven Shapin meant when he referred to "science-in-the-making." Echoing the recommendations made decades earlier by Barber, Shapin advised that the public ought to be aware of what actual scientific processes entail, "warts and all":

> We should find ways of showing and explaining to them such things as...the collective basis of science, which implies that no single

scientist knows all of the knowledge that belongs to his or her field; the ineradicable role of trust in scientific work, and the consequent vulnerability of good science to bad practices; the contingency and revisability of scientific judgment, and thus the likelihood that what is pronounced true today may, without culpability, be judged wrong tomorrow; the interpretative flexibility of scientific evidence, and the normalcy of situations in which different good-faith and competent practitioners may come to different assessments of the same evidence.[21]

Failure to do so, in Shapin's mind, could result in the perpetuation of some unfortunate myths about science. As he explained,

> some fairy-tales would have the public believe there is a universal efficacious scientific method which sorts out good from bad data and confirms or disconfirms scientific theories. To the extent that the public labour under such apprehensions they will have little choice but to pick among these conclusions: (a) that one lot or another of scientists is incompetent, or lying, or in the pay of special interest groups...; or (b) that the area concerned is not science at all.[22]

Shapin thus challenged the plausibility of a formal, orderly, one-size-fits-all objective method that scientists must adhere to for their work to be accepted as science. As he articulated, "the conceptual unification of all the sciences on a hard and rigorous base of materialist reductionism is an old aspiration, but it has never commanded (and does not now command) the assent of all scientists."[23] Indeed, Shapin has not been alone in this line of thinking.

In his pivotal work *Personal Knowledge* and many other contributions, physical chemist and philosopher of science Michael Polanyi cast doubt on the oft-stated notion, so prevalent in the 1950s and '60s, that scientific knowledge, and knowledge in general, evolves objectively. Polanyi maintained that there is a "personal participation of the knower in all acts of understanding," noting that "we must see the universe from a centre lying within ourselves and speak about it in terms of a human language shaped by the exigencies of human intercourse. Any attempt rigorously to eliminate our human perspective from our picture of the world must lead to absurdity."[24]

A thinker who followed closely in Polanyi's footsteps was physicist and historian of science Thomas S. Kuhn. He, too, was extremely skeptical about uncritical empiricism. Like Polanyi before him, Kuhn regularly tested the popular credo, held both within and outside mainstream science, that scientific knowledge is built from a never-ending accumulation of facts and advanced in a dispassionate and piecemeal way. He stressed that exactly this kind of "cumulative" process is portrayed in standard science textbooks when it is usually not the case. He accentuated the "different concept of science that can emerge from the historical record of the research activity itself."[25] In works such as his renowned *The Structure of Scientific Revolutions*, Kuhn presented ideas on how science develops, focusing on what he termed "paradigms": that is, "universally recognized scientific achievements that for a time provide model problems and solutions to a community of practitioners."[26]

Throughout the history of science have been periods of what Kuhn called "normal" science, in which scientists committed to one paradigm or another concentrated on answering questions meant to reaffirm pre-established theories. Although advances were made by this sort of "puzzle-solving," it had its limitations. Kuhn argued that "normal science" seems to have been

> an attempt to force nature into the preformed and relatively inflexible box that the paradigm supplies. No part of the aim of normal science is to call forth new sorts of phenomena; indeed, those that will not fit the box are often not seen at all. Nor do scientists normally aim to invent new theories and...are often intolerant of those invented by others. Instead, normal-scientific research is directed to the articulation of those phenomena and theories that the paradigm already supplies.[27]

Progress in science, as Kuhn stipulated, has also been punctuated by periods of crisis followed by revolutions "in which an older paradigm is replaced in whole or in part by an incompatible new one."[28] Expanding on this concept, he wrote that "scientific revolutions are inaugurated by a growing sense...that an existing paradigm has ceased to function adequately in the exploration of an aspect of nature to which that paradigm itself had previously led the way."[29]

The influence of Kuhnian concepts on how people have come to understand science is undeniable: *The Structure of Scientific Revolutions* continues to be cited often, and its terms (e.g., "paradigm shift") are employed in everyday language. This popularity has led some scientific commentators to suggest that Kuhn has been widely misinterpreted and his concepts incorrectly applied, whereas others contend that his concepts have far-reaching relevance.[30] Psychiatrist Sayeed Nassir Ghaemi has asserted that the concept of paradigms can be useful in examining psychiatric schools of thought and dealing with a range of issues in psychiatric practice.[31] Alan Baumeister and Mike Hawkins, too, in their article on the development of modern psychopharmacology, point to the paradigmatic qualities inherent in psychoanalytical and neurochemical interpretations of mental illness.[32]

Thus, though Kuhn's ideas are normally used in discussions of the natural sciences, they are applicable to the social sciences and even relevant to studies on the altered states of consciousness often experienced with psychedelic drugs. Psychologist Charlie Tart advanced this point: "Kuhn's concept of paradigm has far wider applications than to formal scientific theories guiding the investigatory activities of scientists. We all have paradigms, world views, about different areas of reality. We have personal and cultural paradigms about economics, politics, religion, sexuality and so on."[33]

Kuhn's perspectives, as well as those of Polanyi and other theorists of science, thus provide a conceptual framework within which to view scientific research in general and psychiatric research in particular; such varied perspectives also help the historian to understand why experts disagree on an issue, why there is resistance to certain scientific innovations, and what makes a particular scientific discovery revolutionary.

THE ORGANIZATION OF SCIENTIFIC RESEARCH

Although my overall intention in this book is to focus on the scientific aspects of the psychedelic experiments in Saskatchewan, I do not want to lose sight of the broader social factors at play as the research progressed. A good deal of scientific practice, prior to becoming public, takes place behind closed doors. Science, however, is not completely isolated; as with anything else, it does not occur in a vacuum and is subject to the same winds of social, economic, cultural, and political change. As pointed out by sociologist David Bloor, "features of culture which usually count as non-scientific greatly influence both the creation and the evaluation of scientific theories and findings."[34] Historian of psychiatry Andrew Scull puts it more bluntly when he warns that "the scientific and the social are inextricably intertwined, and the historian ignores either at his peril."[35]

Although great scientific innovations have resulted from individual, or small teams of, scientists following independent lines of inquiry, events or interests outside the scientific setting have often dictated the style of research and the pattern and pace of scientific discovery. The organization and bureaucratization of wider society and scientific activity had greater relevance from the mid-twentieth century on, a trend that did not go unnoticed by observers of the time. Writing of this transformation in his classic social commentary *The Organization Man*, William H. Whyte Jr. identified a radical ideological shift moving society toward increasing collectivization. He referred to this newfound "organization" as "that body of thought which makes morally legitimate the pressures of society against the individual."[36] He outlined the main tenets of this "social ethic" as "a belief in the group as the source of creativity; a belief in 'belongingness' as the ultimate need of the individual; and a belief in the application of science to achieve this belongingness."[37]

Whyte was speaking primarily about the change occurring in the consciousness of Americans, though it was a "bureaucratization of society that [had] affected every Western country."[38] The problem, as he saw it, was not the "organization" per se but "our worship of it" at the expense of individualism. He argued that emphasis on the social ethic was misguided: "People do have to work with others, yes; the well-functioning team is a whole greater than the sum of its parts, yes—all this is indeed true....We need to know how to co-operate with the Organization but, more than ever, so do we need to know how to resist it."[39] In the 1950s, belief in the social ethic was most obvious in large corporations, but it was pervasive in other areas of work as well, whether in government ranks, the suburbs, the classroom, or the laboratory.

For Whyte, once the social ethic was "applied to science," several developments were likely to follow:

(1) Scientists would now concentrate on the practical application of previously discovered ideas rather than the discovery of new ones; (2) they would rarely work by themselves but rather as units of scientific cells; (3) organization loyalty, getting along with other people, etc. would be considered just as important as thinking; (4) well-rounded team players would be more valuable than brilliant men, and a very brilliant man would probably be disruptive.[40]

The danger, Whyte continued, is that the organization tries "to mold the scientist to its own image; indeed, it sees the accomplishment of this metamorphosis as the main task in the management of research."[41] Truly independent research had become endangered; this was unfortunate, Whyte thought, because many of the most important discoveries could be attributed to open inquiry.[42]

Whyte was by no means the only academic warning against this alarming trend in society. Other well-known social critics of the time, such as Paul Goodman (*Growing Up Absurd*) and David Riesman (*The Lonely Crowd*), reiterated many of these sentiments, as did a growing number of scientific theorists.[43] Perhaps the most vocal advocate for the freedom of science was Polanyi. Like Whyte, he argued that ideologies, whether Marxist, positivist, or organizational, could wreak havoc on the practice of science. According to Polanyi, "the primary decisions in the shaping of scientific progress are made by individual investigators when they embark on a particular line of inquiry."[44]

Whether modern, more "organized," forms of scientific research were completely displacing traditional approaches was questionable. Sociologist of science Warren Hagstrom, for one, remained optimistic and cautioned against making generalizations across the entire scientific community, writing that the phenomenon was observable in some scientific fields more than others and that it was occurring at various rates and with varying results in each area of specialization.[45] In post–Second World War psychiatry, psychiatrists and other mental health professionals were embroiled in many of the same debates about science and scientific research, and many of the transformations of which Whyte, Polanyi, and others spoke were taking place in the field.

"SCIENTIFIC PSYCHIATRY" IN POST–SECOND WORLD WAR AMERICA (1945–75)

The post–Second World War era marked a tumultuous period in the history of American psychiatry. Many histories of the time have mentioned controversies and crises, turf battles, revolutions, and paradigm shifts as the mental health sector grew.[46] From the Second World War on, mental health began to assume increasing importance in American society. The ravages of war had clearly been demonstrated by the psychological impacts on soldiers; one of the most visible indicators was the sudden and precipitous rise in the number of psychiatric casualties and the demand for the treatment of "war neuroses" on and off the battlefield.[47] The postwar years saw notable shifts in mental health policy with the establishment of a federal mental health infrastructure (e.g., the National Institute of Mental Health), the rise to prominence of professional associations (e.g., the American Psychiatric Association), and the infusion of private and public funding for training and research in the area.

The impact of the Second World War on the psychiatric profession was considerable. Until the 1950s, psychiatry had been a marginalized medical discipline, its practice mostly confined to the isolated mental hospitals or asylums. Psychiatry was viewed by many as the poor cousin of the medical community, and its influence in society was negligible. War, however, changed all that. With the rapidly increasing demand for mental health services during and after the war, psychiatry began to move from asylums into communities, where practitioners filled the ranks of university departments, medical institutions, and private and community practice settings. The field witnessed phenomenal growth in both numbers and scope as psychiatric expertise was sought on everything from the general health of civilians to an expanding range of social issues.[48] Psychology, whose history during this period was closely interwoven with that of psychiatry, was similarly impacted by the war and the heightened demand for treatment services. Psychology underwent its own transformation, shifting from its more academic and laboratory-based research traditions (psychometrics, personality testing, etc.) to clinical practice. As with psychiatry, psychology experienced an overwhelming professionalization and expansion of its roles. It also received major investments in training and research opportunities, which rivalled those of its psychiatric sibling and led to competition between the two fields.[49]

This period in psychiatry also marked the peak of psychoanalytical power. By the end of the war, Freudian-derived theories and psychotherapies held sway over much of the mental health sector and society in the United States, directly attributable to the war, military psychiatry, and the influence of psychoanalytically trained leaders such as William and Karl Menninger, who promoted psychoanalysis as the answer to mental illness.[50] The war, in fact, gave legitimacy to the argument advanced by many proponents of psychoanalysis that the underlying causes of mental illness were psychogenic and best treated with psychotherapeutic interventions. As N. G. Hale Jr. reported,

> the experience in WWII suggested that environmental stress, far more than premorbid personality, could account for the neuroses and perhaps for mental disorders.[51] Moreover, prompt early treatment had forestalled the development of more serious symptoms among servicemen. Accordingly, it was assumed that the alleviation of stress and early treatment in [the] community could prevent nervous and mental disorders in peacetime.[52]

In the years following the war, psychoanalysis had a dominant influence on "much of the political and educational apparatus of American psychiatry."[53] In psychology, too, psychoanalytical theories and therapies (once the exclusive jurisdiction of medically trained psychiatrists) had an increasingly strong grip on the profession as up-and-coming clinical and research psychologists clamoured to take advantage of the newfound opportunities presented by psychoanalytical training institutes.

The immense influence of psychoanalysis did not go unchallenged, how-ever. Many within psychiatric and psychological communities contested its authority, insisting that it was dogmatic and scientifically bankrupt. The view of psychoanalysis as non-science was common in the wider sci-entific community. As one of the most recognized authorities on scientific knowledge, Popper stated that, when it came to "Freud's epic of the Ego, the Super-ego and the Id, no substantially stronger claim to scientific sta-tus can be made for it than for Homer's collected stories from Olympus."[54] Understandably, many in the ranks of psychiatry came to see psychoanaly-sis as a barrier to the profession's much-desired status as a scientific medical specialty; the growing popularity of psychoanalysis only reinforced the image of psychiatry as an immature science among its mainstream medi-cal counterparts. One faction that regarded the psychoanalytical paradigm as backward and unscientific was the biologically oriented psychiatrists, more disposed to the "medical model" idea that mental illness was pre-dominantly somatic and genetic in origin and should be treated as such.

For psychologists who had more scientific aspirations in mind for their profession, psychoanalysis and its increasing influence in the training of clinical psychologists and psychological research was looked on with dis-dain. Within the wider scientific community, the opinions on psychology were comparable to those on psychiatry; psychology, like other social sciences, it was said, was scientifically deficient in a number of respects. So an increasing number of psychologists sought to emulate the meth-ods of the natural sciences in clinical practice and research. In an effort to counter the psychoanalytical trend in the field, an American Psychological Association–sponsored team of psychologists came together in the late 1940s to begin development of a standardized and more medically based training platform, blending academic psychology, research methodology, and clinical practice into what came to be known as the "scientific-practi-tioner" model.[55]

The move toward a more positivist and objectivist model of science within both psychiatry and psychology was greeted with enthusiasm in many camps, not only because it stemmed the tide of psychoanalysis, but also because it represented a much-needed mark of scientific progress. This view of objectivism as science was entrenched in almost every scientific discipline. From the nineteenth century on, orthodox science established as one of its main tenets that scientific truth arose from the collection of cold, hard facts. This firm belief naturally led to the questioning of anything that could not be "objectively" verified or observed impartially by others. As Polanyi wrote, this "prevailing conception of science, based on the disjunction of subjectivity and objectivity, seeks—and must seek at all costs—to eliminate such emotional appraisals of theories from science, or at least minimize their function to that of negligible byplay."[56]

Objectivism came to guide and shape how many within the psychiatric and psychological communities would see clinical practice and research. For Seymour Kety, a paragon of biological psychiatry, the subjective bias was one of the greatest sources of error in scientific research; consequently, everything possible had to be done to eradicate it. As Kety saw it, "carefully controlled and 'double-blind' experimental designs, which are becoming more widely utilized, can help to minimize this bias."[57] Similarly, among many psychologists, the argument was that psychology "should study human behavior scientifically, and the distinctive feature of science is its method, which implies that there should be controlled observation, and whenever possible, measurement."[58]

Within psychological circles, the objectivist ideal was best represented by the behaviourist approach. The determination of behaviourists to establish a more scientific basis within their field was not in itself a negative thing, for it produced some phenomenal and important findings about human behaviour and mental processes. Criticism began to mount, however, when its proponents held it to be the psychological standard by which all else was measured. What inevitably troubled humanistic thinkers such as Abraham Maslow and Carl R. Rogers was behaviourists' insistence on perceiving human beings simply as automatons who responded mechanically to a wide range of stimuli; such an interpretation made no room for human values, which of course could not be accurately and objectively measured. As Maslow stated, "pious positivists as a group accept the same strict dichotomizing of facts and values that the professional religionists do. Since they exclude values from the realm of science and from the realm of exact, rational, positivistic knowledge, all values are turned over by default to non-scientists and to non-rationalists (i.e., to 'non-knowers') to deal with."[59] Opponents of the objectivist paradigm worried about the overwhelming power of behaviourism in determining future research in the field. This concern led Maslow to issue the following warning:

> The greatest danger of such an extreme institutional position is that the enterprise may finally become functionally autonomous, like a kind of bureaucracy, forgetting its original purposes and goals and becoming a kind of Chinese Wall against innovation, creativeness, revolution, even against new truth itself if it is too upsetting. The bureaucrats may actually become covert enemies to the geniuses, as critics so often have been to poets, as ecclesiastics so often have been to the mystics and seers upon whom their churches were founded.[60]

Clearly, by the 1950s, there were profound differences of opinion within psychiatry and psychology on scientific practice. The counterargument to purely objectivist models of science, advanced by Kuhn, Polanyi, Maslow, and others, was essentially that "all knowledge is experiential knowledge."[61] As explained by Rogers, "this aspect of science...has been greatly ignored in American science. Especially it has been ignored in American psychology, where it has been considered slightly obscene to admit that psychologists feel, have hunches, or passionately pursue unformulated directions."[62]

The other major turning point came in the early 1950s with the introduction of powerful psychotropic drugs (e.g., chlorpromazine), the discovery of which triggered the modern "psychopharmacological revolution" and a period of transformation in psychiatry as "profound and far-reaching" in its consequences as the war had been.[63] The new "tranquilizing" drugs served as an immediate boon to psychiatric therapeutics, enabling mental hospitals to better manage and treat the symptoms of patients with severe mental disorders such as schizophrenia; the new medications also helped to substantially reduce patient numbers in mental hospitals, thus preparing the way for another significant shift in psychiatry, from the age of the asylum to that of deinstitutionalization and "community psychiatry."[64] Outside institutions, these drugs and their successors, which came to include an array of antipsychotic, antidepressant, and anti-anxiety medications, became widely used in the prevention and treatment of minor and major mental illnesses. For biological psychiatrists, the new drugs confirmed their conviction that the roots of mental illness are found in brain chemistry; determining the mechanisms and effects of these drugs soon became psychiatry's "Holy Grail" crusade in understanding and treating what many thought were chemical imbalances.[65]

Psychopharmacology thus helped to construct a firm scientific footing upon which to remedicalize the field. The stunning success of psychoactive substances saw hundreds of millions of dollars channelled into scientific research programs to test their effects and related hypotheses concerning their actions; along with the drug research came numerous advances in understanding brain chemistry and the function of neurotransmitters such as serotonin. Psychiatry also experienced a major refinement of its methods as researchers turned to the randomized controlled testing of therapeutics to scientifically evaluate their efficacy. As biological viewpoints on mental illness began to gain precedence in psychiatry, so did reforms in diagnosis, with incremental efforts made toward a strict medical model of classifying mental illness that harkened back to Kraepelinian psychiatry.[66] No longer would official psychiatric diagnoses be driven by psychoanalytical interpretations of symptoms and unproven etiological theories (as in the first and second versions of the American Psychiatric Association's *Diagnostic and Statistical Manual* or *DSM*[67]); instead, they would rely on a standardized classification of disorders based upon specific patterns of symptoms that could then be targeted by scientifically validated drugs and other therapies.[68] Closely connected to these developments, pharmaceutical companies began to invest heavily in testing, manufacturing, and marketing psychopharmacological agents.

Despite indications in the mid-1960s that scientific progress was finally under way in psychiatry, all was not well in the field. Amid quarrelling within and among psychoanalytical and biological groups, psychiatry entered a period of instability that mirrored the turmoil and civil unrest in American society. In addition to schisms in the profession, psychiatry in the 1960s and '70s confronted a crisis of confidence as a growing number of professional and non-professional dissenters challenged its legitimacy. Many critics questioned the increasing medicalization of psychiatry at the expense of important psychological and social factors, some even asking whether mental illness is truly a medical disease or something more socially constructed and "manufactured" by the psychiatric establishment.[69] Others opposed the involuntary institutionalization of, and experimentation on, mental health patients, bringing the field of biomedical ethics and patient rights to the fore. Added were concerns about the purported efficacy of the new drug treatments as reports of deleterious side effects (e.g., Parkinsonism) became more public. A darker side was also revealed as ties between pharmaceutical corporations and psychiatric research became more apparent. Complicating matters further, psychiatry had to deal with diminishing government support and increasing regulatory and medical insurance regimes in the mental health area. All of these developments created an environment "antipsychiatric" in both tone and substance, in which psychiatry struggled to survive. This vulnerability only hardened the resolve of some within the psychiatric community to steer the field in a more scientific direction, a move that reached fruition with the publication of *DSM-III* in 1980.[70]

Much of the historiography of psychiatry in the post–Second World War era has cast the period as a drawn-out clash between psychoanalytical and biological paradigms, with the latter prevailing in the early 1970s and psychiatry moving out of the ideological dark ages and along a path to scientific maturity. Another popular metaphor has been that of a pendulum swinging between psychoanalytical and biological extremes, with some historians framing the move to the biological end as "a smashing success."[71] Although such overarching descriptions are accurate to some extent, other historians have begun to poke holes in this conventional narrative, criticizing it for concealing the complexity of, not to mention the intraparadigmatic conflict in, psychiatry at the time.[72] As Rasmussen argues, "we must regard psychiatric medicine in the United States as more diverse and eclectic and less polarized...than any scenario placing a long struggle between psychoanalysts and biological psychiatrists on center stage."[73] Similarly, Pickersgill depicts a psychiatry "characterized by diversity, even within particular approaches and discourses; without this heterogeneity the psychiatric profession would not move between dominant professional programmes—without the multiplicity of perspectives the pendulum masks, it would not swing at all."[74]

Alternative accounts of this period have also disputed the fall of Freud in American psychiatry being as definitive as Shorter, Valenstein, and others have suggested. Although the influence of psychoanalysis had certainly waned by the mid-1960s, some historians have stressed that its death at the hands of the psychopharmacological revolution and biological psychiatry has been overblown. Jonathon Metzl's gender-based analysis of the rise of popular drugs such as Miltown (meprobamate) has found a good deal of Freud in contemporary psychiatry's medicine chest, thus revealing the roots of the biological revolution to be firmly intertwined with psychoanalysis.[75] Revisionist histories like those of Metzl and Rasmussen have also judged the more common portrayal of psychotropic drugs in the 1950s and '60s as signalling a paradigm shift to biological psychiatry to be equally problematic.[76]

One of the more attractive points in the standard narrative of psychiatry in this era is the profession's evolution into a scientific specialty. In many of the official histories of this era, the biological revolution has stood apart as an important, if not the most important, chapter in the story of scientific psychiatry. Although the "science as progress motif," as Scull calls it, has been in keeping with modern psychiatry's view of itself as a scientific medical discipline, some interpretations of this period have raised doubts about whether psychiatry has been as scientifically progressive as we have been led to believe. As some see it, biological psychiatry has overextended its reach with theories and treatment claims that have not always been supported by strong scientific evidence.[77] Nevertheless, biological reductionism has had its benefits. As Grob notes, "the advantages were obvious. An emphasis on somatic mechanisms allied psychiatry with the biomedical sciences, whose prestige and funding had been growing at an exponential rate; it also seemed to eliminate the gap between psychiatry and other medical specialities."[78] Others have been much less sanguine, underlining the power that economic and political interests have exerted in this tale of medical progress.[79] Rasmussen has suggested that "historians ought to consider the roles of pharmaceutical companies in shaping medical theories and practices during the entire twentieth century."[80]

Almost every historian of psychiatry will admit downsides to the psychopharmacological era, with instances of drug overuse and misuse, yet the overall assessment has been positive. Psychotropic drugs revolution-ized mental health as we know it and were important in the advancement of a biological, and more scientific, psychiatry. "The other possibility," as proposed by medical journalist Robert Whitaker, "is that psychiatry, eager to have its own magic pills and eager to take its place in mainstream medicine, turned the drugs into something they were not. These first-gen-eration drugs were simply agents that perturbed normal brain function in some way.... It stands to reason that the *long-term* outcomes produced by the drugs might be problematic in kind."[81] This scenario led Whitaker, and a host of other critics both inside and outside psychiatry, to speculate whether there has been a "true scientific revolution or a societal delusion."[82]

One curious thing in many of the historical accounts of post–Second World War psychiatry and the biological and psychopharmacological rev-olutions is the lack of reference to another class of psychotropic drugs: hallucinogens (or what would later become psychedelics). At best, psyche-delics have occupied a minor part in the psychiatric historiography of this era.[83] In some readings, their significance has been downplayed, distorted, or ignored altogether. Yet hallucinogens such as LSD played as much a role as other psychotropic drugs in the foundational years of the biological and psychopharmacological revolutions. In fact, psychedelics constituted a sci-entific revolution in their own right, but one that does not fit neatly into the conventional narrative of this period. In the same ways that Rasmussen and Metzl have used their own research in the history of psychopharma-cology to put forward perspectives that contrast sharply with the reigning narrative, one can scrutinize the history of LSD and other psychedelic drugs to determine just how complex and eclectic psychiatry was at this time.

THE PSYCHEDELIC REVOLUTION

In the ongoing quest to understand mental illness, a growing number of psychiatrists and psychologists were drawn in the 1950s to LSD and other psychedelics, mainly because of their ability to serve as model psychoses in normal, healthy individuals. As early advocates argued, the effects of psychedelics, or psychotomimetics ("madness-mimicking," as they were labelled), mirrored symptoms of naturally occurring mental illnesses, the most enigmatic of which was schizophrenia. According to the theory, the ingestion of psychedelics gave initiates, for a few hours, a way to explore the diseased mind, thereby suggesting new and better methods of treatment. Naturally, such a suggestion provoked an uproar among those in the scientific and psychiatric mainstream who doubted the biochemical connections between these mysterious hallucinogenic drugs and mental illness. The essential question was whether mental illnesses such as schizophrenia could be attributed to a biochemical quirk. If they could, then the discovery had major implications for psychiatric theory. The debate would probably have rested on this matter had it not been for the revelation that psychedelics could do much more than induce psychotic-like states of mind.

In some circumstances, they led to feelings of ecstasy, expanded consciousness, enlarged perception, and experiences that could be classified as mystical and religious. The tremendous variability of many psychedelic drug tests was overwhelming, and by the mid-1950s some researchers heralded the drugs as having far-reaching implications for understanding how the brain, diseased or healthy, functioned. Stanislav Grof, a respected Czechoslovakian psychedelic researcher who immigrated to the United States in the 1960s, commented that "the capacity of LSD and some other psychedelic drugs to exteriorize otherwise invisible phenomena and processes and make them the subject of scientific investigation gives these substances a unique potential as a diagnostic instrument and as research tools for the exploration of the human mind."[84]

The next crucial development involved hallucinogens as an alternative psychotherapeutic treatment for disorders such as alcoholism. Scientific reports began to suggest stunningly high rates of success compared with more traditional therapies. Soon experiments with psychedelic drugs branched into a host of multidisciplinary studies, ranging from religious mysticism to the creative process and parapsychological phenomena (e.g., extrasensory perception or ESP).

By the late 1950s and into the early '60s, reports on psychedelic drugs, notably LSD, inundated the scientific community. Thousands of papers appeared on the subject. More problematically, psychedelics became a sociocultural phenomenon as people from mainstream society began experimenting with them. Soon psychedelics moved from the relatively confined and controlled environments of laboratory and hospital to the street, where they became part of a grand and uncontrolled social experiment. Adopted by the burgeoning youth movement of the '6os, the drugs contributed to the countercultural revolution. In the pandemonium that ensued, researchers found themselves competing with sensationalist media reports, to the point where myth and reality became blurred. This competition inevitably spelled disaster for scientists carrying out legitimate research with LSD, the most promising and powerful of the psychedelic substances. Faced with aggressive clampdowns by governments to bring public hysteria under control, research on, and clinical use of, psychedelics quickly fell into disrepute. From that point on, mainstream psychiatry began to display what David Healy referred to as a "jaundiced view of LSD and hypotheses that went with it."[85] Gradually, the contributions of psychedelics to psychiatric and psychopharmacological history were "flushed down the memory hole."[86]

Although the demise of psychedelic science was predominantly the result of factors outside the scientific environment,[87] it is important to consider what was happening inside the medical and psychiatric communities leading up to and during these external developments. As with other drugs in the psychopharmacological revolution, generalizations about psychedelic research have proven to be extremely difficult. Experimentation with them in psychiatry was widespread in the 1950s and varied greatly. Biological psychiatrists explored them in the hope of locating a biochemical substrate to mental illness. LSD in particular became immensely valuable to early neuroscientific efforts by helping to chart the metabolic pathways in the brain, shedding new light on the function of neurotransmitters such as serotonin and norepinephrine. From a therapeutic angle, the drugs were used as adjuncts to psychotherapies of all types, psychoanalytical ones included. Even the U.S. Army and CIA believed that the drugs held untold potential, which explained why they financed the bulk of scientific research with the drugs (though their purposes were focused on espionage and chemical warfare rather than on improvements to mental health). For all of its popularity, though, psychedelic science did not last.

Contrary to common perceptions, the '6os counterculture and public manifestations of psychedelia were not the sole causes of this downfall. By the early 1960s, there was already a consensus brewing within psychiatry that psychedelics were too dangerous and had no place inside or outside the field. Partly, this reaction stemmed from the fact that aspects of certain psychedelic-related studies were too troublesome for the profession; some methods of using the drugs and some research findings challenged dominant psychiatric theories and treatments in ways that made psychedelics incompatible with transformations in psychiatry. Unlike other psychopharmacological therapies, psychedelic therapy, with its overt mysticism and emphasis on subjective patient experiences and values, did not fit neatly within the new objective world of "scientific psychiatry" and evidence-based medicine.[88] Rather, it stood out as a direct challenge. This led in turn to a scientific backlash and antipsychedelic movement in medical and psychiatric communities. From the mid-1960s on, scientific reports citing results from randomized controlled trials began to disprove the positive claims of earlier research findings. Psychedelic science had suddenly become non-scientific, relegated to the status of psychoanalysis and other defunct fads in psychiatric history.

How, then, can one explain the resurgence of psychedelic drugs in contemporary medicine and psychiatry in Europe and North America? If there is indeed scientific merit to psychedelic research today, then what does it tell us about previous research in places such as Saskatchewan?

DECONSTRUCTING PSYCHEDELIC SASKATCHEWAN

Saskatchewan had one of the first research communities to achieve significant results in psychotomimetic studies. It was also one of *the* places to which other scientists looked as a model for the successful employment of hallucinogens in psychotherapy, particularly in the treatment of chronic alcoholism. The Saskatchewan project, in many respects, was a microcosm of the psychedelic experiments conducted in the rest of the world, yet something about it made it unique.

Hoffer, Osmond, and Blewett were at the forefront of the scientific investigation of psychedelic substances. Their pioneering endeavours set the Saskatchewan research apart from similar experiments elsewhere, and their work guided many interdisciplinary studies with these substances after the mid-1950s. Within a decade, the Saskatchewan research project spawned nearly 200 scientific publications, the findings of which culminated in several innovations in the etiology, diagnosis (e.g., the Hoffer-Osmond Diagnostic Test), and treatment of mental illness. As three of the most ardent proponents of the scientific study of psychedelics, Hoffer, Osmond, and Blewett were revolutionaries. The revolutionary aspect of their work occurred on two levels: in its innovations in understanding the human mind (e.g., mental illness) and in its challenge to dominant assumptions about scientific research.

Saskatchewan's psychedelic drug research, like many other experiments with hallucinogenic substances, began with an exploration of their psychotomimetic properties. When Hoffer and Osmond initiated rigorous testing of these substances, their goal was to acquire more knowledge of and empathy for the schizophrenic experience. This, they hoped, would help them to isolate similar compounds in the human body and lead to more effective treatment of the disease.

When Humphry Osmond left England in the fall of 1951 to become clinical director at the Saskatchewan Hospital–Weyburn, he brought a radical new approach to mental illness. Before coming to the province, Osmond had completed a study with his partner, John Smythies, in which they observed striking similarities between acute schizophrenia and intoxication with the hallucinogenic drug mescaline, a synthesized derivative of the peyote cactus. The paper based upon this study theorized that schizophrenia was the result of a malfunction in one's biochemistry. In particular, it was a "disorder of the adrenals in which a failure of metabolism occurs and a mescaline-like compound or compounds are produced."[89] Their suggestion that a toxic, endogenous substance—the oxidized derivatives of the hormone adrenaline (or epinephrine)—was at least partially, if not wholly, responsible for the onset of schizophrenia and other mental illnesses was not the first such biochemical hypothesis. As early as 1906, Carl Jung had speculated about the existence of a "Toxin X":

The mechanisms of Freud are not comprehensive enough to explain why dementia praecox arises and not hysteria; we must therefore postulate for dementia praecox a specific concomitant of the affect—toxins?—which causes the final fixation of the complex and injures the psychic functions as a whole. The possibility that this "intoxication" might be due primarily to somatic causes, and might then seize upon the last complex which happened to be there and pathologically transform it, should not be dismissed.[90]

Throughout the first half of the twentieth century, a flurry of biochemical hypotheses was advanced to explain schizophrenia, but few were useful in pinpointing an etiology of the disorder and related mental illnesses. For years, researchers scoured the schizophrenic's body for this supposed toxin, and many theories abounded about what was responsible for the disease (e.g., bowel toxins), but none was generally accepted. Whether because of inadequate technological advances or faulty hypotheses, this failure undoubtedly had a lot to do with the predominance of Freudian views as well as those of influential psychiatrist Adolph Meyer on mental illness, especially in North America.

In most psychiatric schools of thought, schizophrenia was seen largely as a result of psychosocial factors. A popular interpretation in the 1950s was that schizophrenics retreated into a fantasy world of delusions and hallucinations because of their inability to cope with early childhood trauma. For psychoanalytical practitioners such as Harry Stack Sullivan, Frieda Fromm-Reichmann, and Theodore Lidz, the answers to questions about schizophrenia would be found not in the patient's biochemistry but in his or her "schizophrenogenic" family environment. Fromm-Reichmann was more forthright when she observed that the schizophrenic individual "is painfully distrustful and resentful of other people…because of the severe early warp and rejection that he has encountered in important people of his infancy and childhood, as a rule mainly in a schizophrenogenic mother."[91] Although Freud never fully subscribed to this view, or believed that schizophrenia could be successfully treated using psychotherapeutic techniques, some of his followers believed otherwise. More often than not, "Freudian" views on schizophrenia were elevated to cult status without having been proven scientifically.[92]

In terms of Meyer's more psychobiological approach, the possibility that biochemistry played a role in schizophrenia was never completely ruled out, but it habitually shied away from giving it too much credence. As Hoffer and Osmond saw it, such a "holistic approach was not particularly favorable to a toxic hypothesis for it tended to lay emphasis on all possible variables rather than any one in particular. The supporters of toxic hypotheses, in contrast, seemed narrow, fanatical and one-eyed, compared with this much broader view."[93] As a general rule, in the 1950s biochemical theories of schizophrenia were almost always greeted with pessimism and some hostility in the more mainstream scientific community. As Manfred Bleuler, son of the eminent Eugen Bleuler, authoritatively put it in his review of works on "the pathological physiology of schizophrenia" in 1955, "these works have failed to bring us even one step closer to the possibility of finding behind the psychological psychosis of schizophrenia a definable, specific, pathological, somatic schizophrenia. We have no evidence of any disturbance which would neatly differentiate it from other psychoses, somatic disorders or from the norm."[94] By the time of the biochemical theory of schizophrenia of Osmond and Smythies, most professionals considered it both outdated and unrealistic. A more likely explanation of the repudiation by the more orthodox psychiatric establishment, however, might be that their theory was a notable exception to the psychiatric status quo.

Aiding Osmond in his quest to better understand and treat schizophrenia was Abram Hoffer, a talented young medical doctor and biochemist and director of psychiatric research for Saskatchewan's Psychiatric Services Branch. Osmond, with his skills as clinician, administrator, and researcher, and Hoffer, with his extensive knowledge of biochemistry and vitamins, complemented one another perfectly. They struck up a lifelong friendship and partnership, and from 1951 on their names were inseparable. Their theories regarding mental illness provoked both enormous curiosity and profound skepticism, and set the stage for a continuous assault on conventional psychiatric practice.

Continuing along the lines first set out by Osmond and Smythies in their hypothesis, Osmond and Hoffer began an extensive research program to deal with the plight of the schizophrenic. At the heart of this program was the model psychosis experimentation with hallucinogenic drugs. Expanding on the original adrenaline-related theory, they concentrated on the relationship between hallucinogens and a number of the hormone's immediate derivatives, such as adrenochrome and adrenolutin, believing them to be promising candidates for the mescaline-like substance. The result was one of the most hotly contested discoveries in modern psychiatric history: namely, the finding that megadoses of vitamins such as niacin and ascorbic acid could alter the drug experience and treat mental illness.

The initial purpose of this testing was to attain greater knowledge of schizophrenia, but by the mid-1950s the experiments were headed in directions that few could have anticipated. Osmond, Hoffer, and their team of researchers were among the first scientists to discover that some people under the influence of these drugs could have experiences that lay well outside the realm of psychosis. They realized that, depending entirely on the frame of mind of the individual and the environment in which he or she was placed ("set and setting"), many people could have "transcendent" experiences. According to William Braden, "the Hoffer-Osmond studies are far from conclusive, and similar theories have been advanced in the past. . . . The line dividing insanity and mysticism has never been too sharply drawn, and the biochemical theory of schizophrenia makes it all the more tenuous."[95] Braden, along with many other psychedelic scholars, questioned whether "insanity, mysticism, and the psychedelic experience [are] in some way related."[96] This unexpected development in the research led both to Osmond's use of the term "psychedelic" and to an array of new interdisciplinary studies of the drugs' applications.

Another individual who became acutely aware of these new possibilities was Duncan Blewett. Drawn to the early work of Osmond and Hoffer, he became an integral part of the experimentation. As supervising psychologist for the Saskatchewan government's Psychiatric Services Branch, Blewett, along with his two colleagues, charted new goals for this research. All three primarily wanted to improve mental health, and each was eager to demonstrate the scientific value of psychedelics. Whereas Hoffer and Osmond focused on schizophrenia and its relationship to these drugs and human biochemistry, Blewett became enamoured with the potential of these drugs, namely LSD, as psychological instruments to increase spiritual awareness and self-understanding, and he worked to have this potential acknowledged within psychology and the wider scientific community. As one of the Saskatchewan project's leading proponents, Blewett argued that psychedelics "are to psychology what the microscope is to biology or the telescope to astronomy."[97]

Although model psychoses remained the core of the experiments, psychedelics were fast becoming known among early practitioners for their therapeutic possibilities, predominantly with people suffering from severe addictions. Here, too, Saskatchewan led the way. Hoffer, Osmond, and Blewett were pivotal in the co-creation of "psychedelic therapy" and myriad other important inroads made with these drugs, in areas such as architectural design, perceptual studies, and parapsychological phenomena. By the end of the 1950s, all eyes in the psychiatric community were fixed on Saskatchewan, a situation that led Healy to remark that "Weyburn and Regina were almost as important lights in the psychopharmacological firmament as Paris and Basel."[98] As he pointed out, Hoffer and Osmond's adrenaline-based hypothesis and treatment theories for schizophrenia were "at the heart of biological psychiatry," placing the two psychiatrists "at the forefront of biological speculation."[99] In the same way, Saskatchewan's successful use of LSD to treat alcoholism, and the spectacular claims attributed to psychedelic therapy, reverberated throughout the psychiatric world and beyond. But the "heady buzz," as Healy referred to the state created by the province's psychedelic innovations, was evanescent.[100]

By the time Saskatchewan's psychedelic research program concluded in the early 1960s, its claims had led to a number of major scientific controversies that, if left unresolved, put many in psychiatry at risk. Because they tested the supremacy of psychoanalytical frameworks as well as competing branches in biological psychiatry, the Saskatchewan claims were attacked on several fronts. There was a prolonged period of challenges from the U.S. National Institute of Mental Health (NIMH), the American Psychiatric Association (APA), the Canadian Mental Health Association (CMHA), the Addictions Research Foundation (ARF), and others, whose replication studies produced results refuting many of the Saskatchewan claims. By the end of the 1960s, the scientific evidence against Saskatchewan's psychedelic research was significant enough to destroy any scientific merit that it, and similar psychedelic drug research, had.

Much of the resistance did not originate solely from the claims or the use of psychedelics (though they were partly to blame for it) but had more to do with the "style" of science practised. One of the most controversial issues in the work of Hoffer, Osmond, and Blewett was that much of it was grounded in what many in the scientific community labelled "subjectivity" and "emotionalism." Many of the hypotheses coming out of Saskatchewan had, as their fundamental bases, the personal experiences of the investigators and, more importantly, their patients. Whether this was a weakness or failure continues to be a matter of debate.[101] A frequent criticism of the work of Hoffer and Osmond is that they relied too much on personal observations in developing their hypotheses. To many scientists involved in the debate in the 1950s, the fact that these two psychiatrists tested the hallucinogenic drugs on themselves disqualified their research from being objective and made their claims overly "simplistic" and unable to stand up to "sober research."[102]

Blewett faced similar resistance. The idea of spiritual values, his main preoccupation, was out of step with traditional Freudian psychoanalysis and especially with the reigning behaviourist philosophy, which, like orthodox science itself, emphasized objective, observable, and strictly empirical data. As Maslow wrote of the behaviourist tendency,

> the pure positivist rejects any inner experience of any kind as being "unscientific," as not in the realm of human knowledge, as not susceptible of study by scientific method, because such data are not objective, that is to say, public and shared. This is a kind of "reduction to the concrete," to the tangible, the visible, the audible, to that which can be recorded by a machine, to behavior.[103]

In many ways, Blewett's insights on psychedelics and spirituality are representative of the fledgling humanistic movement in psychology in the 1950s, led by Maslow and Rogers, to make science less dependent on mechanistic, reductionist principles and more inclusive of all perspectives, even the most personal ones.

In spite of the many achievements of Saskatchewan's psychedelic research, scientific assessments of it from the 1970s on were decisively negative. Reviewing the adrenaline-based theories of, and alternative biological treatments for, schizophrenia (which came to be grouped under the umbrella of orthomolecular psychiatry and megavitamin therapy), a *British Medical Journal* editorial spoke for much of the psychiatric community when it declared the following: "Sadly, orthomolecular psychiatrists have generally not followed scientific methods of proof; when others have had poor results with scientific methods they have tended to reply by verbal argument instead of with the argument of renewed experiment and fresh fact. The controversy is sterile."[104] Similar appraisals were made of claims regarding LSD and alcoholism. Summarizing the history of LSD treatment of alcoholism, psychiatrist Frederik Baekeland wrote that it "starts with sensational anecdotal reports, followed by enthusiastic and almost always successful uncontrolled studies and, finally, by consistently unsuccessful controlled trials. Some of the latter trials showed transient short-term effects but they...like the others were not able to demonstrate any long-term efficacy of LSD."[105] With the controversies surrounding the Saskatchewan research supposedly dead, the work faded into obscurity. At least this has become the standard narrative in the official annals of scientific psychiatry.

Several decades have now passed, and historians are finally revisiting psychedelic Saskatchewan. The biggest contributors to the subject have been medical historian Erika Dyck and psychologist John Mills (though both give limited attention to Blewett and his role in the psychedelic research program). Dyck has taken the broadest view of the experiments and their overall place in psychiatric and psychopharmacological history. Her examinations of the various aspects of the research, from the work with schizophrenia and alcoholism to the use of LSD in architectural design, have come, like my own, to view it as significant on many counts. She contends that, "far from being fringe medical research, the LSD trials represented a fruitful, and indeed encouraging, branch of psychiatric research occurring alongside more famous and successful trials of the first generation of psychopharmacological agents, such as chlorpromazine and imipramine."[106] Much like my analyses, hers have concentrated on the research on schizophrenia and alcoholism and the resulting new therapeutic paradigm for understanding and treating mental disorders, what

Dyck has labelled "psychedelic psychiatry." The new approach "challenged researchers to reconceptualize mental illness in terms of an individual's experiences and self-awareness, thus abandoning rigid systems of symptom classification based on interpretations from an observer, in favour of interdisciplinary collaboration involving the patient....It promoted a different kind of medical discourse for advancing clinical assessments based on connections among individuals' experiences, beliefs, and behaviours."[107] The main practitioners of the approach, the "psychedelic psychiatrists," as Dyck calls them, "designed a therapy that concentrated on empowering patients to play a more active role in their recovery, instead of passively accepting treatments doled out by psychiatrists. Far from being simply another competitor in the growing pharmaceutical industry, LSD threatened to undermine it."[108]

A major part of Dyck's assessment of the research involves the influence of Saskatchewan's geopolitical landscape and "relatively unbureaucratized state" in shaping the psychedelic experiments.[109] As Dyck emphasizes, the province's "political culture provided an ideological sanctuary for LSD experimentation," a point that I agree with to a certain degree, though I focus more on the scientific freedom accorded to the Saskatchewan researchers than on any particular political ideology. Scientific freedom is integral to understanding how the psychedelic drug experiments were carried out in Saskatchewan.[110] It is also crucial in understanding why independent-minded and iconoclastic thinkers such as Hoffer, Osmond, and Blewett flourished. All three were fiercely independent and produced some extremely interesting and productive research findings, but this independence also caused them a great deal of trouble.

Although scientific research in Saskatchewan in the 1950s was headed in the same direction that Whyte and others like him criticized, special circumstances permitted the free and open research that he championed. As Hoffer recounted, in 1950 a "unique constellation of events" in Saskatchewan enabled the research to begin.[111] There was no College of Medicine, no Department of Psychiatry, no official Board of Ethics, and no Research Board. This absence of organizational infrastructure in the province's mental health system allowed researchers to invest time in individual pursuits. In addition, the social democratic government of the Cooperative

Commonwealth Federation (CCF) and its leader, Tommy Douglas, encouraged bold, innovative ideas in health care. Finally, Saskatchewan was relatively isolated geographically, which made locales such as Weyburn ideal for conducting independent research and testing new theories. All of the above factors coalesced to create an environment unique for its time.

Absent from Dyck's analyses of the experiments is a fuller discussion of the latter stages of Hoffer and Osmond's theories and treatment of schizophrenia, what the two psychiatrists considered the most important things to come out of their research with psychedelic drugs.[112] This gap is filled to some extent by the writings of Mills. In his assessment, he grants that the research provoked some novel theories and findings, and led to some significant milestones in psychiatric history, but he believes that the success of the research was short-lived because of its inability to meet acceptable scientific standards and methods. In centring his critique on the adrenochrome hypothesis (and the resulting megavitamin/orthomolecular treatment), Mills acknowledges that it was the "world's first theoretically based and empirically testable theory of the biological origins of schizophrenia, and therefore it attracted substantial research funding," yet he concludes that the claims "could not be substantiated and [that] the theory fell into disrepute," and he heavily emphasizes the "conclusive" findings of the NIMH and CMHA replication studies to support this argument.[113]

Concerning the scientific conclusions that can be made today about Saskatchewan's research with psychedelic drugs, and the theories and experiments of Hoffer, Osmond, and Blewett, what can be said? I aim in this book to provide a more complete picture of these men and their research. I have tried to avoid making definitive conclusions about the scientific validity of their theories and research, yet the more I read about their work, and the more I sought to understand these men and their activities in the wider context of the scientific community, it became difficult not to make a strong argument for the revolutionary nature of their psychedelic experiments.

According to Kuhn's scientific framework, major paradigm shifts in history were the transition from the Ptolemaic system to the Copernican system, Newtonian mechanics, Galilean astronomy, Darwinian evolution, and Einsteinian relativity. Although it would be an exaggeration to place Saskatchewan's psychedelic discoveries on the same level, I argue that the theories and findings of Hoffer, Osmond, and Blewett fall well within the category of what Kuhn considered "minor revolutions," those "associated with the assimilation of a new sort of phenomenon, like oxygen or x-rays."[114] Much of the Saskatchewan psychedelic work qualified as revolutionary in the Kuhnian sense. Hoffer and Osmond's biochemical theories, for instance, can be seen as an attempt to establish a biological paradigm for approaching schizophrenia; even within the biological paradigm that dominates psychiatry today, their theories continue to challenge the reigning views on how to understand and treat the disease. The same can be said of the contributions of Hoffer, Osmond, and Blewett to psychedelic science and therapy. In bridging medical/biological and mystical perspectives, they helped to establish a hybrid style of scientific practice (with psychedelics) that some are now labelling "mystical materialism."[115] Certainly, much of the Saskatchewan psychedelic work has not resulted in paradigm shifts in psychiatry and other fields, but this in no way diminishes its revolutionary qualities. As Kuhnian scholar Joseph Rouse noted, "revolutions succeed by giving renewed impetus to research," and this is precisely what has been happening since the 1990s.[116] As progenitors of the psychedelic revolution, Hoffer, Osmond, and Blewett helped to lay much of the groundwork upon which today's promising scientific studies rest.

STRUCTURE OF THE BOOK

I have divided the book into two parts. Part 1 deals with Saskatchewan's psychedelic research program from 1951 to 1961. Chapter 1 analyzes the origins of Hoffer and Osmond's biochemical theories and their adoption of hallucinogens for their psychotomimetic properties as an alternative approach to dealing with schizophrenia. The chapter covers the two psychiatrists' research during its embryonic stages (1951–54), from the gradual expansion of the initial Osmond-Smythies hypothesis to what became a fully fledged multidisciplinary attack on the schizophrenic disorder. Blewett is also introduced. Chapter 2 expands on Chapter 1, discussing the refinement of Hoffer and Osmond's early findings and their adrenaline-based hunches, and how they were received in the wider scientific community. Chapters 3 and 4 concentrate on the second major phase of the province's hallucinogenic drug experiments, the evolution of psychedelics as a concept and form of therapy, and the various contributions of Hoffer, Osmond, and Blewett. The biggest breakthrough in Saskatchewan at this stage came with the therapeutic use of psychedelics, notably in the treatment of alcoholism. Chapter 5 is a more detailed examination of important contributions made by the Saskatchewan research team in other areas: architectural design, perceptual studies, and parapsychology. By the mid-1950s, many researchers around the world had abandoned the model psychosis side of research in favour of psychedelics used in therapy and newer forms of scientific exploration. Saskatchewan was one of the few places that continued to use and further test and develop both types of hallucinogenic drug experiences, the psychotomimetic and the psychedelic. At the same time, Hoffer, Osmond, and Blewett further dissected the psychedelic properties of the drugs, for instance as instruments in exploring and identifying parapsychological phenomena.

Part 2 concentrates on the period after the research (1961–75) and examines the fallout in the scientific community regarding whether Saskatchewan's psychedelic research has scientific value or not. The mid-1960s were a significant turning point in scientific culture, especially for how future research would be funded, conducted, and judged. This shift was largely responsible for shaping and controlling both the response to and the perception of Saskatchewan's psychedelic research. Published laboratory and clinical test results began to question the validity of earlier positive findings with LSD and similar drugs. The authors of these reports, many of them experts in their own right, claimed to possess incontrovertible evidence of unscientific methodology in research like that done in Saskatchewan. Chapters 6 and 7 examine the opponents, their arguments, and how they affected assessment of the research, and focus respectively on the results of the two main components of the research: the studies of schizophrenia (Chapter 6) and the use of LSD to treat alcoholism (Chapter 7). Chapter 8 tackles some of the myths and misconceptions about the Saskatchewan research directly following its conclusion, largely as a result of the 1960s counterculture and the CIA conspiracy theory. It also provides a clearer view of where Hoffer, Osmond, and Blewett weighed in on the social and political debates that arose over LSD.

I conclude the book with an assessment of where their work stands today, how it has been portrayed both within and outside the scientific community, and what can and cannot be said definitively about these controversial experiments five decades later.

PART ONE

Psychedelic Science: The Saskatchewan Experiments (1951–61)

Like the earth of a hundred years ago, our mind still has its darkest Africas, its unmapped Borneos and Amazonian basins. In relation to the fauna of these regions we are not yet zoologists, we are mere naturalists and collectors of specimens. The fact is unfortunate; but we have to accept it, we have to make the best of it. However lowly, the work of the collector must be done, before we can proceed to the higher scientific tasks of classification, analysis, experiment and theory making.
—Aldous Huxley, *The Doors of Perception and Heaven and Hell* (69)

Post–Second World War Saskatchewan was the site of a host of innovative experiments—socio-economic, political, and scientific. It was a place of firsts in North America: the first socialist government, the first system of public medical insurance (i.e., Medicare), and the first scientifically testable biochemical hypothesis for schizophrenia. The province was heralded, in fact, as a world leader in mental health in the 1950s, noted for its cutting-edge psychiatric research. Hallucinogenic drugs figured centrally in this research.

The years 1951–61 witnessed the rise to scientific prominence of Saskatchewan's hallucinogenic drug research and the work of three of its key architects, Abram Hoffer, Humphry Osmond, and Duncan Blewett. Beginning with explorations of the psychotomimetic effects of hallucinogens, Hoffer and Osmond focused on the drugs' chemical relationships to adrenaline, a result of the experiments of Osmond and his first collaborator, John Smythies, with the structurally similar hallucinogen mescaline. From there, Hoffer and Osmond mapped out a unique hypothesis for schizophrenia that became the basis for a decade-long, multifaceted research program. Using the Saskatchewan Committee on Schizophrenia Research as their base, the team carried out a wide range of studies into the psychological, perceptual, and biochemical world of schizophrenia.

The discoveries resulting from these "model psychosis" experiments attracted major scientific and psychiatric interest, drawing both praise and derision. For Osmond, who characterized the period in military terms as a build-up to the attack on schizophrenia, 1961 was a pivotal year. Writing to Hoffer, he remarked that by that time they had begun "to show in a variety of ways that our general hypothesis, our initial discoveries and our fundamental work had reached a point where it could and was being applied in diagnosis, prognosis and treatment."[1] Their use of hallucinogens was critical to this success. In Osmond's opinion, "our greater understanding of the experience of schizophrenic patients, derived from studying the madness-mimicking effects of [psychotomimetic] substances, has enabled us to do things which might have been otherwise impossible."[2]

Offshoots of the schizophrenia work led the Saskatchewan researchers into some uncharted territories and unexpected discoveries. Along with Blewett, Osmond and Hoffer expanded the research and prepared the ground for "psychedelic" as a concept and unique form of drug-assisted therapy. Here, too, the results were promising, in particular those achieved with the psychedelic treatment of alcoholism. As a whole, the hallucinogenic experiments in Saskatchewan had groundbreaking scientific implications for research on not only the diseased mind but also the healthy mind.

But not everyone saw the Saskatchewan research in this way. The experiments occurred at a time when psychiatry was struggling to find its scientific feet. That the Saskatchewan discoveries challenged many of the established (psychodynamic) and competing (biomedical) theories of and approaches to mental illness, while promoting an alternative (and sometimes hallucinogenic) approach to scientific methods, guaranteed controversy. Hoffer and Osmond summed up the difficulty in what they were doing:

> The traveler who is expert in finding his way by means of maps will not necessarily survive when exploring unmapped country. It is a mistake to let researchers believe that they can depend on the "maps" of the methodologists, when their task is to go where there are no maps. There is a danger that those giving monies for research may become so preoccupied with the need for "sound research design" that they forget there is a world of difference between demonstrating what is known already and exploring the unknown.[3]

Working in a field with so many unknowns, the Saskatchewan researchers were like the explorers described above by Huxley. Undaunted, they explored the "darkest" and "unmapped" regions of the human mind.

CHAPTER 1

MODEL PSYCHOSES AND THE ADRENOCHROME HYPOTHESIS

umphry Osmond and Abram Hoffer's findings with hallucinogenic drugs and their adrenaline-centred hypotheses elicited extremely mixed reactions in the psychiatric world in the early 1950s. When the two Saskatchewan psychiatrists established their research program in 1951, schizophrenia persisted as the most crippling and mysterious mental health affliction. Generations of psychiatrists had been rather unsuccessful in uncovering an all-embracing answer to the enigmatic disease, and few of them thought a cure possible. Practitioner after practitioner arrived at the sombre conclusion that schizophrenia would forever remain an irreparable way of life and that the best hope for the patient stricken with it was palliative care. This attitude manifested itself in continued dependency on electroconvulsive therapy (ECT) and enthusiastic adoption of new wonder drugs, such as chlorpromazine and reserpine, that seemingly inhibited many schizophrenic symptoms. Amid many competing psychological and sociocultural theories was the suggestion by a couple of unknown researchers that the disease, along with similar mental illnesses, was possibly the consequence of an error in adrenaline metabolism.

Hoffer and Osmond's idea that hallucinogenic-like derivatives of the adrenaline hormone, namely adrenochrome or adrenolutin, were somehow responsible for the onset of schizophrenia spawned widespread criticism throughout the psychiatric profession. By 1950, as noted in the introduction, biochemical hypotheses for schizophrenia were extremely unpopular in most spheres of psychiatric practice. The model psychosis work with hallucinogenic drugs such as LSD did renew psychiatric interest in biochemical hypotheses; nevertheless, though many experts thought the uncanny resemblance between intoxication with these chemical substances and the schizophrenic condition interesting, most refused to take the research seriously. Few believed that these drugs would be of any consequence in locating the cause of or determining a treatment for schizophrenia.[1] The Saskatchewan experiments, however, challenged many people to rethink their assumptions about the disease. The research under Hoffer and Osmond once again raised the prospect of a biochemical factor in schizophrenia. The results of their studies made Saskatchewan a frontrunner in the hunt for Carl Jung's legendary "Toxin X."[2]

The earliest scientists who performed model psychosis tests with hallucinogens, or psychotomimetics, spent much time in the preliminary stages collecting data on what was then a relatively obscure group of substances. In this respect, the Saskatchewan research project was not all that different, but it was one of the few such projects to try to answer a specific question: does a mescaline-like substance derived from adrenaline contain the missing link to schizophrenia? This premise, originally put forward by Osmond and Smythies, became the catalyst for the project in Saskatchewan and the foundation for much of the subsequent research in the province on schizophrenia.

Any examination of Hoffer and Osmond has to take into account psychiatry in Saskatchewan in the post–Second World War period because it was within this intellectually fertile environment that their theories took root. The CCF government of T. C. Douglas, touted as the first socialist government in North America, had a humanitarian disaster looming in provincial mental hospitals; the overcrowding and unsanitary conditions were so bad, in fact, that it necessitated an overhaul of the mental health system. Part of the government's strategy consisted of allowing new ideas to be tested, and this enabled the two young psychiatrists to begin their model psychosis trials and search for the elusive mescaline-like factor, or M-factor, in the schizophrenic body.[3] Their research also secured approval of and financial sponsorship from the federal government and leading benefactors such as the Rockefeller Foundation. This opportunity for Hoffer and Osmond to engage in what William H. Whyte Jr. described as "idle curiosity" became a determining factor in the ultimate success of Saskatchewan's hallucinogenic drug experiments. Citing some American financers of research (e.g., Bell Labs), Whyte wrote that "instead of demanding of the scientists that they apply themselves to a practical problem, they let the scientists follow the basic problems they want to follow. If the scientists come up with something then they look around to see what practical problem the finding might apply to. The patience is rewarded."[4] Although Hoffer and Osmond began with no guarantee of success, their research on schizophrenia culminated in a long list of accomplishments and a number of medical firsts.

The years 1951–54 marked the developmental period of their research program on schizophrenia, during which many of their basic ideas were devised and much of the rudimentary testing with hallucinogens was initiated. After Hoffer and Osmond secured the necessary funding and resources, they took the pioneering Osmond-Smythies hypothesis as their starting point and commenced their inquiry into hallucinogenic drugs to determine if they might be related to schizophrenia.

Hoffer and Osmond became the first scientists to perform human tests with adrenochrome and the first to show that this substance has psychotomimetic characteristics like the other hallucinogens. They also spearheaded an alternative approach to schizophrenia and other mental illnesses with megadoses of common vitamins such as niacin and ascorbic acid, an innovation that eventually formed the pillars of a new branch in medicine known as orthomolecular psychiatry.[5]

Closely interwoven with these accomplishments were the Saskatchewan researchers' model psychosis experiments with drugs such as mescaline and LSD, noteworthy for a number of reasons. This research catapulted the province onto the world stage as one of the pivotal centres for LSD studies, with Hoffer, Osmond, and their colleagues among the earliest researchers to have their findings on the effects of LSD in humans published, in both scientific journals and the lay press. They were also the first to document how agents such as niacin could be used to modify hallucinogenic drug experiences. These initial discoveries helped to bring about greater understanding of and empathy for schizophrenics and sufferers of other mental diseases, and a significant improvement to mental health in Saskatchewan.

Transformation of the province's mental health system began under the CCF government of Douglas in the mid-1940s. Previously, the system in Saskatchewan, like those of many other places in the world, was in a hopelessly depressing and appalling state. Traditionally, individuals suffering from mental illnesses faced segregation from their communities and families, and confinement in one of the two large rural psychiatric institutions constructed in North Battleford (1914) and Weyburn (1921). This policy endured for decades as a convenient means for successive governments and a fearful public to deal with something about which they understood little and with which they cared not to be associated on a daily basis. Years of neglect and mismanagement, however, did not make the problem magically disappear. Instead, it got progressively worse. Severe overcrowding at these institutions led to a massive drain on government resources and became a public eyesore. In a 1944 study commissioned by Douglas, Dr. Clarence Hincks of the National Committee for Mental Hygiene reported that "the North Battleford Hospital with a normal capacity of 1,174 patients is caring for 1,716; and the Weyburn hospital with a normal capacity for 1,040 patients is sheltering 2,485. In other words, at both institutions there are 4,201 patients with adequate accommodations for 2,214. This represents overcrowding to the extent of 89%."[6] Although first conceived as the panacea to handling the more troublesome members of society, the mental hospitals wound up being a disgrace. Other deficiencies included a lack of skilled personnel, extremely low staff morale, corrupt management practices (e.g., racketeering and patronage), and a high turnover rate of the few professionally trained psychiatrists in the province.

Not surprisingly, this atmosphere did not bode well for mental patients. The institutions were not only overcrowded but also overconcentrated, with hundreds of patients of all sorts—from the intellectually disabled and epileptic to the schizophrenic—lumped together in small spaces. Dr.

F. S. Lawson, a former superintendent of both the North Battleford and the Weyburn hospitals and a one-time director of psychiatric services, was well aware, like many others, of the herculean task for doctors and staff in treating their patients, and he frequently commented on the dismal circumstances. The Weyburn building, he said, "stunk like something out of this world. It was crowded and the whole basement area was a shambles, naked people lying all over the place, lying around incontinent. There was a lot of seclusion in use, and mechanical restraints."[7] Although the North Battleford hospital might have avoided such restraints, it had one ward "of 175 patients who didn't wear clothes, who slept on straw ticks on the floor, and who were roused at 2 a.m. to go to the toilet and then hosed down with cold showers to clean them off. The verandah had a sawdust dam on the doorway to the stairs to keep the urine from running down the stairs. These were mostly chronic schizophrenics."[8]

In addition to the unbelievably unsanitary conditions, patients often had to endure a painful process of desocialization, and the longer the patient was isolated from mainstream society, and the more adapted he or she became to institutional life, the more remote the likelihood of his or her ever being successfully reintegrated into society. There were also many disadvantages to the locations of these institutions. Long distances separated many patients from their families and therefore made visits few and far between. In journalist-author Fannie Kahan's assessment, the patient could only "sit and wait": "That's what most of Saskatchewan's 3,500 patients in North Battleford and Weyburn hospitals are doing because we failed to heed humanitarian voices of long ago, and because we give more credence to our superstitious fears and ignorance than the voices of experts today."[9]

For years, professionals such as Drs. David Low, Henry Sigerest, Clarence Hincks, and others had proposed more appropriate, and in theory more humanitarian, means of coping with mental illness in Saskatchewan.[10] They called for greater specialization in mental health services (e.g., training nurses and recruiting professionals), administration of services by a Department of Public Health (DPH, with the assistance of a Psychiatric Services Branch),[11] and a network of mental health clinics operating either as separate units or as add-ons to general hospitals. They hoped that out-patient services would form the basis of a new preventative approach to mental illness emphasizing treatment and rehabilitation, as opposed to the outdated and inhumane practice of containment.

By the late 1940s, it appeared that honest efforts were again being made to improve the conditions of the mentally ill and to correct the social and financial catastrophe created by the mental hospitals. Following the Second World War, mental health in Western society underwent significant changes, notably increasing bureaucratization and professionalization of mental health services and widespread acceptance of previously unheard of and incredibly potent tranquilizing drugs.

A significant transformation (in both Saskatchewan and abroad) was the switch from a policy of containment to a system in which the large mental hospitals were gradually scaled down and modified and the mentally ill placed back in the community to be treated. Newfound empathy for the mental patient was exemplified in what came to be known as the Saskatchewan Plan, a main tenet of which was "a genuine respect for the individual who is mentally ill...and a recognition that he can best be helped by being treated in the community with as little interference with his normal living as possible."[12]

By 1950, the CCF government was implementing most, if not all, of the recommendations in reports like those of Sigerest and Hincks to fix the problems of the decrepit mental hospitals and raise the standards of care for the patient. Assisting Premier Douglas in this effort was Dr. Griff McKerracher, appointed as commissioner of mental health services in 1946 and later as director of the Psychiatric Services Branch (PSB) of the DPH formed in 1950. McKerracher helped to set up educational programs for training psychiatric nurses, develop psychiatric wards in the general hospitals (e.g., the Munroe Wing of the Regina General Hospital), initiate training programs for up-and-coming psychiatrists (e.g., Hoffer), and establish mental hygiene clinics (staffed with a psychiatrist, psychologist, and social worker). Douglas wanted the hiring of new workers to go through the Public Service Commission, a move that he hoped would clean up the past practice of patronage, and he took significant steps to unionize hospital workers so that they would have collective bargaining rights. With the assistance of federal funding in 1950, the first provincial chapter of the Canadian Mental Health Association was organized in Saskatchewan. The CMHA was effective in educating the public about the conditions of mental hospitals and their patients, thereby helping to reduce the stigma attached to the illness. The government also began a massive recruitment drive, both within and outside the country, to attract the brightest minds in the field of mental health. Douglas wanted the best in the field so that his government could fulfill its mandate for meaningful social change. To this end, the CCF actively encouraged and sponsored innovative ideas and approaches, like those of Osmond, giving researchers almost limitless freedom to pursue unorthodox lines of inquiry. In many respects, the changes signified a truly golden age for mental health in the province.

HUMPHRY OSMOND: PSYCHEDELIC GRANDMASTER

Born on July 1, 1917, in Surrey, England, Humphry Fortescue Osmond was the second-born son of Dorothy and George William Forbes Osmond, both of whom descended from strong seafaring roots. His grandfather on his mother's side was a Peterhead whaling captain, and his grandfather on his father's side, and Humphry's father too, were both lieutenant commanders in the Royal Navy. Sadly, his mother passed away in 1919, shortly after giving birth to his younger sister, leaving them to be raised by their father, who managed with a shore job near Plymouth while the children were cared for by a nanny and her daughter. Soon the family made its way back to Surrey to live with a grandmother and two aunts, a happy time for the young Osmond.

When he was seven, Humphry attended the Amesbury boarding school, or a preparatory as it was then called, in the Surrey village of Hindhead. Following that, he went to a public school in nearby Haileybury that he apparently detested. Had it not been for good friends such as Christopher Mayhew, who became a guinea pig many years later for Osmond's mescaline experiments and a well-known member of Parliament in the Labour government, Osmond would likely have found his public school years unbearable. At this time, he became a voracious reader, spending any spare money on poetry anthologies and Shakespearean works. His reputation as an "intellectual force" was acknowledged by many who knew him and became evident in the countless literary references and quotations in his later writing and lectures.

For a time, it looked as though Osmond might pursue a livelihood in banking. Upon his matriculation, he was offered a job with the Hong Kong and Shanghai Bank, but because of delays in transportation he arrived for his scheduled interview in an exasperated state and did not pass the medical. He then had a stint as a building inspector with a financier and property development firm in Chitton. Although this work became useful in his later years in designing mental hospitals, it, too, was an occupation for which Osmond was ill suited. Fate, it seemed, had other plans. His father recognized his burgeoning interest at a young age in medicine, and it was an automatic decision to apply for the Kitchner scholarship, a perquisite reserved for sons of retired naval officers, so that Humphry could attend medical school. From that point on, his life took a decidedly different course.

Osmond started his medical journey at Guy's Hospital in London in 1937. By the end of the decade, most English doctors had been pushed into front-line service in the war, leaving Osmond and other medical students to deal with the many casualties from the German Blitz and bombed air raid shelters in London between 1940 and 1941. When he finally received his papers in 1942, he was once again thrown into service, this time to Fairborough, where the staff of Guy's were evacuated for a gruelling six-month period to act as on-call doctors. Osmond subsequently signed up for duty in January 1943 aboard the HMS *Volunteer*, part of a naval convoy that provided support to the United States, as a surgeon lieutenant in the Royal Navy. As Osmond's sister recounted, British losses at sea were enormous, and the grim experience must have had a profound impact on her brother, for the typically gregarious doctor shared little about it with family members.[13]

These two episodes, his hospital experience and his time in the navy, inevitably led Osmond to psychiatry, his life-consuming interest. Following brief psychiatric residencies in the Mediterranean, on the island of Malta, Osmond ventured back to London in 1946, where he married Jane, a young nurse, and took up a post as the first assistant in the Department of Psychological Medicine at St. George's Hospital and Medical School. Around this time, Osmond and some of his fellow doctors began to develop an interest in the powerful mind-altering effects of the relatively unknown drug LSD and similar substances. As his sister humorously recalled of these early experiments, Humphy's wife, instrumental in her husband's decision to come to Saskatchewan, provided "calm and dignity to some odd episodes."

As the 1940s drew to a close, Osmond, along with his friend and psychiatric colleague Smythies, reported on a number of intriguing observations regarding schizophrenia that would lead Osmond down a startling new path of inquiry and eventually take him to Saskatchewan Hospital–Weyburn in 1951. In a paper written with Smythies titled "Schizophrenia: A New Approach," they noted the striking similarities between acute schizophrenia and intoxication with mescaline, a synthesized derivative of the American cactus plant otherwise known as peyote. Osmond and Smythies highlighted many of the characteristics of schizophrenia, including "disturbances of association, thought, mood, and behaviour without dementia. Delusions and hallucinations occur frequently and disturbances of bodily function lumped together as catatonia are important. The illness may sweep down in a few hours or may be very insidious. It may last a few days or a whole lifetime."[14] They also underscored the immense variability of schizophrenic experiences. Such observations did not provoke a great deal of disagreement, but contention did arise over locating the cause, or etiology, of such symptoms and their treatment (e.g., deep-insulin and electroconvulsive therapies).

Osmond and Smythies briefly reviewed the research on schizophrenia, mentioning the various treatments used and what the future might hold for further biochemical approaches to the problem. That the effects of mescaline intoxication in a normal person mirrored in many ways the schizophrenic experience had been noted long before the two British psychiatrists came along. They were well aware that drugs such as mescaline and LSD had been used to induce "model psychosis" in test subjects and that such drugs had been "taken for years by psychiatrists who wish[ed] to experience schizophrenic symptoms and to investigate synaesthesia."[15] Osmond and Smythies went on to list the common features of mescaline poisoning and schizophrenia. It was also known that the chemical structures of mescaline and adrenaline closely resembled one another. The researchers put these facts together as the basis of a radically "new approach" to schizophrenia, later known in professional circles as the "transmethylation hypothesis." According to them,

> the striking implications of the relationship between the bizarre Mexican cactus drug and the common hormone have, so far as we know, never been recorded before. Psychiatrists appear to have been unaware of its biochemical structure, biochemists to have been uninterested in its production of an artificial psychosis. The close clinical connections between schizophrenia, anxiety states and stress have been known for a long time, and the process involved may be that in certain people when the adrenals are overworked…the process of methylation becomes disturbed and highly toxic substances are produced. This M-substance, once produced, would naturally set up a vicious circle, since one of the prominent features of both the mescal "psychosis" and of many cases of schizophrenia is the stress and terror experienced by the victim.[16]

Furthermore, Osmond and Smythies argued that

> the result of these endogenous disturbances in sensation, feeling, and thought upon the sick person would depend on the age of onset, the rate at which M-substance is produced, the patient's previous personality and special mode of reaction to M-substance, and the cultural setting in which the illness occurs. These variables allow for the enormous range of different reactions such as is actually observed in schizophrenia, which could thus appear to have many and various "precipitating causes."[17]

The principal task for the researcher wanting to follow this line of inquiry necessitated a "pharmacological investigation of mescaline, its breakdown products and the range of compounds between it and adrenaline….The attempt would have to be made to isolate these substances from the body fluids of acute schizophrenics."[18]

Such an undertaking would require a significant level of funding for human resources, equipment, and space to conduct the experiments. Osmond and Smythies could find no one in England willing to back their research proposal. As Abram Hoffer recalled, when Osmond presented the idea to the Maudsley Institute of Psychiatry, one of the premier research institutes in England at the time, the directors there "literally laughed at him," which only made the young doctor all the more determined to "get as far away from this sort of attitude as possible."[19] Fortunately for Osmond, his wife had spotted an ad in the *Lancet* for a job as the clinical director of a mental hospital in a far-off place called Weyburn, Saskatchewan. He jumped at the opportunity, arriving in the province in the fall of 1951.

DR. HOFFER, I PRESUME?

When Osmond landed in Saskatchewan, he met a young medical doctor and biochemist by the name of Abram Hoffer. Hoffer was born on November 11, 1917, in the small Jewish Hungarian farming community of Hoffer, Saskatchewan, a settlement founded by his Austrian immigrant parents. Firm believers in the value of education, they ensured that their children had all the curricular opportunities available to them. It was not long before Hoffer began to excel academically, developing a special interest in chemistry.

In 1934, he enrolled at the University of Saskatchewan's College of Agriculture, obtaining his BSA in 1938 and his MSA two years later. He recalled these early university years as interesting ones indeed, "analyzing wheat from around the province for its protein content" and writing his master's thesis on root rot, a fungus that attacked wheat.[20] He also remembered Professor Roger Manning, who sparked his interest in the adrenaline derivative, adrenochrome, and niacin, more popularly known as vitamin B3. However, medicine, specifically psychiatric research, was not at the top of his mind at this time. Although working on the farm did not appeal to Hoffer, he entered the field of agricultural chemistry intending to continue in a related career; his schooling, he noted, "represented a compromise between my wish and my father's desire that I remain on the farm."[21]

Following his graduation from the University of Saskatchewan, Hoffer headed south to the University of Minnesota, where he earned a PhD in cereal chemistry in 1944. While pursuing his PhD studies, Hoffer also worked as a chemist for Purity Flour Mills in Winnipeg, married his sweetheart Rose Miller, and had their first of three children. His curiosity about nutritional research and vitamins reached the point where a career in medicine seemed to be the next logical route. Thanks largely to the encouragement and support of his wife, family, and friends, he began his medical studies back at the University of Saskatchewan in 1945, culminating in 1949 with an MD from the University of Toronto and successful completion of the Canadian Medical Council exams. With these accomplishments in hand, Hoffer returned home to work as an intern at the City Hospital in Saskatoon.

When finally ushered into the Psychiatric Services Branch of the Department of Public Health by McKerracher in 1950, Hoffer was presented with the opportunity of a lifetime. Heading the PSB might seem odd since Hoffer was not a fully qualified psychiatrist, yet he received the posting under the conditions that he take up a position in the Munroe Wing at the Regina General Hospital and complete the four-year psychiatric training program. He was also required to carry out an extensive tour of research centres in both Canada and the United States. As Hoffer later explained, McKerracher started the research program with two goals in mind: "to obtain new information about the psychoses, the defectives and the mentally disturbed geriatric patients, and to introduce a critical attitude into the psychiatric thought of his department."[22]

Writing to McKerracher in April 1950, Hoffer noted that he had "been able to accomplish a moderate amount of research in the field of vitamin methodology and bread baking" and expressed the hope that he would be able to adapt these skills to psychiatry.[23] Hoffer quickly applied his background in biochemical research so that he could help to make McKerracher's vision a reality. In the months ahead, the Saskatchewan Committee on Schizophrenic Research was formed, with the young Hoffer acting as chair. It was in this climate that the research with psychedelic drugs began in earnest. In their first meeting in 1951, in McKerracher's office in Regina, it became apparent that Hoffer and Osmond were headed down the same path. In fact, with their profound interest in schizophrenia and the role that biochemistry might play in it, they were ideally suited for one another. Hoffer thought Osmond's research ideas worthy of further exploration. Osmond, operating out of his base in Weyburn, and Hoffer, from the Munroe Wing, formed a close working alliance, corresponding daily, swapping journal articles, and expanding on many of the initial hunches of Osmond and Smythies.

When they ultimately decided to concentrate on searching for the elusive M-substance, Hoffer and Osmond realized that the next step was to secure appropriate funding. In November 1951, they prepared a "Research Plan for the Investigation of the Biochemical and Clinical Importance of the Compounds Lying between Mescaline and Adrenaline and Their Relationship to Schizophrenia" and sent it to the DPH and, through it, the Department of Health and Welfare (DHW) in Ottawa. In their proposal, they laid out a detailed research program for extensive biochemical, physiological, and psychological testing with mescaline and other known hallucinogens. These studies would be complemented by examinations of the body fluids and tissues of schizophrenic patients. Osmond insisted that the

> one great difference...previous biochemical investigations of schizophrenia always faced was that the structure of the possible toxic substances was very uncertain. The body produces a huge number of chemicals any one of which might be responsible for schizophrenia. For the moment we are only concerned with a single question: "Is schizophrenia caused by a mescaline-like substance lying between adrenaline and mescaline?"[24]

Another important implication of their proposal was that, if their theory was correct and schizophrenia is indeed caused by an error in adrenaline metabolism, then new forms of treatment would likely follow. As Osmond stressed, "the current physical therapies for the functional psychoses are all based on theories for which there is very little supporting evidence."[25] His assertion contained an important truth. The dominant modes of treatment, such as ECT and insulin coma, produced sketchy results at best. In some people with schizophrenia, they brought about a temporary reprieve of symptoms. In others, the symptoms became worse as they experienced many unpleasant side effects. Hoffer and Osmond thought it worth looking into anti-adrenaline substances, such as ascorbic acid (vitamin C) and niacin (vitamin B₃), to see if they might shed any valuable clues. Hoffer, a specialist in vitamin research, knew of ascorbic acid's properties as an anti-oxidant that could cut down the conversion of adrenaline and of the successful use of niacin in cases of pellagra psychoses. This kind of treatment met two other key criteria: it was affordable and relatively safe in large doses. By incorporating this method, the two doctors saw the possibility of validating much of what they were saying before they actually isolated the so-called M-substance.

A proposal of this magnitude required a substantial amount of money to provide for the necessary staff (e.g., nurses and psychologists), volunteers, drugs, and equipment, but Osmond stood firm on the matter: "It certainly would be initiated for little more than the cost which the community pays for the average schizophrenic during his or her lifetime."[26] Things soon began to fall into place. Within the next year, an amazing chain of events led to one of the first and most important discoveries of their research. By early spring 1952, their proposal had been approved, and they would receive the $5,200 initially requested to get the project under way.[27] The money came in the form of a national public health grant through the DPH. More funding from both the provincial government and the federal government soon followed, as did generous donations of drugs and vitamins from pharmaceutical companies. Within a couple of years, their research also received financial assistance from outside foundations (e.g., the Rockefeller Foundation) that saw merit in it.

Hoffer and Osmond were by no means the only scientists interested in studying hallucinogens for their ability to serve as model psychoses. American psychiatrist Oscar Janiger recalled in a 1959 paper that the original idea for model psychosis "was inferred more than 100 years ago by Moreau, who attempted to correlate the effects of hashish with the manifestations of mental illness."[28] Because these hallucinogenic drugs, in effect, "mimicked madness," they came to be more frequently known by the term "psychotomimetic." From a strictly scientific point of view, little was known about them. Detailed accounts of substances such as mescaline began to appear in the late nineteenth century in the works of psycho-pharmacological pioneers such as Louis Lewin (e.g., *Phantastica*). Lewin, as Janiger wrote,

> brought peyote to the attention of the scientific world. He predicted its possible use in psychiatric research and was among the first to underscore its unique activity on the psyche. Several eminent physicians of this period, notably Emil Kraepelin, Weir Mitchell and Havelock Ellis, experimented with peyote, and its active ingredient, mescaline, and were greatly intrigued by its possibilities. Their interest provided a stimulus to many of the subsequent investigators who gave exhaustive but primarily descriptive accounts of the effects of mescaline on a number and variety of subjects. Lacking a unifying principle or guiding hypothesis, research on the hallucinogens became gradually attenuated.[29]

Long before Hoffer and Osmond began their own experiments with hallucinogens, there was much speculation about the effectiveness and reliability of model psychoses. Among experts in the field of schizophrenia, the debate led to confusing and often contradictory rhetoric. A central question was whether psychotomimetics merely produced a toxic condition and/or delirium, or actually produced a condition resembling functional psychoses such as schizophrenia. Elaborating on this conundrum, Janiger noted that

> many agents, one of the most common being alcohol,...produce toxic psychotic reactions when taken in the body. Closely related are those psychotic reactions produced when the body itself manufactures noxious agents as a consequence of disordered metabolism, as in specific liver, thyroid or kidney conditions. These are referred to as endogenous toxic psychoses, while the former are characterized as exogenous reactions. The distinction between the two modes of origin is poorly and arbitrarily delimited.... If the drug reaction is of the order of a functional psychosis, one without any clearly defined

toxic or any known organic cause, then we are indeed interested. Schizophrenia, in this category, is probably the most obscure and malignant of the mental afflictions—and the most widespread.[30]

Many experts charged that advocates of model psychosis were claiming that the illness and the psychotomimetic experience were identical. Critics reasoned that the differences between them were too substantial to ignore and that any valid connection between actual illness and model psychosis was therefore inappropriate. Hoffer and Osmond argued repeatedly that they were not demonstrating an exact replica of schizophrenia:

> A model is not a reproduction of an original. A model is required to clarify certain aspects of an original. This the psychotomimetic substances do reasonably well. There undoubtedly is more similarity between the mescaline experience and schizophrenia than between a simple deteriorated schizophrenic without hallucinations and delusions and a vividly excited catatonic schizophrenic with vivid visual and auditory hallucinations. There is less similarity between toxic confusional states and schizophrenia.[31]

When the two Saskatchewan psychiatrists began mapping out their studies, they had a difficult time deciding precisely where to begin. The challenge was formidable.[32] For the purposes of their research, the hallucinogenic substances had to demonstrate, as mescaline did earlier, pronounced similarities to schizophrenia. In normal volunteers, therefore, such substances had to "cause profound changes in perception, affect, thinking and behaviour, without clouding of consciousness, confusion or gross physiological disturbances."[33]

Prior to the announcement of funding, Hoffer was busy surveying the medical literature for information on commonly known hallucinogens, which at that time probably amounted to five (mescaline, hashish, ibogaine, harmine, and LSD). During this process, he made a vital observation about their chemical properties: each contained an indole nucleus.[34] This was an exciting finding for the two doctors because now "instead of any one of fifty thousand compounds [they] had it narrowed down to three."[35]

Was the hypothetical toxin in the schizophrenic body hallucinogenic? Was it an indole? Did it derive from adrenaline?

A curious observation had originally attracted Osmond and Smythies to adrenaline. An asthmatic historian, upon hearing a recording of Osmond's mescaline experience, made numerous comparisons between the psychotomimetic experience and his own repeated use of adrenaline.[36] This was later combined with an account by Regina anaesthetist Eric Asquith on how discoloured adrenaline used during the war had triggered psychotic effects in soldiers whom he had treated. Was adrenaline or something close to it a hallucinogen? The main problem was that Osmond and Hoffer were comparing hallucinogens derived from plants with something in the human body. Osmond remained adamant that "while it is true that one cannot argue directly from an assortment of exotic vegetable alkaloids to a hypothetical M-substance, it is folly not to concern oneself with these puzzling hallucinogens whose action sometimes resembles schizophrenia so closely that casuistry has been needed to avoid this unwelcome conclusion."[37]

The two partners proceeded to outline their research program at a meeting of medical associates, who eventually made up the Schizophrenia Committee, at the University of Saskatchewan in the spring of 1952. When they mentioned in passing the substance pink adrenaline, one of the people present, a young pharmacologist by the name of Duncan Hutcheon, drew everyone's attention to it:

> Hutcheon pointed out that "pink adrenaline" certainly contained among other things adrenochrome. In the exciting ten minute discussion which followed after Hutcheon drew the spatial formula of adrenochrome, it was shown that this substance was related chemically to every hallucinogen whose chemical composition has been determined. It is evident that we had stumbled upon a compound which has an indole nucleus in common with the hallucinogens, which is readily derived from adrenaline in the body, and which can be fitted into a logical scheme relating to stress. Under stress the quantity of adrenalin in the body will increase and this might be turned into adrenochrome in the schizophrenic individual.[38]

Hoffer and Osmond then set out to have adrenochrome synthesized and to carry out the necessary tests. They also began to explore more seriously the effects of all known hallucinogens, notably mescaline and the more revolutionary LSD-25.

LSD COMES TO SASKATCHEWAN

It was only in 1943 that the effects of the most powerful hallucinogenic compound, d-lysergic acid diethylamide (LSD-25), were brought to light by Swiss chemist Albert Hofmann. Working for the Sandoz Pharmaceutical Company in Basel, he had been investigating the properties of the rye fungus, ergot, as a circulatory stimulant, and in synthesizing ergot he had inadvertently come across LSD. While working in the lab, Hofmann accidentally absorbed an infinitesimal amount of the drug through his skin, and he soon lost touch with reality. Recounting this first experience with LSD-25 as an unbelievable transformation of his consciousness, he mentioned how he had witnessed "an uninterrupted stream of fantastic pictures, extraordinary shapes with intense, kaleidoscopic play of colors."[39] More controlled studies of the drug followed by Hofmann and another Sandoz researcher, W. A. Stoll, and with them some of the first scientific publications on the mysterious ergot derivative. That LSD bore many of the same features of other psychotomimetics became clear. In one instance, LSD might produce moments of profound beauty and significance, accompanied by spectacular auditory and visual images; in another instance, it could occasion horrific Kafkaesque hallucinations and feelings of extreme paranoia and terror. Like other psychotomimetics, LSD possessed incredible variability, making it difficult, if not impossible, to reach generalizations about it. This made it an ideal candidate for model psychosis research.

One predicament of the early research with LSD was the number of unknowns. Scores of tests had been carried out on animals, from spiders to Siamese fighting fish, but little was understood, even a decade after its discovery, about its effects on the psyche and its metabolism in the human body. What made LSD stand out from the rest of the psychotomimetics, as researcher after researcher soon discovered to their astonishment, was its unbelievable potency. Whereas someone would have to take the other hallucinogens in dosages of grams and/or milligrams to provoke a reaction, LSD could accomplish the same or even better results using mere micrograms (one millionth of a gram). Hofmann's experiences confirmed that "the novelty of LSD as opposed to mescaline was its high activity, lying in a different order of magnitude. The active dose of mescaline, 0.2 to 0.5 g, is comparable to 0.00002 to 0.0001 g of LSD; in other words, LSD is some 5,000 to 10,000 times more active than mescaline."[40]

That there existed a substance that could alter a normal person's consciousness and personality to such an extreme degree piqued the interest of many psychiatrists. By 1952, more and more scientists clamoured to use LSD, initially supplied free of charge by Sandoz, as a model psychosis. Hoffer and Osmond wondered whether the drug affected the adrenal system in any way and how an agent such as niacin might react to it. On April 2, 1952, Osmond wrote to Hoffer:

> Do we know where LSD acts anyway? And what it does? I rather think not. If it got fixed in the suprarenals it might act in relatively high concentration. I thought that we might take the blood and urine from people who had LSD and test them with Rheineck salts. If it does work by producing abnormal suprarenal functions then we would find M substance present.[41]

Osmond, aided by the young psychiatrist Ben Stefaniuk and Smythies, who immigrated from England to assist with the project for a year, initiated pilot studies cataloguing the effects of LSD in twenty normal volunteers. The study yielded nearly 500 pages of unpublished data summarizing the twenty volunteers' reactions to the 200 microgram (or gamma) dose of LSD that each had been administered. These reactions included physiological changes (e.g., temperature, hearing, and taste), problems in communication and thought processes (e.g., perception of time), changes in perceptions of reality (e.g., depersonalization or loss of ego), as well as mood alterations (e.g., depression, fear, and euphoria). The study also tried to gauge participants' performance on a variety of mental tests under the influence of the drug (e.g., Rorschach, ESP, and mathematical questions).

The doctors never equated the LSD experience with schizophrenia no matter how many similarities between them could be shown. As Hoffer and Osmond stated, the former

> is a single episode of short duration occurring in a healthy volunteer, who usually knows that he may have an experience of this sort and how long it is likely to last. Most experiments are conducted in a friendly atmosphere and the volunteer is encouraged to report his experiences to others who are often keenly interested in what is happening to him. Contrasted with this the schizophrenic patient may have been ill for months or even years, has no rational explanation for his experiences; the friends who remain to him are not interested in his experience...and may even refuse to believe what he tells them. In addition he cannot tell whether he will endure this for the rest of his life or even for eternity.[42]

An intriguing trait of this early testing was that it gave the normal person who had taken LSD greater empathy for the schizophrenic patient. The psychotomimetic experience forced normal people to view mental illness from a whole new perspective. As one of the Saskatchewan volunteers recalled about his LSD adventure, "not once have I since looked at the patient without experiencing a strange pang, something I feel sure is akin to understanding and to brotherhood, more than anything I have ever before felt for anyone, outside the usual circle of family and close friends."[43]

Since his first encounter with mescaline, Osmond saw that psychotomimetics offered a perfect training tool to educate psychiatrists, nurses, and other mental health professionals about schizophrenia. If these drugs provided the healthy individual with a facsimile of the naturally occurring illness, as Osmond believed, then one might approach the patient in an entirely different manner. As he viewed it, "our business [was] to help when we can...or at least harm the sick person as little as possible."[44] Furthermore, society ought to

> listen seriously to mad people, for, in phrases which are usually clumsy, ill-constructed, and even banal, they try to tell us of voyages

of the human soul…[to] another world than this; but mostly we don't hear because we are talking at them to reassure them that they are mistaken.…The least we can do for these far voyagers…is to hear them courteously…and try to do them no harm.[45]

Many in the field interpreted this assessment of the drugs as potential educational tools as utterly ridiculous, but some were intrigued by the idea and thought that it deserved closer examination. Psychologist Peter McKellar admitted that "not all that happens in a model psychosis experiment is relevant to the study of psychosis," but in areas where overlap did occur these experiments "permit[ted] detailed observational, introspective, and psychometric studies of a kind not otherwise possible."[46]

While the LSD tests were being conducted at Weyburn, Hoffer and another associate, research psychologist Neil Agnew, began tests of their own with volunteers at the Munroe Wing of the Regina General Hospital. These experiments, like those in Weyburn, were designed to find out more about model psychosis as induced by LSD in normal volunteers. They were chosen according to strict criteria. They had to have a health checkup as well as an assessment of their medical background and level of education. Reports on LSD research had indicated that people with extreme anxiety, depression, and tension usually had prolonged and adverse reactions to the drug, as did people with past histories of liver disease; for patients with schizophrenia who were administered LSD, the drug either exacerbated the psychotic experience or had no effect at all.[47] As for the preference that subjects be of average or above-average intelligence, the researchers thought that they could get more precise accounts of the LSD experience. It became obvious to Hoffer and Osmond that "a Ph.D. physicist will not react the same way as a ward aid. Not only is the experience different but the ability to describe what has happened is different."[48]

The two studies—those of Osmond-Stefaniuk and Hoffer-Agnew—reached remarkably similar findings. Chief among them was the great variability in the LSD experiences. As Hoffer and Agnew remarked, "the variability of response to LSD, whether the dosage be small or large, has made it exceedingly difficult to make generalizations."[49] This variability could have resulted from any number of factors, such as the roles of investigators in the process and their relationships with subjects, the environment in which the experiment occurred, and the personality of the individual taking the drug. Researchers were intrigued by just how much an experience could be altered when such variables were considered. By paying closer attention to the structure of the experiment, or the "set and setting" as it later became known, researchers could better determine its outcome. These factors became crucial in the use of LSD for treating illnesses such as alcoholism and behavioural problems (see Chapters 3 and 4).

There were also several important differences in the early Saskatchewan LSD projects. The study of Hoffer and Agnew was unique in that it incorporated niacin (or nicotinic acid) to see if and how it might modify the LSD experience.[50] Early testing at the Munroe Wing resulted in one of the earliest scientific publications in North America on the effects of LSD in humans. Based upon previous observations that LSD somehow interfered with and/or inhibited carbohydrate metabolism, Hoffer and Agnew theorized that a compound that could facilitate such metabolism would counteract the LSD effects. Succinic acid provided one possibility. It had been used to negate the effects of mescaline intoxication. Niacin, it seemed, did likewise. At the beginning of this study, Hoffer pointed out to Osmond that

it is quite likely that we are even closer now to an experimental schizophrenia than we were with lysergic acid alone. You will remember that the lysergic acid molecule contains a nicotinic acid group so that these can be considered competitive antagonists. The administration of LSD alone appears to block the pyridine enzymes and produces at least three types of disturbances.[51]

Hoffer went on to list these "disturbances" as changes in perception, changes in affect, and changes in thinking, noting that niacin had reduced most of them, with the exception of affect. Hoffer and Agnew later admitted that they could not determine "why the affective disturbances were more resistant to modification."[52]

The setup of the Agnew-Hoffer study was as follows. A group of ten male volunteers was randomly divided into two subgroups, one group treated with nicotinic acid at the height of the LSD experience, the other group pretreated with three grams of nicotinic acid three days before the LSD experience. Hoffer and Agnew hoped that in this way "nicotinic acid could have a preventive action, that is modifying the effects of LSD when administered prior to it, and also that it could have a treatment effect, that is modifying the effects of LSD when given during the model psychotic experience."[53] The results of both groups confirmed their hypothesis that niacin would modify the LSD experience. However, the modification did not appear to result from any direct biological antagonism, as they had initially thought: "We should expect the modification of the LSD experience to be at least as pronounced in those subjects receiving nicotinic acid prior to the LSD which was not the case. It may be that it is acting in an indirect way, for instance, by interaction with one or more compounds produced by the introduction of LSD into the organism."[54] Data collected from the early LSD projects shed more light on the relationship between model psychoses and schizophrenia and allowed researchers to predict more accurately what to expect in certain situations in which the drug was used. They also provided a rich foundation for further inquiry. The Saskatchewan schizophrenia research team was joined in this pursuit by another independent-minded researcher.

DUNCAN BLEWETT: PSYCHEDELIC TRICKSTER

Born on October 28, 1920, in Edmonton, Duncan Basset Blewett spent much of his childhood and teen years in various locales in Canada and the United States because of his father's fluctuating employment, which included stints as an employee with the Tennessee Valley Authority in Muscle Shoals, Alabama; an entrepreneur in Grand Prairie, Alberta; a farmer in eastern Ontario; and an Indian agent for the Canadian government.

In the spring of 1941, Blewett left high school to enlist in the army as an infantryman on the front line in Europe.[55] Joining the 1st Battalion Kent Scottish Regiment, based in Chatham, Ontario, Blewett served as an artillery gunman, gradually winding his way through France, Italy, Holland, and Germany. In one of his tours of duty, he was part of a small delegation to Westende, Belgium, forcing the surrender of German forces. Like

Osmond's time on the front line, Blewett's combat experiences exacted a heavy physical and emotional toll, and Blewett was not inclined to discuss them with others. From what can be gleaned from some of his wartime correspondence, the degradation, misery, and poverty that he witnessed while in service affected him greatly. Even at this early stage in his life, his concern for the human condition and its betterment was apparent. Commenting on what he had seen in a letter to his first wife, Irene, Blewett revealed glimpses of his budding psychological and philosophical leanings, empathic nature, and characteristic rebelliousness:

> As soon as fascism and its associated evils are buried away in a deep grave people will be able to live like humans again and not like a lot of savage dogs fighting over which one is the toughest....Why can't people understand how much cooperation among themselves is going to mean?...All social progress is halted when the fight begins and security becomes as extinct as the dodo....Poverty is a sin not of the poverty stricken but of the society which permits it to flourish....Snobbery of race or color or creed is a sin...[as are] greediness and indifference to the sufferings of another. They all have no place in the world.[56]

After his demobilization in 1945, Blewett returned to Canadian soil. Taking advantage of the educational funding available to veterans, he registered as a provisional student at the University of British Columbia, where he went on to major in psychology and philosophy and eventually complete his BA and MA (1947 and 1950, respectively), penning his thesis on a "Comparison of Certain Measures of Interest in Armed Service Trade Selection." This introduction to the field inspired Blewett to pursue a career in psychology, and following his graduation from UBC he enrolled at the University of London, England's, Maudsley Institute of Psychiatry to undertake his PhD studies under the tutelage of famous behavioural psychologist and personality theorist Hans J. Eysenck. Eysenck was one of the earliest advocates of what became known as the factor analytic approach, a psychometric/statistical method employed in psychology, education, and other fields. Upon the successful completion of his doctoral studies in 1953, Blewett received an offer to work at the University of Illinois at Urbana-Champaign with another leading proponent of the factor analytic approach, psychologist Raymond Cattel, who went on to psychological fame for his 16 Personality Traits. It is an irony that Blewett studied under some of the foremost empirical, behavioural psychologists of the twentieth century, whose efforts were geared toward making psychology more scientific through objective measurement of human personality traits, only to break rank and devote his energies to advancing a more humanistic, spiritual path in the profession. The transformation occurred as a direct result of his move to Saskatchewan in 1954 to become involved in the schizophrenia research of Hoffer and Osmond and, not long after that, the development of the psychedelic treatment approach.

When Blewett arrived in the province to take up duties as the PSB's supervising psychologist, a new position that entailed coordination and maintenance of clinical and research services in the provincial mental hospitals, as well as in the psychiatric wards of general hospitals and outpatient clinics, he met with Hoffer and Osmond and quickly immersed himself in the research program. As part of his duties on the Saskatchewan Committee on Schizophrenia Research, Blewett began looking into issues and gaps in psychological testing for schizophrenia, spending time in the Saskatchewan Hospital–Weyburn and applying his factor analytic skills, working alongside Ben Stefaniuk, on the development of what became known as the Weyburn Assessment Scale (WAS), an instrument designed to assess the degrees of illness in hospitalized patients with schizophrenia, measure behavioural change, and objectively assess the success or failure of therapeutic procedures. Around this time, Blewett also put his name forward to chair some of the model psychosis studies with hallucinogens, attempting to lay out criteria and definitions of the model that might help to differentiate between psychotomimetic drugs and other substances.

Blewett brought a trickster element to Saskatchewan's psychedelic research, eventually developing a reputation for blurring the boundaries of acceptable scientific practices. In mythology and folklore, the trickster is commonly seen breaking rules, cultural values, and societal norms and tends toward mischief and both intelligence and foolishness. In seeking new experiences and ideas, the trickster sometimes acts impulsively and irresponsibly, causing his actions to backfire, yet his actions also bring new knowledge and understanding. In his psychedelic research exploits, Blewett exhibited such qualities. As he became more involved in the drug research, his novel approach often teetered between science and spirituality and therefore became the subject of intense scrutiny both within and outside the scientific community. But his approach led to discoveries that might not have occurred otherwise, and he is thus important in Saskatchewan's psychedelic story. In his subsequent work with hallucinogenic drugs in treatment settings and what came to be known as psychedelic therapy, Blewett, alongside Hoffer and Osmond, would take centre stage in turning Saskatchewan into a pioneering force.

THE ADRENOCHROME HYPOTHESIS UNFOLDS

By the first part of 1953, Hoffer and Osmond and the rest of the Saskatchewan schizophrenia research team were simultaneously pursuing several different research ends, including the adrenochrome hypothesis, still in its infancy. Had they located the M-substance? Was it a hallucinogen capable of provoking the same psychological changes as mescaline and LSD? Weighing all of the evidence accumulated to that point, it seemed plausible to reason "that since microgram quantities of LSD could produce such a vivid psychological change it was not improbable that similar quantities of a chemical could be produced within the body and produce schizophrenia."[57]

Adrenochrome had been looked at as far back as 1937 (Green and Richter), but it had never been tied to any comprehensive theory of schizophrenia, nor had it been seriously tested to see how it affected the human body. To Hoffer and Osmond, there remained only one way to find answers to their questions—testing the compound on humans. So, with synthesized adrenochrome in hand, they and their wives volunteered for the first experiments ever done with adrenochrome on human subjects. These pilot tests were the first to show that adrenochrome could properly be classified as a psychotomimetic that might occur in the human body, thereby marking a milestone, albeit a controversial one, in psychiatric history.

Hoffer and Osmond began by administering small doses of adrenochrome to themselves to verify if anything would happen. They were not
disappointed. Osmond wrote of his first experience in his journal:

> After the purple red liquid was injected into my right forearm I had
> a good deal of pain. I did not expect that we would get any results
> from a preliminary trial and so was not, as far as I can judge, in a
> state of heightened expectancy. The fact that my blood pressure did
> not rise suggests that I was not unduly tense. After about 10 min-
> utes, while I was lying on the couch looking up at the ceiling, I found
> that it had changed colour. It seemed that the lighting had become
> brighter. I asked Abe and Neil [Agnew] if they had noticed anything
> but they had not. I looked across the room and it seemed to have
> changed in some not easily definable way. I wondered if I could have
> suggested things to myself. I closed my eyes and a brightly coloured
> pattern of dots appeared. The colours were not as brilliant as those
> which I have seen under mescal, but were of the same type. The pat-
> terns of dots gradually resolved themselves into fish-like shapes. I
> felt that I was at the bottom of the sea or in an aquarium among a
> shoal of brilliant fishes. At one moment I concluded that I was a sea
> anemone in this pool. Abe and Neil kept pestering me to tell them
> what was happening, which annoyed me. They brought me a Van
> Gogh self-portrait to look at. I have never seen a picture so plas-
> tic and alive. Van Gogh gazed at me from the paper, crop-headed,
> with hurt, mad eyes and seemed to be three dimensional. I felt that
> I could stroke the cloth of his coat and that he might turn around
> in his frame. Neil showed me some Rorschach cards. Their texture,
> their base relief appearance, and the strange and amusing shapes
> which I had never before seen in the cards were extra-ordinary.[58]

Agnew and Hoffer noticed a severe transformation of Osmond's regular
behaviour during his second adrenochrome trial "marked by strong preoccupation with inanimate objects, by a marked refusal to communicate with
us and by a strong resistance to our requests."[59] The research team's subsequent experiments with adrenochrome, and soon after another of the

oxidized derivatives of adrenaline, adrenolutin, appeared to exhibit many of the hallmark characteristics of both schizophrenia and various model psychoses. The researchers were able to relate adrenochrome to other psychotomimetics on a number of levels, pointing to many of the same changes in perception, affect, and thinking. However, Hoffer, Osmond, and Smythies cautioned other experimenters that adrenochrome "is more insidious than [mescaline and LSD], its effects last longer and possibly, in consequence of this, its administration is accompanied by loss of insight."[60]

Apart from the psychological experiments with the compound, other tests had shown that adrenochrome caused abnormalities on the electroencephalograph (or EEG) machine (in normals and epileptics), that it was similar to LSD in that it inhibited intermediary carbohydrate metabolism, and that it lowered body temperature, an immediate sign that it could "cross the blood brain barrier in contrast to adrenaline which is not able to do so."[61] These findings seemed to bolster support for a biochemical hypothesis for schizophrenia but did not deter pessimistic critics, who said that the work was non-sensical and no more than a product of the researchers' fanciful thinking. When Hoffer and Osmond began the adrenochrome research, they already had two strikes against them. The presence of adrenochrome in the human body had never been proven, and few scientists believed that it ever could be. Others contested that the body could produce enough adrenaline in the first place to form adrenochrome.

One's response to the findings depended, of course, on one's interpretation of schizophrenia. If a person tended to believe that the disease likely has a biochemical root, then the research being done in Saskatchewan offered great hope. Another example of the early success of the research was that it spurred a resurgence of research on other possible neurotoxins, such as taraxein and serotonin.[62] If a person viewed the disease from an opposite standpoint, however, say from a psychosocial or psychobiological perspective, then the research was a threat to the prevailing teachings on mental illness. As one of the earliest theorists to emphasize the importance of "seeing resistance by scientists themselves as a constant phenomenon with specifiable cultural and social sources," Barber called attention to how "substantive concepts and theories held by scientists at any given time become a source of resistance to new ideas."[63]

Intriguingly, some of the harshest detractions of Hoffer and Osmond's discoveries came from other biological psychiatric researchers, many of whom were equally committed to a medical model of schizophrenia. But it was not the biochemical theory so much as the scientific methodology by which Hoffer and Osmond arrived at their discoveries that many biological competitors found problematic.[64] One of the staunchest critics, Seymour Kety of the NIMH, noted years later that the two merely "prepared some adrenochrome, gave it to each other in a rather uncontrolled experiment and convinced themselves that it did produce hallucinations."[65] This was reason enough to regard the adrenochrome hypothesis of schizophrenia as "only one of many simplistic and premature experiments that sober research was unable to support."[66]

But as Nicolas Langlitz and others have pointed out, scientific self-experimentation was once regarded as respectable and distinguishing: "In a competitive field, both the self-experimenter's heroism and the fact that he had experienced certain phenomena firsthand with which his colleagues were personally unfamiliar served as sources of social distinction."[67] With the emphasis on objectivity in scientific research, however, what constituted appropriate scientific practice and behaviour changed considerably:

> Objectivity was born out of a deep-seated distrust, even fear, of the subjective and its inclination to defile an impartial perspective on the world.... In the case of hallucinogen research in particular, the ardor with which a few vocal individuals from the previous generation of researchers working in this field had come to advocate drug use had raised grave concerns whether drug experiences did not corrupt the dispassionate outlook expected from scientists.[68]

Despite early criticisms of the Saskatchewan work as simplistic, Hoffer and Osmond never asserted that what they were doing was simple. That they had to "search for a fairly unstable compound in a concentration of about one part in five million" attested to the complexity of their task.[69] And "if one agrees that M substance could exist then there is no reason why it should not produce a wide variety of clinical pictures."[70]

Only two years into the research program, the findings of Hoffer and Osmond and the rest of their team had much of the world's attention focused on Saskatchewan. The sudden interest played out in a wide-ranging debate within the psychiatric community. Moreover, the work started to attract public attention. In the summer of 1953, Canada's national magazine, *Maclean's*, sent one of its editors, Sidney Katz, to do a feature story on what was happening in the province's mental hospitals. Katz volunteered for an experiment at the Weyburn hospital, which provided the material for one of the first articles of its kind in the lay press. As the title suggests, "My 12 Hours as a Madman" recounted his LSD experience in vivid detail, replete with photographs of Katz and drawings of what he witnessed while under the powerful drug. Katz began his article using fantastic terms:

> On the morning of Thursday, June 18, 1953, I swallowed a drug which, for twelve unforgettable hours, turned me into a madman....I experienced the torments of hell and the ecstasies of heaven. I will never be able to describe fully what happened to me during my excursion into madness. There are no words in the English language designed to convey the sensations I felt or the visions, illusions, hallucinations, colors, patterns and dimensions which my disordered mind revealed. I saw the faces of familiar friends turn into fleshless skulls and the heads of menacing witches, pigs and weasels. The gaily patterned carpet at my feet was transformed into a fabulous heaving mass of living matter, part vegetable, part animal. An ordinary sketch of a woman's head and shoulders suddenly sprang to life. She moved her head from side to side, eyeing me critically, changing back and forth from woman into man. Her hair and her neckpiece became the nest of a thousand famished serpents who leaped out to devour me. The texture of my skin changed several times. After handling a painted card I could feel my body suffocating for want of air because my skin had turned to enamel. As I patted a black dog, my arm grew heavy and sprouted a thick coat of glossy black fur.[71]

Katz departed Weyburn with a sense of urgency:

Half of all our hospital beds are filled with mental patients. Half of these again suffer from schizophrenia. We don't know yet the cause of this disease but there is good reason to suspect that it is due to an error in body chemistry. A few specks of a drug changed me, a normal person, into a madman. Is it therefore...not entirely possible that the schizophrenic is a person whose body constantly manufactures minute particles of a similarly poisonous substance?[72]

The report created a storm of controversy. To many outside observers, it was a well-orchestrated attempt by Hoffer and Osmond to sensationalize the topic and arouse a positive response to their theories and research. Whether this was the case or not is debatable. By this point, Hoffer and Osmond had already made great strides and did not need to resort to such tactics to advance their findings. Their research was producing some incredible results and becoming accepted in respected scientific journals, encouraging other researchers in the field to pursue a similar course. As much as can be gleaned from the massive collection of personal notes that the two left behind, such grandstanding and proselytizing were never their intention. In commenting on the pathetic situation in Weyburn, Osmond stated to Hoffer early on that "the government must understand that as doctors we don't want publicity of a favorable or scandalous sort....We just want something done."[73] To this end, they were instrumental in educating both the professionals and the public through lectures and visits to the mental hospitals.

By 1954, the work of Osmond and Hoffer had come a long way in a short time with a modest budget and limited resources. This was an impressive feat. In his new position as superintendent of the Saskatchewan Hospital–Weyburn, Osmond had begun implementing extensive changes in patient care.[74] He had managed to imbue the staff with a new spirit and transform the hospital into an efficiently run unit that could be used for teaching, training, and research. There remained much to be done, but many observers were impressed and even taken aback by the speed with which improvements had been made in the hospital's conditions. That Osmond was able to do this says a lot about his skill not only as a clinician and researcher but also as an administrator. As Hoffer confessed to his friend and partner in December 1953, when Griff [McKerracher] told me he was bringing a Clinical Director from England we were all hopeful but cautious. Not even in our wildest fantasies did we think you would become a brilliant scientist in a most difficult field, a brilliant administrator in the most miserable hospital in Canada, [or]...excel in both areas where other men had failed in either one alone.[75]

Hoffer soon found himself occupied with other responsibilities, too, accepting professorial duties at the newly opened Department of Psychiatry at the University of Saskatchewan's College of Medicine in Saskatoon.

The research of the two psychiatrists was firmly established by this time in the province and well on its way to becoming increasingly known abroad, yet controversy regularly followed on the heels of their findings. Many within the mainstream psychiatric establishment were incensed at what was happening in Saskatchewan. In their view, not only was the doctors' research moving too fast and in too many directions, but also Katz's LSD article, the first report on LSD to appear in the lay press, was a source of scandal. Not everyone, of course, viewed the research in this light. Others noted the positive turnaround in Saskatchewan's mental hospitals. The Rockefeller Foundation apparently saw enough merit in what the Hoffer-Osmond team was doing that it granted their request for nearly $300,000 in research funding.

It remained to be seen whether the theories could be verified. At the least, the findings paved the way to an alternative understanding of schizophrenia and helped to relaunch the biochemical debate during a period when others had sounded the death knell for such theories. Hoffer and Osmond accomplished this by using an interdisciplinary approach to studying the disease, an approach that drew some very talented and imaginative professionals in the mental health field to Saskatchewan. The work also led to a new course in treating the disease in its incipient stages, at a time when the market was inundated with the new tranquilizing drugs (e.g., phenothiazines) advocated by many as a revolutionary means of treating mental illness.

CHAPTER 2

PSYCHIATRIC PARADIGM CLASH

From the mid-1950s on, the studies of Humphry Osmond and Abram Hoffer on a biochemical understanding and treatment of schizophrenia and other mental disorders continued to generate an overwhelming response, much of it negative, in scientific circles outside Saskatchewan, and as their findings reached a wider audience the volume of criticism grew.

During the next and final stage of their work on schizophrenia in the province (1955–61), Hoffer and Osmond spent much time refining their earlier experiments and carrying out new ones. In expanding on the adrenochrome hypothesis, they added another toxic derivative of adrenaline, adrenolutin, to their list of potential M-substances, highlighting its psychotomimetic traits, as they had done with adrenochrome. Much of the psychiatric profession downplayed these discoveries as well as other statements about niacin as a viable treatment for schizophrenia. Widespread disbelief in their research findings eventually led Hoffer, Osmond, and some of their associates to produce the first pure samples of adrenochrome and adrenolutin in the lab in the late 1950s, a feat that finally permitted other scientists to test and confirm their claims. In the last years of their studies, Hoffer and Osmond also created two new diagnostic tests (the Hoffer-Osmond Diagnostic [HOD] Test and mauve factor test), which, in their opinion, made the diagnosis, prognosis, and treatment of mental illness much simpler. They also added important information on how LSD works in the human body by pointing to a strong connection between its reaction and that of the various adrenaline metabolites (e.g., LSD causes a rise in adrenochrome levels).

Despite these apparent successes, orthodox psychiatrists remained largely unconvinced and unimpressed. Their refusal to even consider what came to be referred to more regularly as the "adrenaline metabolite theory" should perhaps come as no surprise. As shown throughout history, resistance has frequently been the typical response to any radically new scientific discovery. Rarely have new theories in science been easily accepted, especially if they have gone against an ingrained tradition.[1] Nonetheless,

> good scientists spend all their time betting their lives, bit by bit, on one personal belief after another. The moment discovery is claimed, the lonely belief, now made public, and the evidence produced in its favor, evoke a response among scientists which is another belief, a public belief, that can range over all grades of acceptance or rejection. Whether any particular discovery is recognized and developed further, or is discouraged and perhaps even smothered at birth, will depend on the kind of belief or disbelief which it evokes among scientific opinion.[2]

By positioning themselves in direct opposition to contemporary psychiatric wisdom, therefore, Hoffer and Osmond more or less ensured that their theories would encounter resistance.

A principal revolutionary quality of their hypotheses on schizophrenia, then, was that they marked a decisive split with accepted views. Hoffer and Osmond constantly argued that psychiatry, by not being held up to the same standard as most other scientific disciplines, possessed an inherent weakness—hence their insistence on application of the "scientific method." As a rule by which most scientists operated, this method required that they produce hypotheses that could be subjected to rigorous testing. According to Osmond and Hoffer, this was severely lacking in most psychiatric research of the day. The overwhelming control of psychoanalysis and its advocates in North America hindered the profession from being recognized by other medical experts as a true science. This was mainly because "the doctrines of Freud were interpreted as immutable principles rather than as hypotheses to be tested and discarded if faulty."[3] Another well-known Saskatchewan psychiatrist, Colin Smith, offered a like-minded estimation of his field when, in 1960, he noted that "the dogmas of psychoanalysis are perhaps to be compared to the theories of physics current before Galileo, thoroughly rationalistic and grounded in insufficient attention to experimentation."[4] Hoffer and Osmond gave a similar critique of the empiricists, who, though attempting to remedy the scientific deficiencies of psychiatry with a ceaseless collection of facts, remained largely guilty of the same offence as the psychoanalytical theorists. Many empirical procedures, too, suffered from the absence of working hypotheses. The adrenaline-based hypotheses coming out of Saskatchewan claimed to be scientifically based, and symbolized a new paradigm or style of scientific research on schizophrenia in particular and mental illness in general.

By the mid-1950s, the adrenochrome work stood apart as the first specific biochemical hypothesis for schizophrenia, leaving other psychiatric researchers to either confirm or refute the findings. Experimental psychologist Heinrich Kluver (University of Chicago—Culver Hall), who had encouraged both Hoffer and Osmond on separate occasions to look more intently at the mescaline phenomenon, greatly admired the Saskatchewan research, stating that it had made adrenochrome "one of the most interesting [compounds] in the world."[5] Accusations of fraud came from those still wont to criticize Hoffer and Osmond for failing to substantiate their claims.

One of the biggest problems was the instability of adrenochrome. As they later explained,

> we lost our source of adrenochrome in 1953. At first we thought this merely a temporary setback because we had many kind offers to supply us with it. We soon discovered that the generosity of our benefactors was not matched by an ability to make catechol indoles.[6] Batch after batch of adrenochrome either started to deteriorate while in transit or arrived as a mess of sticky, black amorphous melanin compounds which were too insoluble to be tested for psychotomimetic

properties. The deteriorating samples could be used a few times but were unsuitable for a series of cases.[7]

This made it next to impossible to corroborate the earlier finding of Hoffer, Osmond, and Smythies that adrenochrome was a psychotomimetic that might appear in the human body. Following their claims, numerous unpublished reports disputed the adrenochrome hypothesis, but it was hard to determine if these "counterexperiments" used real adrenochrome or not. Questions remained about the purity of the substance used, the means of administration (e.g., intravenous or inhaled), and the types of subjects on whom it was tested.[8] As Hoffer and Osmond contended, "the early workers used adrenochrome semicarbazide, or poor preparations of commercial adrenochrome that contained large quantities of many impurities including some of the silver used to catalyze the oxidation of adrenaline to adrenochrome. The commercial preparations were very unstable because of these impurities."[9]

The two carried on, however, and in the absence of pure adrenochrome they began to study samples of an equally promising candidate for the M-substance, adrenolutin. It was another of the oxidized derivatives of adrenaline and a close relative of adrenochrome. It "consisted of smooth running, shining golden crystals which had a greenish sheen when looked at askew. They dissolved easily in water and were very stable in solution. Indeed, this could be boiled without decomposing. We felt this would be much easier to work with than fugitive adrenochrome."[10]

Once again using themselves as test subjects, the two sought to find out if adrenolutin had psychological effects comparable to those of adrenochrome, which it did, though its effects were not nearly as pronounced as those of LSD or mescaline (e.g., perceptual anomalies), as Hoffer and Osmond had hoped they would be. The effects of adrenolutin also proved to be much more subtle and prolonged than those associated with adrenochrome. The essential point was that adrenolutin was shown to be a psychotomimetic. As usual, the self-experiments got Osmond and Hoffer into trouble, leaving them wide open to criticism that their pilot tests were wholly subjective and therefore lacked the proof needed to give adrenolutin any serious attention. In an attempt to satisfy their detractors, the two doctors later performed a series of double-blind tests with adrenolutin using university students as paid volunteers.[11] They persistently defended their earlier actions, however, insisting that "no one ever told [them] how one could carry out preliminary experiments any other way" and that "double-blind experiments on new drugs," as far as they were concerned, "would be worse than useless since one would not know what activity to look for."[12] There were similar accusations regarding many of the other Saskatchewan psychotomimetic drug tests, for example when Osmond, using himself as a guinea pig, wrote the first medical paper on oliloqui, the ancient drug used by the Aztec Indians.[13]

With adrenolutin, Hoffer and Osmond encountered the same road-block as with adrenochrome. After they used up their first supply of the drug, no pure adrenolutin could be found, and none could be synthesized. This situation changed only when Saskatchewan researchers discovered a way to do so in the late 1950s. It was difficult to say with certainty whether Hoffer and Osmond had located the M-substance, and even they were hesitant to make too hasty a conclusion on the matter. In the 1950s, there were still many uncertainties about the metabolism of adrenaline in the human body, with all of the intricate enzymatic processes involved; and even after thousands of papers had been written on the hormone many questions remained unanswered.

Despite the progress of the Saskatchewan research, both Hoffer and Osmond realized that their scientific hunches had limitations. Osmond stated that there was "no reason why derivatives of serotonin, acetylcholine and dopa should not be psychosis-mimicking."[14] Even Hoffer, with his knowledge of biochemistry, admitted ignorance:

> The hypothesis that epinephrine metabolites have some relevance in the production of schizophrenia does not indicate that they are the sole cause of the disease. No metabolic abnormality per se can be more than a mediator. The final clinical disorganization must be expressed in terms of personality reaction to disturbances of internal and then external function. According to this hypothesis, these metabolites are necessary to produce the disease, but this hypothesis does not offer a complete explanation. In order to show

a relationship between epinephrine metabolites and schizophrenia, it is essential to know what the metabolites are, whether they have psychological effects, whether substrate from which they may be formed is present in the body, and finally...that enzyme systems are present which can effect the transformation.[15]

The elusive M-substance had not yet been isolated; however, some positive developments lent further credence to the Saskatchewan research. In 1955, a U.S.-based research team led by psychiatrist Robert Heath isolated a toxic protein in schizophrenic blood, known as ceruloplasmin, and from it another substance that they called taraxein. Taraxein was later shown to have hallucinogenic qualities in both human and animal testing.[16] The Saskatchewan studies began to lean more heavily on the possibility that adrenochrome somehow bound itself within red blood cells (or erythrocytes), which might have given some clue about why the percentage of adrenaline in serum was so low. Hoffer hypothesized that adrenaline metabolism was therefore primarily an intracellular process.

As for the LSD studies, by the late 1950s the amount of psychological data on the drug had, as the Saskatchewan research team conceded, "far outdistanced the understanding of the meaning of the biochemical and physiological responses to taking it."[17] Further complicating the situation was the fact that these two categories, the psychological and the biochemical/physiological, had yet to be connected "in some comprehensible manner."[18] It was well established by the work of a number of others in the field, and accepted by most, that LSD interfered with the transmission of stimuli or messages along the brain's synapses. Exactly what caused this interference was still the subject of intense scrutiny. One mystery was that LSD was excreted from the body long before the psychological symptoms disappeared. Researchers thus began to hypothesize that the drug somehow antagonized enzyme metabolism in the body, particularly in the brain, and set off a chain reaction that inevitably resulted in severe psychological transformations. Although there was some agreement on this point among researchers conducting biochemical studies, they often disagreed on the specific enzymic process involved.

The work of Osmond and Hoffer, of course, concentrated on adrenaline metabolism. Two main theories that ran counter to the adrenaline-related ones were that LSD was a powerful inhibitor of serotonin, a mildly psychotomimetic neurohormone derived from the amino acid tryptophane, and that LSD raised the level of acetylcholine, another neurohormone. These catalytic reactions were believed to be the primary factors responsible for psychological manifestations induced by LSD. Although Hoffer and Osmond did not deny the importance of these findings, they did argue against their primacy in the LSD reaction. One finding of the Saskatchewan experiments was that Brom-LSD, another ergot derivative, did almost the same things to serotonin and acetylcholine as LSD. Yet Brom-LSD was not psychotomimetic and therefore did not produce psychological changes when given to people. This suggested that the main reaction lay elsewhere. In focusing on LSD's influence on adrenaline metabolism, the Saskatchewan studies were about to reveal some fairly startling evidence that inevitably had significant ramifications for the research on schizophrenia and other mental illnesses, and the therapy for alcoholism.

Hoffer and Osmond made considerable headway in their search for the M-substance. Their research had proven adrenochrome and adrenolutin to be psychotomimetic compounds.[19] This was confirmed by other scientists in a number of animal and human experiments, first by Americans and then by Germans.[20] Perhaps the most convincing studies were those completed by Stanislav Grof and his Czech colleagues, who verified, in 1961, adrenochrome's psychotomimetic qualities using double-blind tests on human subjects.[21] In 1957, Saskatchewan researchers also became the first to produce pure and stable samples of these two substances, some of which they distributed to outside investigators to test for themselves. The noisiest critics of the Saskatchewan research, such as Seymour Kety, Julius Axelrod, and John Benjamin, remained completely unconvinced, raising once again the fact that Hoffer and Osmond had not isolated these toxins in the human body nor crystallized them from living tissue.

The two psychiatrists were never intimidated by criticism that their findings were "far from conclusive" or blown out of proportion. On the contrary, they welcomed such fault-finding. Yet some of the most vicious criticism of their biochemical theories over the years came from "propagandists" who lashed out at their theories simply because, if true, they would have exposed many inadequacies of orthodox theories on schizophrenia based upon psychoanalysis. Hoffer and Osmond thought that scientists, in the words of renowned renaissance scholar Francis Bacon, had a "duty to risk being wrong."[22] They presumed that a hypothesis had to meet the following criteria:

> First: It must account both inclusively and economically for what is already known: an hypothesis which fails to do this would be automatically disqualified. Second: It must do this better than any previous hypothesis. Third: It must be testable in a way which will readily lead to its refutation should it be false, using methods available to the science under scrutiny. An hypothesis which depends for testing upon methods or principles not yet developed, while not absolutely inadmissible, would be less useful than one which can be tested immediately.[23]

What distinguished Hoffer and Osmond's early work with model psychoses from the work of so many of their contemporaries, and of previous contributors, was that it was in fact "guided" by a carefully reasoned hypothesis that could be retested by experiment. As the two doctors saw it, many existing theories on schizophrenia failed to do this. That they often equated their research method with those of great scientists such as Galileo might have sounded arrogant, but their claim was accurate nevertheless. As Michael Polanyi reported, "modern science must continue to be defined as the search for truth on the lines set out by the examples of Galileo and his contemporaries. No pioneer of science, however revolutionary—neither Pasteur, Darwin, Freud, nor Einstein—has denied the validity of that tradition or even relaxed it in the least."[24]

As the 1950s drew to a close, the Saskatchewan research on the biochemical factors of the LSD experience fused some of the more disparate elements of the program. One peculiarity that the researchers came across in their therapeutic treatment of alcoholism was that patients who displayed excessively high anxiety and tension usually responded less well to treatment than others. In these circumstances, LSD caused their tension levels to escalate, thereby hindering the psychedelic experience toward which therapists had geared their sessions. At this point, Osmond and Hoffer began to question more seriously whether the fault was with a specific biochemical abnormality in the adrenaline metabolism of the alcoholic. In examining the issue further, Hoffer and his colleagues began to understand more about the relationship between LSD and the adrenal system:

> Anxiety, which is a function of the adrenaline secretion, may become very intense, remain moderate, or disappear within the first hour after taking an adequate quantity of LSD. During this period, plasma adrenaline levels showed fluctuations. Physiologically, autonomic changes are marked. A very constant finding is papillary dilation. This in itself is a useful index of LSD activity. One of the major sites for the production of adrenaline, the adrenal medulla, shows an increase in metabolic activity, i.e., the radio phosphorus uptake is markedly increased. This is specific for the medulla since there is no comparable change in the cortex.[25]

Once Hoffer and Osmond had pure adrenochrome and adrenolutin to work with, a number of new chemical assays were devised, and several new lines of experimentation became possible. Thanks to the ingenuity of Hoffer and some of his researchers (e.g., Payza and Mahon), a method was formulated in 1958 by which adrenochrome levels could be measured in bodily fluids, both in vitro and in vivo. This was accomplished using a type of fluorescence test on blood samples drawn from various patients and normal, healthy volunteers. This assay proved to be a great boon to the research project, for it allowed Hoffer and Osmond to make a stronger case for the relationship between the LSD reaction and adrenaline metabolism. They set out on a series of different experiments, alternating among those that used adrenochrome, those that used LSD, and those that used a combination of them. They also tested other lysergide derivatives (e.g., LSM, or d-lysergic acid morpholide, and Brom-LSD) and continued to apply a variety of modifiers (e.g., ascorbic acid and penicillamine), as they had earlier with niacin, to see how they would affect the LSD experience. They related one of these experiments as follows:

The clinical response after the 35 gamma of LSD-25 was slight and considered mainly of an increase in tension. There was no alteration in adrenochrome levels. While the adrenochrome was being injected, a remarkable feeling of muscular weakness set in, there was some constriction of the chest and a need for deep breathing. This continued for some minutes. A few minutes later, a typical adrenochrome or adrenolutin experience begins. But this is sometimes as intense as that seen with 100–200 micrograms of LSD-25 and continues from 5 to 9 hours. The experience is not an intensification of the usual LSD-25 experience but is an intensification of the adrenochrome experience.[26]

Other tests revealed that adrenaline in plasma could be converted in vitro into adrenochrome and adrenolutin after LSD was added, and that LSD stabilized adrenochrome in blood. Perhaps one of the more significant discoveries to surface from this recent chain of developments was that LSD caused some subjects' adrenochrome levels to rise. This tended to be the case with individuals who had more intense responses to LSD and in whom tension remained relatively low or non-existent. This immediately led Hoffer and his colleagues to hypothesize that, "since the plasma adrenochrome value is clearly related to the psychological experience after LSD, it seems likely that it is the intermediate by which LSD acts."[27]

Further adrenochrome work outside the province demonstrated that ascorbic acid added to adrenochrome diminished the latter's psychotomimetic features in animals. In subsequent studies of this reaction in lab tests and humans, Hoffer and Osmond realized that adrenochrome was partially converted into leuco-adrenochrome, another indole that they found to be non-toxic and actually to have some tension-reducing qualities. When ascorbic acid was combined with LSD, "perceptual changes were not altered,...disturbances in thought were less marked, and suspicion was less evident. The ability to concentrate and appreciate the perceptual changes was heightened."[28]

Another substance that the Saskatchewan researchers used in concert with LSD was penicillamine. It was similar to ascorbic acid in that it converted adrenochrome into leuco-adrenochrome. It was also notable in that it eliminated all of the affective symptoms associated with the drug experience. Because lack of affect, or emotion, was such a common feature in so many forms of schizophrenia, the combination of penicillamine and LSD constituted yet another possibility for model psychosis. For a while, Osmond and Hoffer even thought that these drugs together might be of considerable benefit in treating alcoholics who suffered from extreme tension, but it was ultimately concluded that the "LSD experience without affect is of little value therapeutically for alcoholics."[29]

After sifting through all of their data, Hoffer and Osmond began to chart a hypothetical course for adrenochrome in the human body, which would become the basis for their adrenochrome-adrenolutin hypothesis for schizophrenia. According to this theoretical map, whether adrenochrome was converted into the harmless leuco substance or the poisonous adrenolutin depended on the biochemical nuances in each individual. This was what supposedly separated healthy individuals, whose biochemistry functioned normally, and schizophrenics, whose biochemical apparatus was impaired. In essence, LSD poisoned the enzyme that caused adrenochrome to turn into leuco-adrenochrome, the result being a number of possibilities: "(1) an increase in adrenochrome since it is not destroyed so quickly and the excess appears in blood and urine, (2) some diversion of the adrenochrome into adrenolutin, and (3) consequently a decrease in the rate of destruction of injected adrenochrome. This inhibition probably occurs in the brain as well as peripherally."[30]

Two other unique tests developed in Saskatchewan bolstered the biochemical theories of Osmond and Hoffer and aided them in the diagnosis, prognosis, and treatment of schizophrenia: the mauve factor test and what came to be known simply as the Hoffer-Osmond Diagnostic Test. Here, too, the LSD work played a guiding role. Building upon the initial idea of Osmond and Hoffer that LSD might replicate the chemical changes taking place in the schizophrenic body, Saskatchewan researchers (Payza, Irvine, et al.) created a test to identify chemical abnormalities in urine samples using refined paper chromatographic techniques.[31] Alcoholic patients who provided urine samples before and at the height of the LSD experience produced mauve spots in the post-LSD specimens, Hoffer and Osmond discovered. They followed the same procedure with schizophrenic patients

and normal individuals. The results indicated that many of the schizophrenic patients, who had received no treatment and were in the early
stages of the disease, had the mauve factor present, whereas the normal
volunteers did not. Moreover, in a significant proportion of schizophrenics treated with niacin, the mauve spots no longer appeared. This was
further verification of a biochemical difference between schizophrenics
and normal people. One difficulty was that researchers could not properly
identify the chemical; thus, it was referred to as an unknown substance.[32]
Hoffer and Osmond would coin the term "malvaria" to describe anyone
who excreted this mauve factor.

In the early 1960s, they put together all of the information collected
over the past decade of research and fashioned an objective card-sorting
test (the HOD Test) that they thought would assist them in psychiatric
diagnosis, notably in schizophrenia. As Hoffer described the test, it

> consisted of a set of 145 cards each with a question printed on one
> side and a number on the other. Questions were selected so that
> schizophrenic patients would be more apt to find the statement
> true than patients with some other psychiatric disease or a normal
> comparison group. Both normals and non-schizophrenic patients
> are likely to consider the statements, as applied to themselves,
> false. The questions were obtained by examining many mental
> descriptions of schizophrenics and others, by reading several hun-
> dred accounts written by normal subjects who described their own
> experiences with LSD or mescaline, by studying several biographies
> written by schizophrenic patients when ill or after they had recov-
> ered, and by referring to workers such as Lewis and Potrowski who
> leaned heavily on perceptual changes in developing diagnostic crite-
> ria for schizophrenia.[33]

Interestingly, individuals who had mauve spots on the test also tended
to produce higher scores on the HOD Test, so the two tests could be
correlated. These new tests were not received enthusiastically in the psychiatric world. The suggestion that schizophrenia could be diagnosed with
a test that required little or absolutely no psychiatric and/or psychological

expertise was abhorrent to many, and years were spent arguing over the usefulness of such a test.

The controversy over Hoffer and Osmond's research project, in all of its facets, ultimately led to its demise in Saskatchewan in the early 1960s. Understanding the reasons for its cancellation, however, is a complicated task, especially with the multiple factors involved, both internal and external to the experiments themselves. The work on schizophrenia represented only one aspect—albeit one of the most important—of the entire program.

As early as 1953, there was a dramatic rise in interest in the use of hallucinogens. Curiosity was no longer confined to psychiatrists preoccupied with the drugs' psychotomimetic effects but had gradually spread into the social sciences and mainstream culture, lured to the drugs' apparent ability to foster and/or replicate mystical experiences. This aspect became central in another phase of the Saskatchewan research, the adoption of LSD in therapeutic situations, notably in the treatment of chronic alcoholism. Thus, in addition to the studies on schizophrenia, Hoffer and Osmond, together with Duncan Blewett, pioneered what came to be known throughout the world as "psychedelic therapy."

What conclusions can be drawn, then, about the hypotheses of Hoffer and Osmond regarding schizophrenia? In 1961, their tireless search for a biochemical precursor to schizophrenia remained an unsolved mystery of sorts but had shown promising leads. Their findings with adrenaline metabolites and the connections made with other hallucinogenic drugs, though debatable and inconclusive on some counts, were impressive in many respects, including the synthesis of pure adrenochrome and adrenolutin and their classification as psychotomimetic substances. These major feats cannot be discounted.[34] More importantly, from these discoveries and related findings arose many other innovations in the diagnosis and treatment of schizophrenia.

Did all of this work qualify as revolutionary in scientific terms? Hoffer certainly thought that it did. Writing to Osmond about the hostility that the two and their theories faced, he noted that it had

little to do with adrenochrome. I am certain the serotonin people are receiving [an] equal blend of criticism. We are attacked because we have spearheaded a counterrevolution in psychiatry which may destroy what analysts have so slowly acquired. They have rightly ascribed to us the leader's role in this movement. I am not worried about this type of attack. The scientific method has flourished in spite of attack because scientists read papers, not in crowds where crowd psychology prevails, but alone. Mob hysteria may slow down progress but cannot halt it.[35]

Hoffer's views predated those of Kuhn in *The Structure of Scientific Revolutions*. Nevertheless, it is interesting to note the parallels between the two, for example the kind of political strife that could occur within scientific communities when a new discovery challenges a reigning paradigm. Referring to this "mob psychology," Feyerabend commented that "revolutions lead to quarrels between opposing schools. The one school wants to abandon the orthodox program, the other school wants to retain it....The fight between opposing schools is a power struggle, pure and simple."[36]

Hoffer and Osmond took a unique hypothesis and developed and tested it. They disseminated their theories, had them published and in some instances verified by outside observers, yet many specialists steadfastly refused to acknowledge the value of their work. Although Manfred Bleuler did not address their findings specifically, he made it clear at the first International Congress of Neuro-Pharmacology in 1958 that work like that being done in Saskatchewan could never be taken seriously. Bleuler spoke for many when he said that "the psychopathological pictures resulting from the administration of [LSD] and other phantastic drugs do not correspond to the usual pictures presented by schizophrenia."[37] To add insult to injury, he noted that "it is not only incorrect to assume that drugs influencing the 5-hydroxy-tryptamine [serotonin] and noradrenaline metabolism in one direction produce schizophrenia; it is equally erroneous to affirm that drugs acting in a contrary sense on this metabolism could be specific remedies for schizophrenia."[38]

By 1961, Hoffer and Osmond's work, like that of Woolley, had restimulated interest in biochemical theories of mental illness and helped to kick-start the second coming of biological psychiatry. But it would be another decade and a half before it could be seen how their adrenochrome hypothesis (and controversial megavitamin treatment for schizophrenia) would play out in the psychiatric community (investigated more fully in Chapter 6). The adrenochrome hypothesis thus bore the markings of prerevolutionary science. It was followed by a period of intense crisis, much like other important discoveries in the history of science.[39] It rapidly evolved into a candidate for a biological paradigm in psychiatry, and schizophrenia research in particular, but "the reception of a new paradigm often necessitates a redefinition of the corresponding science.... The normal-scientific tradition that emerges from a scientific revolution is not only incompatible but often actually incommensurable with that which has gone before."[40] Although one might hesitate to view Hoffer and Osmond's discoveries in the same light as those, say, of Galileo, Newton, or Einstein, the comparisons, however minor, are there. There "can be small revolutions as well as large ones,... some revolutions affect only the members of a professional subspecialty, and... for such groups even the discovery of a new and unexpected phenomenon may be revolutionary."[41] What remained to be seen in the years after Osmond and Hoffer discontinued their studies in Saskatchewan was whether their theories would die or be "sufficiently unprecedented to attract an enduring group of adherents away from competing modes of scientific activity."[42]

CHAPTER 3

BEGINNING HALLUCINOGENIC THERAPY

The Saskatchewan research with psychotomimetic drugs was headed, by the mid-1950s, in some startling directions. Humphry Osmond acknowledged that the strange, drug-induced experiences offered much more to the researcher than a means for modelling psychoses such as schizophrenia. As he put it, they were "of more than medical significance" and pertained "to psychology, sociology, philosophy, art and even...religion."[1] Abram Hoffer, too, realized from the start that the drugs they were studying presented researchers with a glimpse of the workings of the human mind. While carrying out his initial studies with LSD, he triumphantly remarked to Osmond in a letter that "we appear to have stumbled upon a way of chemically dissecting a person's psyche," and this had "vast philosophical and psychological implications."[2]

Osmond and Hoffer continued to concentrate on schizophrenia, but they did not necessarily restrict their examination of hallucinogens to their psychotomimetic properties. They were prepared to explore many other avenues of research. A prominent turning point came when hallucinogens began to be used as adjuncts in psychotherapy and, especially in Saskatchewan, the treatment of alcoholism.

In 1953, Hoffer and Osmond began experimenting more seriously with LSD and mescaline in an effort to tap the hidden potential of hallucinogens as a psychiatric treatment. When they inaugurated this phase of their research, a few other professionals were engaged in similar studies. Prior to the Saskatchewan work, teams of scientists in both Europe and the United States had called attention to hallucinogens as aids in the psychotherapeutic process. Many, in fact, went on to publish numerous findings on incorporating these drugs in treating various mental disorders (e.g., anxiety). This speculation eventually caught the attention of Hoffer and Osmond, and they soon added their names to the growing body of experts working in this area. Unlike their contemporaries, though, the two Saskatchewan psychiatrists decided to investigate a problem that, until then, had remained more or less unexplored by other scientists: the use of hallucinogens in the treatment of alcoholism. They believed that the drugs also afforded an ideal model for the toxic psychosis, or delirium tremens (DTs), that sometimes afflicted alcoholic individuals.

The treatment stage of hallucinogenic experimentation in Saskatchewan, like the first part of the research on schizophrenia, represented an evolutionary period in which Osmond and Hoffer conceived and tested some of their earliest hypotheses on the matter. They started with the simple idea that the psychotomimetic experience could be used, in effect, to frighten the alcoholic into becoming sober. By using hallucinogens to imitate aspects of naturally occurring DTs, which caused some alcoholics to quit drinking altogether, Hoffer and Osmond thought that they could successfully treat the disease. Intriguing as this notion was, it was not what led to the positive findings of their pilot tests. In a serendipitous development, the Saskatchewan researchers found that the experience of mysticism and/or spirituality accounted for many of their patients' sudden conversions.

All that Hoffer, Osmond, and Duncan Blewett (who joined this phase of research) could do at the outset was make observations and collect relevant data. What separated them from many of their contemporaries was that, relatively quickly, they "proceeded to the higher tasks of classification, analysis, experiment and theory making."[3] Their work was significant in that it served as a beacon for many other workers in this area. In their own studies, Hoffer, Osmond, and Blewett gradually became attuned to the possibilities of drugs such as LSD not only for the scientific world but also for society as a whole. Blewett, in particular, was drawn to the spiritual dimension of the drug experience. To him, the study of these drugs should not be limited to applications for the mentally ill; it should also be open to those with healthy frames of mind. Blewett therefore focused on LSD for its role in promoting human awareness and self-understanding. Although Hoffer and Osmond remained sympathetic to a degree to this objective, they preferred to concentrate on the biochemical reactions of hallucinogens in the human body, specifically those relating to the adrenaline hormone and its derivatives. Differences aside, the three men shared many common interests and goals, and they pooled their efforts to make Saskatchewan a leading place in carrying out research of this kind. They arguably became authorities on the effects and uses of what would come to be known as psychedelic drugs.

HALLUCINOGENS AND PSYCHOTHERAPY

By the early 1950s, as noted above, a significant body of literature indicated the efficacy of using LSD and mescaline in psychotherapeutic settings. Authors such as Anthony K. Busch, Warren C. Johnson, Walter Frederking, and Ronald Sandison, to name but a few, wrote extensively on the value of these drugs in various clinical trials and how effective they were in helping to resolve patients' issues.[4] The precise nature of the healing process remained obscure. As Hoffer and Osmond remarked, "when a chemical is used as an integral portion of the therapy another variable is added to the complexity of factors already operating."[5] Reflecting on the subject, the founder of LSD, Albert Hofmann, noted that its

apparent benefits as a drug auxiliary in psychoanalysis and psychotherapy are derived from properties diametrically opposed to the effects of tranquilizer-type pharmaceuticals. Whereas tranquilizers tend to cover up the patient's problems and conflicts, reducing their apparent gravity and importance, LSD, on the contrary, makes them more exposed and more intensely experienced. This clearer recognition of problems and conflicts makes them, in turn, more susceptible to psychotherapeutic treatment.[6]

The characteristic effect to which Hofmann referred, known as "abreaction," enabled some patients while under the drug to gain access to a wealth of buried or repressed material in their unconscious minds. This process sometimes permitted patients, with the help of an experienced therapist, to thoroughly reassess themselves and their patterns of behaviour, at which point it was thought that they could take the next step to leading more healthy, well-adjusted lives. The controversy over this type of drug therapy was that it was carried out in considerably less time, and in many instances with astoundingly greater success, than more traditional therapeutic measures. Czech researcher Stanislav Grof revealed from his own studies that "many clinicians, knowing how difficult it is to change deep-rooted psycho-pathological symptoms, not to mention the personality structure, were incredulous of the dramatic effects achieved in a matter of days or weeks."[7] This was not to say that the drug was a panacea for all that ailed the patient. Most researchers at the time—Hoffer, Osmond, and Blewett included—stressed that LSD was only a part, albeit a vital part, of an overall treatment program. Another member of the Saskatchewan team, psychiatrist Colin Smith, voiced a similar sentiment:

> The treatment is not simply a "chemical" one but rather a "psychological" one in which the personalities of patient and therapist interact in complex ways. The psychological situation created will be regarded differently by different patients and therapists. To one therapist the hallucinogenic experience may be a form of aversion therapy, to another an experience interpretable only by Jungian concepts, to yet another a religious-mystical experience and so on. Moreover, in this kind of treatment the personal factor looms large and goes beyond mere differences of technique.[8]

The ultimate success or failure of treatment hinged not on the drug itself but on the kind of experience that resulted from its use, and this in turn was reliant on a number of variables, which came to be referred to as "set and setting." Administered by a trained and compassionate therapist who empathized with what the patient was experiencing, the drug could work wonders. Conversely, given by an inept physician who failed to establish mutual trust or strike rapport with the patient, the drug could cause inestimable damage. Osmond had but one stipulation for those working with these mysterious substances: "One should start with oneself."[9] As he saw it, "unless this is done one cannot expect to make sense of someone else's communications and consequently the value of the work is greatly reduced."[10] Many who worked in the field wholeheartedly agreed. Another Canadian team of researchers underscored the need for the therapist "to understand these experiences himself, else it would be a matter of the blind leading the blind. This understanding can only be gained by taking the drug and learning how to control a successful experience. Osmond's Golden Rule, 'You start with yourself,' is of utmost importance in utilizing the psychedelic experience for therapeutic ends."[11]

A predominant form of hallucinogenic drug therapy in the early 1950s was psycholytic therapy, in which low doses of LSD were administered in single cases over an extended period of time (e.g., five to seven sessions over three months). Evaluating data from this technique, Grof noted that,

rather than being unrelated and random, the experiential content seemed to represent a successive unfolding of deeper and deeper levels of the unconscious. It was quite common that identical or very similar clusters of visions, emotions, and physical symptoms occurred in several consecutive LSD sessions. Patients often had the feeling that they were returning again and again to a specific experiential area and each time could get deeper into it. After several sessions, such clusters would then converge into a complex reliving of traumatic memories. When these memories were relived and integrated, the previously recurring phenomena never reappeared in subsequent sessions and were replaced by others. It soon became clear that this observation might have important implications for the practice and theory of dynamic psychotherapy. The use of repeated LSD sessions in a limited number of subjects appeared to be much more promising than the study of single sessions in large groups of individuals.[12]

When the Saskatchewan researchers began to look at using hallucinogens in psychotherapy, they departed from the psycholytic method. They gave patients "overwhelming" doses (e.g., 200–500 gamma or micrograms) with the primary goal of engendering a mystical, conversion-like experience. This approach evolved from their efforts to treat alcoholics.

When Osmond and Hoffer embarked on this project in 1953, the provincial government had just established its Bureau of Alcoholism as a branch of the Department of Social Welfare. Historically, attitudes toward alcoholism had paralleled those toward insanity. Alcoholism was regarded as an inherent weakness or moral imperfection in an individual rather than a disease or illness. Alcoholics were ridiculed and ostracized, often imprisoned for their unruly behaviour, and denied admittance into hospitals for help and rehabilitation. As one of the bureau's first counsellors, Angus Campbell, recalled, many medical professionals of the day "shunned inebriates as one would a leper," but a few "pursued serious scientific avenues to find answers."[13] As Osmond and Hoffer began to examine the issue more closely, they soon saw that predominant views in society and the medical community had made the alcoholic the "most misunderstood, most rejected and most neglected medical phenomenon known."[14]

By the 1950s, the CCF government had become sensitized to the destructive nature of alcohol abuse, as evidenced in the family and workplace, and to its social costs, for example in law enforcement.[15] It concluded that a new approach was necessary if any improvement was to occur. In this regard, Saskatchewan was merely following in the footsteps of other provinces and states that had confronted the devastating consequences of alcoholism. From the outset, the Bureau of Alcoholism, under Executive Director Jake Calder, strove to establish strong connections with various agencies. The bureau approached those in the professional health services (both psychiatric and general health practitioners), prominent members of the clergy, and Alcoholics Anonymous (AA), which until then was perhaps the only organization that had given serious attention to alcoholism. Following these consultations, the bureau came up with basic criteria, which Campbell summarized as follows:

1.   Alcoholism is a disease and the alcoholic is a sick person.

2. The alcoholic can be helped and is worth helping.
3. Alcoholism is a public health problem and therefore a public responsibility.
4. . . . A coordinated program of comprehensive community services is essential for the prevention, treatment and control of the disease (cooperation and teamwork needed).
5. The myths and attitudes around alcoholism and drinking in general have resulted in closing the doors of physicians, hospitals, social agencies, employment and rehabilitation services.
6. . . . AA is not the only pioneering effort to do something constructive about alcoholism. It was, however, in the forefront of the trend to regard alcoholism as a treatable condition. Rather than moral condemnation or therapeutic hopelessness, AA has shown beyond the slightest doubt. . . that an alcoholic can be rehabilitated.
7. . . . Special education will be necessary to bring the subject of alcoholism into the consideration of all of us in the public and helping professions.[16]

Hoffer and Osmond's decision to treat alcoholics with hallucinogens came about in a serendipitous way. The first plans originated during a conversation while they were in Ottawa on an unrelated mission—to seek support from the Department of Health and Welfare for their niacin treatment for schizophrenia.[17] As Hoffer recalled,

> While discussing LSD reaction and its similarity to delirium tremens it occurred to us that LSD might be used for giving alcoholics controlled delirium tremens, or giving them the "hitting bottom" experience with complete safety. . . .
>
> This idea, at 4:00 a.m., seemed so bizarre that we laughed uproariously. But when our laughter subsided, the question seemed less comical and we formed our hypothesis or question: would a controlled LSD-produced delirium help alcoholics stay sober?
>
> We were, of course, well aware of the difficulties, both theoretical and practical. Some of these were: (1) that many alcoholics had

experienced delirium tremens repeatedly; (2) the LSD experience modeled delirium but could not produce the identical condition; (3) many physicians would only see the crazy part of our idea and not its potential value. On closer examination it was evident that these objections were not sufficiently well based to dissuade us.[18]

Hoffer and Osmond pointed to AA co-founder Bill Wilson's DTs experience and its crucial role in putting Wilson on the road to sobriety. Additionally, the writings of philosopher/psychologist William James and psychiatrist Henry Tiebout, among others, highlighted the transformative power of the conversion experience and the act of self-surrender.[19]

In 1953, the first attempt to treat alcoholics began at the Saskatchewan Hospital–Weyburn with two patients given 200 gamma of LSD. After this first trial, Osmond and Hoffer believed that they had achieved some success in assisting the subjects to maintain sobriety. They had not been completely cured, but their experience with LSD forced them to reconsider their behaviour and resulted in a noticeable change in their drinking routines. This success seemed to warrant more extensive testing with the drug. In planning for the next LSD experiments, Hoffer and Osmond thought that double-blinds would be practically useless. Given the intensity of the LSD experience, both the investigator and the patient could figure out whether the drug or a placebo had been administered. To make up for what many perceived to be a shortcoming, Hoffer and Osmond decided to take only the worst-case scenarios into their program, alcoholics who had consistently failed in every other recourse to treatment and been given the most negative prognoses. The first major pilot project was assigned to psychiatrist Colin Smith when he joined the research team in 1955. He subjected twenty-four refractory alcoholics to a combination of psychotherapy and hallucinogens. He tried to establish psychotherapeutic relationships with them over periods of two to four weeks to identify underlying problems that might have led to their alcoholism. He also emphasized the importance of establishing a close rapport with the individual prior to administration of a single dose of LSD or mescaline. Smith further described the project:

Early in the study, it was noted that alcoholics tend to be resistant to these drugs and doses of 200 to 400 micrograms of LSD or .5 g of mescaline were used. A prolonged interview was carried out with the patient, who was never left alone while under the influence of the drug. In addition to discussing with the patient problems leading to and arising out of his drinking, strong suggestions were made to the effect that he discontinue the use of alcohol. No attempt was made to arouse fear. The material which emerged was discussed during the next few days and the patient was discharged. In some cases, follow-up was possible; in most, however, further contact was made through AA, which provided much valuable and objective information. Disulfiram was not employed in the study, nor were tranquilizers.[20]

After follow-up periods, which ran from two months to three years, Smith published his account of the trials. The subjects were labelled according to diagnostic category (character disorder, psychopathy, and borderline and actual psychosis[21]), type of reaction to the drug (intense or mild), and response to treatment: (1) much improved (complete abstinence or drinking only very small quantities), (2) improved (definite reduction in alcohol consumption), and (3) unchanged (no significant change in drinking pattern).[22]

On the basis of the results (of the twenty-four patients, six improved, six were much improved, and twelve were unchanged), Smith arrived at a number of conclusions. He found that, though none of the patients became worse as a result of the treatment, borderline or actual psychotics received little benefit from it. He recommended, therefore, that they be excluded from future testing. The most promising cases were individuals suffering from character disorder, those "exhibit[ing] chronic interpersonal difficulties arising on a neurotic basis."[23] Of those patients who obtained the most favourable outcomes, Smith wrote that the experience had enabled them "to gain an enhanced self-understanding. . .and [that it] appeared to influence their subsequent behaviour."[24] He added that the results in the psychopathic category "were not unimpressive in view of the great difficulty in treating patients of this type."[25] Smith noted the many variables at work throughout the experiments, such as therapeutic technique, and how modifications could lead to better results in future endeavours with hallucinogens. He ended his report by saying that "the phenomenon is of great interest and should provoke further study of both physiological and psychological aspects."[26] This was indeed what happened. However, as the experiments continued to expand, something strange and rather unexpected occurred. Many patients did not have the nightmarish psychotomimetic reactions that Hoffer and Osmond believed they would have. Many had opposite experiences.

As was the case at other research centres, the post-1955 years signified a transition for the psychotomimetic drug experiments in Saskatchewan in which LSD, mescaline, and other drugs began to take on religious/spiritual overtones. The results from the 100 or so people who had volunteered for the Saskatchewan drug tests in the first half of the decade indicated that the personality of the subject, and the setting in which he or she took the drug, could evoke multiple responses. Whereas many had typically psychotomimetic reactions, others had what could be described as transcendental experiences.

One of the more intriguing figures drawn to the Saskatchewan research was the famous novelist and social commentator Aldous Huxley. After familiarizing himself with the earlier mescaline writings of Osmond and Smythies, he expressed a profound interest in using the drug to explore the human psyche. In the spring of 1953, Osmond connected with Huxley at an American Psychiatric Association convention in Los Angeles, following which the psychiatrist administered the author's first taste of mescaline. The result was documented in what would become one of the most eloquent descriptions of the mescaline experience, *The Doors of Perception*. Huxley began by citing the work of the Saskatchewan doctors and went on to make a plea for the future use of mescaline and substances like it:

How can the sane get to know what it actually feels like to be mad? Or, short of being born again as a visionary, a medium, or a musical genius, how can we ever visit the worlds which, to Blake, to Swedenborg, to Johann Sebastian Bach, were home?...For those who theoretically believe what in practice they know to be true— namely, that there is an inside to experience as well as an outside—the problems posed are real problems.... By taking the appropriate drug, I might so change my ordinary mode of consciousness as to be able to know, from the inside, what the visionary, the medium, even the mystic were talking about.[27]

Although Hoffer had some difficulty following parts of Huxley's essay, largely because of his lack of knowledge of philosophy, both he and Osmond placed it in the "out of the ordinary category."[28] As Osmond wrote to Hoffer on July 21, 1953, "I think there are one or two alterations that we should make in Aldous' essay, in biochemistry in particular he seems to have gotten hormone and enzyme mixed up but apart from very minor blemishes it is clearly the work of a great craftsman and is full of stimulating and lively ideas."[29] From the time of their first encounter, Osmond and Huxley would spend countless hours discussing the possibilities of hallucinogenic drugs, remaining close confidants until the latter's death from throat cancer in 1963.

Even before he became involved with hallucinogenic drugs, Huxley had devoted a great deal of attention to mysticism and visionary experiences. One work that addressed this subject was his definitive 1944 text on comparative religion, *The Perennial Philosophy*. This philosophy is "immemorial and universal. Rudiments of the Perennial Philosophy may be found among the traditional lore of primitive peoples in every region of the world, and in its fully developed forms it has a place in every one of the higher religions."[30] Clearly, mystical experience can be defined in many ways, but it was the commonality of the experience that attracted Huxley. His thoughts had lasting influences on both Hoffer and Osmond, and even more so on Blewett.

DUNCAN BLEWETT: SCIENTIST TRANSFORMED

From the start of his role as the PSB's supervising psychologist in 1954, Blewett, with his skills as a psychological researcher and psychometrician, was valuable to Saskatchewan's evolving research program on schizophrenia. Writing to Hoffer in September 1956, Osmond welcomed the addition of Blewett to the team:

I like Blewett's attack on the psychological test problem. The fact is that there is no psychological test which is really any good in schizophrenia....I believe we have been tackling it the wrong way round. We should ask ourselves what is the characteristic of schizophrenia and test that....[We] must not allow our chaps to stagnate in "routine research."...[It] won't and can't lead anywhere....All evidence suggests that schizophrenia is not one of the fields where this is likely to succeed....Our work has shown that new hypotheses alone are really fruitful.[31]

Blewett's involvement in the model psychosis work inevitably led to an interest in the startling psychological effects of hallucinogenic drugs and LSD in particular. Right from the time that Blewett joined the service, he was fascinated by the LSD work of Hoffer and Osmond. He pestered Osmond for a guided excursion into the other world, and Osmond finally relented in 1956 and provided the psychologist with an individual session, which Blewett later recalled as "the most profound experience I had ever had or could have imagined."[32] He was a transformed man. Blewett soon began a close collaboration with Nicholas Chwelos (Hoffer's resident in psychiatry at the University of Saskatchewan), and the two mapped out the LSD experience and tested therapeutic applications of the drug. Blewett was especially struck by its potential for understanding the human mind. To him, hallucinogens were "tools as valuable to the psychologist as the microscope to the biologist or the telescope to the astronomer" and the "lens that magnifies, clarifies and makes discernible the structure and processes of personality."[33]

Blewett's thoughts on mysticism were similar to Huxley's. Commenting on the mystical experience, Blewett referred to

a feeling of some direct contact with the infinite and the eternal, a feeling of being at one with the world's Author and all His Works, a complete abandonment of the self and all its motives. Also frequently noted are tremendous infusions of love, feelings of remarkably enhanced understanding, and an awareness of very great beauty all about one. There is a feeling of being outside the bounds of time, space, and the self. There is an enhanced feeling of reality and authenticity which accompanies the experience, at the same time as the realization that the individual has always known the things he is experiencing.[34]

Humans have relied, from time immemorial, on drugs and/or demanding physical and mental regimens as means of bringing about visionary states of mind. It was not until the 1950s, however, that the scientific world viewed such experiences with more than passing interest. As Osmond noted, from

the perspective of history, our psychiatric and pathological bias is the unusual one. By means of a variety of techniques, from dervish dancing to prayerful contemplation, from solitary confinement in darkness to sniffing the carbonated air at the Delphic oracle, from chewing peyote to prolonged starvation, men have pursued, down the centuries, certain experiences that they considered valuable above all others.[35]

With the advance of science, technology, and secularism, especially since the nineteenth century, the scientific community, perhaps more than any other, abandoned, or at least conveniently cast aside, the traditional religious strictures that had held it in check for so many generations. Jung remarked that, by the second half of the nineteenth century, people had become intent on developing a "psychology without the soul" and that, "under the influence of scientific materialism, everything that could not be seen with the eyes or touched with the hands was held in doubt; such things were even laughed at because of their affinity with metaphysics. Nothing was considered 'scientific' or admitted to be true unless it could be perceived by the senses or traced back to physical causes."[36] For Blewett, this scientific tendency to slough off all things religious appeared to have reaped as much harm as benefit: "The discard of the soul was but an inevitable step toward science's freedom from religious domination. . . . Science assuredly gained much by establishing this freedom; yet in throwing off the fetters of imposed belief systems, it has created a spiritual vacuum."[37]

In a curious way, Saskatchewan's studies of hallucinogenic substances heralded a spiritual renaissance in scientific research. The mystical experience came to form an integral part of the therapeutic use of LSD and mescaline. This fostered a new controversy that can be viewed separately from the other controversies in which Hoffer and Osmond were already enmeshed because of their biochemical theories. In the beginning, their work with psychotomimetics was primarily confined to the scientific world as a compelling argument within psychiatric and psychological circles. This atmosphere changed irrevocably once the drugs (especially LSD) began to be explored for their therapeutic benefits, because then the research became subject to scrutiny at all levels in society. As philosopher Alan Watts argued,

> the idea of mystical experiences resulting from drug use is not readily accepted in Western societies. Western culture has, historically, a particular fascination with the value and virtue of man as an individual, self-determining, responsible ego, controlling himself and his world by the power of conscious effort and will. Nothing, then, could be more repugnant to this cultural tradition than the notion of spiritual or psychological growth through the use of drugs. A "drugged" person is by definition dimmed in consciousness, fogged in judgment, and deprived of will.[38]

Opponents of LSD use attached to it the derogatory epithet "instant mysticism," implying that the drug was a cheap, unfair, and undeserved shortcut to realizing the visionary experience. In the same way that critics cast doubt on comparisons between schizophrenia and the psychotomimetic experience, they questioned the legitimacy and genuineness of the transcendental experience that some subjects reported after taking LSD or mescaline. Much of the criticism came from people not versed in the scientific facts and unfamiliar with the drugs and their effects. Osmond and other researchers closest to the debate considered their uninformed comments to be reprehensible. As Osmond stated, "those who have had these experiences know, and those who have not had them cannot know and...the latter are in no position to offer a useful explanation."[39]

THE NATIVE AMERICAN CHURCH EPISODE

A prime example of public and professional misinterpretation of the religious use of hallucinogenic drugs was the sociopolitical response to the Native American Church (NAC) and its sacramental use of peyote.[40] Appearing in the American Southwest in the early 1900s, the NAC had arisen largely in response to colonialism and the encroachment of the settlers' system of values and the subsequent disappearance of traditional Indigenous ways of life. By fusing elements of Native American culture and Christianity with symbolic consumption of the peyote cactus, thought to be God incarnate, the highly ritualized ceremony would—it was hoped—assist Indigenous peoples in adapting to their new environment. Early reports in mainstream society of Native Americans' use of peyote were almost always sensationalized, with outlandish claims of bacchanalian-like orgies, devil worship, and drug-induced murders. From a strictly Indigenous perspective, peyote was not simply a drug but also a sacred medicine. Reflecting on the origins of the sacramental use of peyote, Sioux medicine man and NAC "roadman" Leonard Crow Dog explained that "peyote is holy. It's not a chemical gimmick but a sacred herb—our sacrament....It's opening three doors to me: makes me recognize myself, makes me understand the people around me, makes me understand the world."[41]

Unfortunately, most of Western society failed to understand or appreciate the NAC from an Indigenous point of view. Reactions of Canadian politicians and law enforcement officials to the appearance of a NAC chapter in Saskatchewan were, like those of their American counterparts, based upon cultural ignorance and fear. When the peyote issue took centre stage in debates in the House of Commons, CCF North Battleford Member of Parliament Max Campbell spoke to the matter on January 24, 1956, seeking an answer from the federal government on whether it had made any decision regarding the prohibition of peyote buttons:

> There have been religious organizations formed in my part of the country within the last year or two, and during what you might call religious celebrations these organizations have been importing from Mexico and the southern states what they call peyote

buttons,...part of the cactus plant....Not yet classified as a narcotic it has somewhat the same effect....These people all get around in a big circle, chew these buttons for a while then become sick. After they become sick they are supposed to vomit up their sins. The next procedure is that they again chew these buttons but the reaction is different the second time. They go into a deep sleep which I am told lasts for 20 to 24 hours, and in that sleep they have beautiful dreams....They are a form of narcotic and they are demoralizing to anyone who uses them.[42]

Campbell pressed for an answer again three months later, asking Minister of National Health and Welfare Paul Martin Sr. whether his department would "consider peyote buttons to be a menace to the health of anyone chewing them or drinking any concoction made from them[.] If dangerous, what action will be taken to prevent their entry into Canada?"[43] Hoffer later found out that Campbell had in fact been

given a planted question by [Martin] who was trying at that time to suppress the use of peyote by the NAC in Saskatchewan. When the statement from the Minister of Health hit the public news, the press called me in Regina and asked me what I thought; I replied that it was all nonsense and that was quoted all across Canada....Later on, I complained to Tommy Douglas, who at the time was in Ottawa, about the performance of his member, and about a year later I met Max Campbell at the office in Saskatoon. We got along really well, became close friends, because Max by then realized what had happened and he had turned.[44]

As the religion continued to spread and prosper, various government agencies pressed to have peyote prohibited, fearing that it had become a widespread menace. Interestingly, First Nations' ceremonial use of peyote placed the federal government in the uncomfortable position of trying to prevent its spread while protecting such use because of amendments to the Indian Act in 1951 ensuring that First Nations could carry on traditional spiritual practices. As Erika Dyck and Tolly Bradford have written, the NAC "created an awkward situation for bureaucrats charged with upholding this new policy. While federal bureaucrats felt obliged to allow peyote when used in the context of the NAC ritual, they were constantly challenged by police and Indian agents who retained their paternalistic tendencies and sought to restrict peyote use on reserves."[45] Such use also raised an intriguing question. With officials beginning to concentrate their efforts on criminalizing the drug, "did the state, having adopted an Indian policy shaped by multiculturalism, have the right to interfere in a culturally legitimate use of a hallucinogenic substance?"[46] In the end, the federal government had little choice but to opt for a policy of non-interference in line with the amended legislation.

That peyote ingestion led to rampant abuse and narcotic-like addiction was not the case at all, and when such rumours reached Hoffer, Osmond, and Blewett, they immediately tried to set the matter straight. In the fall of 1956, the three men, along with Weyburn's new research psychiatrist, Teodoro (Teddy) Weckowicz, and a Saskatoon *Star-Phoenix* reporter and photographer, travelled to the outskirts of North Battleford to examine the ceremony for themselves.[47] This was one of the first documented times that non-Indigenous men had the opportunity to witness the ceremony in Canada. As Hoffer and Osmond discovered, they were not the first ones in the province to use hallucinogens to treat alcoholism. Their North Battleford hosts reported several instances of successful treatment of alcoholics using peyote-based ritual healing. Of course, many parallels would later be drawn between NAC rituals and the psychedelic experience employed by therapists; the peyote ceremony also later inspired the doctors' research endeavours in parapsychology and some interesting group studies that included the use of hallucinogens. Osmond recalled that night in the tepee:

> The Indians have been very skillful in structuring their ceremony so that it best meets their needs. They are such masters of symbol, ceremony and ritual, that this is hardly surprising. It would be unwise and impertinent to ape their religion, which developed from their agony when they lost their hunting grounds at the end of the nineteenth century. Our needs are very different from theirs. So we must follow a different route.[48]

Upon further reflection on the ritual, Weckowicz wrote that much of it could be understood in terms of Jungian thought. As he reasoned, Plains Indians placed great importance on achieving a visionary, mystical state, be it through self-imposed fasting, isolation, the sacred sun dance, or the peyote ceremony:

> Having [an] unusual experience in a highly ritualistic setting whether with peyotl or without it, has always been a central motif in the culture of the Plains Indians. The rituals used...for a mystical experience...enacted allegorically the dramas in which the actors and the props personified archetypes from the collective unconscious, described by Jung. The purpose of the drama was to present symbolically the problem of individuation, finding selfhood and creation of man as different from animal. In the peyotl ceremony witnessed by the author, many Jungian Archetypes are discernible.[49]

Whether NAC practitioners would have concurred with a mainstream psychotherapeutic assessment, Jungian or otherwise, of the peyote ceremony is unlikely. Joseph Calabrese recently stressed that

> psychotherapeutic activity...was a basic human activity long before clinical psychologists and psychotherapeutic office sessions existed. It is not owned by any particular cultural group or professional organization but is a generic activity of humankind. Claims that Freud or whoever else "invented" psychotherapeutic intervention are...insulting to members of other cultural traditions who have discovered the phenomenon in question for themselves.[50]

The North Battleford peyote experience had a profound impact on Blewett. Along with Osmond, he participated in the peyote ceremony while Hoffer, Weckowicz, and others observed, and he summarized his impressions of the ceremonial use of the hallucinogen as follows:

> Peyotism, because of its nature, has frequently been called to question by various agencies and authorities. Its life and growth have

met persistent opposition which has developed almost entirely out of ignorance or bigotry. The encroachment of white culture has brought disaster for the Indian....He was left with the choice of becoming a second class citizen or remaining penned up like a head of livestock on the pasture of a reservation. A remote government dictated his rights and offered him the solace of five dollars a year in treaty money....He is left in a situation in which he must develop a new and satisfying value system which can provide him with self-respect and with a sense of purpose. It is in this situation that peyotism has developed.[51]

Blewett also speculated about the social value of peyote and other hallucinogens when used in certain settings: "Can there be any carry-over of what is realized in the experience into more mundane activities and relationships of day-to-day living?...Indians have gone beyond the scientists of today in their use and control of the psychedelic experience."[52]

In her analysis of the encounter of the Saskatchewan researchers with the NAC ceremony, Dyck contends that they were "defending an Indigenous cultural practice but they were simultaneously caught up defending their own credibility as scientists whose careers were intimately connected with psychedelic approaches to healing."[53] It was precisely this openness to blending spiritual modes of thinking and ways of being with science that would complicate the efforts of Hoffer, Osmond, and Blewett to make a scientific case for their therapeutic work with hallucinogens. But in his own defence of the importance of non-scientific elements in the scientific enterprise, Feyerabend reasoned that "the separation of science and non-science is not only artificial but also detrimental to the advancement of knowledge. If we want to understand nature, if we want to master our physical surroundings, then we must use *all ideas, all methods, and not just a small selection of them*."[54] The sort of methodological pluralism of which Feyerabend spoke can also be read as science being inclusive of Indigenous approaches (even hallucinogenic ones) to health and healing. "True," he wrote,

there were no collective excursions to the moon, but single individuals, disregarding great dangers to their soul and to their sanity, rose from sphere to sphere until they finally faced God himself in all His splendor while others changed into animals and back into humans again. At all times man approached his surroundings with wide open senses and a fertile intelligence, at all times he made incredible discoveries, at all times we can learn from his ideas.[55]

By 1956, hallucinogenic drug therapy in Saskatchewan had advanced several steps. Beginning with their hypothesis that the psychotomimetic experience could be used in the treatment of alcoholism, Osmond and Hoffer initiated some of the earliest known pilot trials to treat the disorder with drugs such as LSD and mescaline. The outcomes of these early therapeutic attempts, however, were far from what they had originally anticipated. When they realized that spiritual and/or mystical elements, and not psychotomimesis, were responsible for the successful "conversions" of their first alcoholic cases, Hoffer and Osmond adopted an entirely different line of inquiry. It would become the catalyst for an alternative and highly controversial psychotherapeutic technique that resulted, incidentally, in a new term to describe hallucinogenic drugs: "psychedelic."

In succeeding years, Hoffer and Osmond, along with Blewett and other members of the Saskatchewan research team, had their therapeutic findings published in numerous journals, stirring up much debate in scientific, social, and political circles. Saskatchewan became one of the first places in the world to develop a full-scale treatment program for alcoholism using psychedelics. This achievement led some experts to name the province as "the only place on earth where LSD had a favorable public image."[56] All of these developments, of course, added to the already complex hallucinogenic drug research under way in Saskatchewan, the dimensions of which are further explored in the next chapter.

CHAPTER 4

THE OTHER WORLD: PSYCHEDELIC THERAPY

From 1956 on, Abram Hoffer, Humphry Osmond, Duncan Blewett, and their research group relied more heavily on LSD for treating alcoholism and other mental disorders. Because of the strength of the early experiments with this drug as a treatment aid (an estimated 50 percent success rate), the Saskatchewan program expanded. Between 1956 and 1961, the group tested this therapy to its full potential, and eventually it became accepted as the method of choice in treating chronic alcoholics in Saskatchewan.

At the heart of this radical therapeutic approach was the psychedelic concept. Osmond introduced the term "psychedelic," and the Saskatchewan researchers, with the assistance of a few other experts outside the province, created a unique form of drug therapy that incorporated religion (the transcendental experience, to be more precise) as a dominant feature. Blewett, along with his close friend and research partner Nicholas Chwelos (a research psychiatrist in Saskatoon), went further than anyone else had in mapping out the LSD world. Their *Handbook for the Therapeutic*

Use of Lysergic Acid Diethylamide-25 attracted widespread attention and is still regarded by many experienced LSD therapists as "one of the most informed documents on LSD therapy."[1]

When word of these achievements reached the larger scientific community, it was neither quick nor generous with its accolades for the Saskatchewan scientists. As with the earlier findings of Hoffer and Osmond, the discoveries from this phase of experimentation met with disdain in many scientific quarters. Inclusion of the therapeutic element in the hallucinogenic drug project, in addition to the multiple studies of schizophrenia, led to a convoluted state of affairs, as Osmond noted: "Even among our close colleagues, there are few who have a wide view of our research, for the obvious reason that it is spread over many hundreds of miles and covers many disciplines which at best have very little communication...and at worst can hardly find a suitable language for discussion."[2]

Within the field of psychology, Blewett's findings with psychedelics led to the same kind of skepticism that the theories of Osmond and Hoffer had evoked in psychiatric circles. In his attempt to persuade others that psychedelics were valuable tools, Blewett attracted much ridicule from conservative peers. Like psychiatric professionals, psychological researchers in the 1950s found themselves pulled in opposite directions, with a "pseudoscientific" Freudianism at one end and an "ultrascientific" positivism or behaviourism at the other. Seated uncomfortably in the middle were researchers such as Blewett. Like other humanistic psychologists, he advocated that psychology should allow for a conception of science different from the more objective natural sciences. As Amedeo Giorgi wrote, the reason behind this desire was that humanistic psychologists

> refused to consider many of the everyday phenomena of human life—such as experiences, feelings, laughter, meanings, misunderstandings, etc.—to be psychologically irrelevant....Implicitly...they were saying that in order to be faithful to the phenomena of man, either a new type of science had to be invented or the meaning of science as it was understood had to be considerably broadened.[3]

Although behaviourists and psychoanalysts differed on many things, they did share a disregard for religious beliefs and/or spiritual values, humanists' special area of interest.[4]

In analyzing the work of Hoffer, Osmond, and Blewett more closely, it is interesting to note the types of scientists they were. They operated from very different frameworks of interpretation, or paradigms, on multiple issues. From another perspective, however, they shared some characteristics. They were committed to hallucinogenic research and fought persistently to have its value recognized by others. They also exhibited a spirit of freedom and independence in their pursuit of scientific truth. Hypothetically speaking, if one were to apply sociologist David Riesman's famous character typology to the 1950s scientific community, Hoffer, Osmond, and Blewett would likely belong to "the inner-directed type" (as opposed to "the tradition-directed" and "other-directed" types).[5] The trend in Western society in the 1950s was toward increasing bureaucratization and "organization." Many of the sociological critiques from this decade indicate that people from all sectors of society were becoming more other-directed in much of what they did. In this sense, the Saskatchewan researchers went against the trend. As Riesman saw it, the distinguishing marks of the inner-directed person were the "capacity to go it alone"[6] and a lack of concern about "continuously obtaining from contemporaries...a flow of guidance, expectation, and approbation."[7] Although William H. Whyte Jr. used different wording, he offered a strikingly parallel appraisal of 1950s society. Referring directly to the scientific community, he said that "in the outstanding scientist, in short, we have almost the direct antithesis of the company-oriented man....What he asks is the freedom to do what he wants to do."[8] As demonstrated in their research, the rebellious streak in Hoffer, Osmond, and Blewett, their frequent tendency to "go it alone," was both to their advantage and to their detriment.

By 1956, the Saskatchewan research with hallucinogens had reached a point where the term "psychotomimetic" seemed too generic, and far too limited, to apply to substances that caused such radical variations in how people saw, felt, and thought. Even the term "hallucinogen" was a misnomer because it automatically connoted the drug's hallucinatory qualities as the primary characteristic of the experience, but not all those who tried

LSD or mescaline had perceptual hallucinations. Colin Smith wrote that the term "hallucinogen" "is somewhat unfortunate since it draws attention to the perceptual changes produced by the drug: changes which may be slight or absent and which are probably of little importance from the therapeutic standpoint."[9] As emphasized earlier, these drugs could induce a model of madness if that was the desired purpose of the investigator; however, as Hoffer and Osmond noted, "there seem[ed to be] no reason why psychotomimetic agents should not be developed to mimic deliria, mood swing illness or even dementias rather than schizophrenia."[10] One of the more important insights of the Saskatchewan researchers was that the drugs could also facilitate, as Osmond phrased it, "enlargements, burgeonings of reality."[11] He regularly borrowed William James's term "unhabitual perception" to illustrate how the drugs allowed people to perceive things in unaccustomed ways and the value of alternative perspectives.

Osmond believed that a new word was required; otherwise, people would forever associate hallucinogens with their madness-mimicking properties. He mentioned to Hoffer that it was incumbent on them, as scientific investigators, to

> establish that peyote, mescaline etc. properly used may be of great value and not harmful at all. I have already introduced some new names for your inspection which you will see in the paper that is coming out: Psychodelic-mind-manifesting, Psychozymic-mind-fermenting [and] Psycholytic-mind-releasing are possibilities. Dunc Blewett prefers "mind releasing." I am rather in favor of mind-fermenting, though "psychorheczic" is rather a good one. It means "mind bursting forth."[12]

Hoffer replied that, of the three, he preferred "the first, since -ozymic has a chemical connotation referring to some enzyme system and psycholithic has a pharmacological connotation referring to release or destruction of something."[13] Aldous Huxley provided his own thoughts on the conundrum in a March 30, 1956, letter to Osmond:

> About a name for these drugs—what a problem! I have looked into Liddel and Scott and find there is a verb phaneroein, "to make visible

or manifest," and an adjective phaneros, meaning "manifest, open to sight, evident." The word is used in botany—phanerogram as opposed to cryptogram. Psychodetic is something—I don't quite get the hang of it. Is it an analogue of geodetic, geodesy? If so, it would mean mind-dividing, as geodesy means earth-dividing, from ge and daiein. Could you call these drugs psychophans? Or phaneropsychic drugs? Or what about phanerothymes? Thymos means soul, in its primary usage, and is the equivalent of the Latin animus. The word is euphonious and easy to pronounce; besides it has relatives in the jargon of psychology—e.g. cyclothyme. On the whole I think this is better than psychophan or phaneropsychic.[14]

Huxley then ended his correspondence with a rhyme—"to make this world sublime, take a half a gramme of phanerothyme"—to which Osmond replied with an oft-quoted rhyme of his own—"to fathom hell or soar angelic, just take a pinch of psychedelic."[15] Osmond officially introduced the word in 1957 at a meeting of the New York Academy of Sciences, and it has remained with us ever since.

With the new understanding of these drugs, Saskatchewan researchers began to make some necessary modifications to the set and setting of their therapeutic trials with psychedelics. No longer was the emphasis on inducing a psychotomimetic experience in patients undergoing therapy. The principal aim was now to induce a transcendental or psychedelic experience, the method that had proven the most successful in treating alcoholic patients.

At the same time, Al Hubbard was carrying out parallel projects with alcoholic patients at Hollywood Hospital in New Westminster, British Columbia. With his history of bootlegging, espionage, power politics, and big business, he became well known throughout the LSD world; he has also been recognized as one of the major innovators, along with Hoffer, Osmond, and Blewett, of psychedelic therapy. Shortly after Hubbard, an "unlikely combination of mystic and redneck," came upon LSD, he took up a mission, which he pursued with great religious vigour, to have the benefits of psychedelics spread throughout the world.[16] He thus became known as "the Johnny Appleseed of LSD."[17]

Despite his colourful and somewhat shady background, Hubbard was a master of psychedelic therapy, showing the utmost care and respect for his patients. After first meeting with him in Regina in 1955, Hoffer conveyed how impressed he was with his "great warmth and understanding of the subject who is undergoing the experiment."[18] That Hubbard had a bogus medical degree alarmed and infuriated most physicians, but it did not stop Osmond, Blewett, and Hoffer from associating with him.[19] Their mutual interest in psychedelics became the foundation of a close working alliance. Together with Huxley, his friend Gerald Heard, and a small band of others, they formed the Commission for the Study of Creative Imagination to foster creative studies with psychedelics and further studies on schizophrenia.

It was difficult for anyone to deny Hubbard's impressive track record when it came to LSD therapy. His attention to detail in his experiments arguably ranked next to none. As Blewett conveyed, "among the original investigators who studied the psychedelics, Hubbard was the first to grasp the full meaning of the discovery. He shared his vision with the few whom he could guide toward his understanding. Over the decade of the 1950s this group worked to produce research that would make the value of the psychedelic experience apparent to the scientific community."[20]

Hubbard's main contribution to Saskatchewan's psychedelic therapy was his use of elaborate stage props. Hubbard incorporated everything from religious iconography to stereophonic music, paintings, mirrors, and stroboscopic lights, which, if correctly used, prompted visionary reactions from patients. In the winter of 1958, at Hoffer's behest, Hubbard came to Saskatoon to perform a demonstration of his therapeutic technique with LSD on three chronic alcoholics.[21] Although the Saskatchewan group had been pursuing a similar course with its own LSD methods, Hubbard displayed an adeptness in his sessions that, as Hoffer and Osmond admitted, "seemed most apt to lead to the transcendental or psychedelic experience."[22] The Saskatchewan researchers were so intrigued by his style, and his consummate skill with visual and auditory aids, that they decided to adapt some of his techniques to their own program. Blewett and Chwelos, adding some modifications of their own, carried out experiments that led to the publication of a detailed manual containing what was surely "the only available codified information regarding methodology which ha[d] yet been made available to readers."[23]

As with peyote earlier, concerns began to be raised in 1957 about LSD's addictive potential despite no accounts of withdrawal symptoms in the hundreds of scientific papers published. Blewett and Chwelos decided to settle the issue. As Blewett recalled, "if we could show addiction was not a problem, [LSD] would be introduced into treatment programs of every hospital, clinic and jail by [the] early 1960s; however, to my mind, it was not only in the clinic that LSD would prove remarkably useful. It seemed to offer absolutely fantastic possibilities for psychological research."[24] The two researchers spent much of the next year administering LSD to them- selves and treating patients with the new approach. They refined their techniques considerably and learned a great deal about how the drug expe- rience unfolded and how it could be directed to produce the most beneficial results. From their own experiences and information culled from thousands of other documented experiences, Blewett and Chwelos noted the common thread of the LSD state. Blewett cited the following common elements:

1. A feeling of being one with the Universe.
2. Experience of being able to see oneself objectively or a feeling that one has two identities.
3. Change in usual concept with concomitant change in perceived body.
4. Changes in perception of time and space.
5. Enhancement in the sensory fields.
6. Change in thinking and understanding so the subject feels he develops a profound understanding in the field of philosophy or religion. Associations of ideas are much more rapid and clear and one tends to see many alternate solutions to each problem.
7. A wider range of emotions with rapid fluctuations.
8. Increased sensitivity to the feeling[s] of others.
9. Psychotic changes—these include illusions, hallucinations, paranoid delusions of reference, influence, persecution and grandeur, thought disorder, perceptual distortion, severe anxiety to others which have been described in many reports on the psychotomimetic aspects of these drugs.[25]

Scanning the data, Blewett and Chwelos discerned a characteristic pattern of reaction that Blewett broke down into six levels. The first two levels represent an attempt by the subject to escape from the effects of the drug. In a "flight into ideas or activity," the individual tries to ward off the effects by "concentrating either upon concepts or things outside the self or upon some activity which can absorb his full attention."[26] At the next level, the "flight into symptoms," the reaction "seems to be correlated with an inability or unwillingness to direct one's attention to things outside one-self."[27] The focus remains on physiological sensations, sometimes including migraines, nausea, and other illnesses. At these escape levels, "the self concept is maintained despite the action of the drug," and the person can "minimize the psychological effect of the drug by developing an *idée fixe* and by clinging desperately to it in a battle against the drug's effects."[28] This was the main reason that Blewett and Chwelos rejected the psycholy-tic approach of using low doses, which they saw as perpetuating the first two levels and preventing an experience with lasting therapeutic value.

The last four levels made up what had come to be known as the psy-chotomimetic and psychedelic experiences. The psychotomimetic levels were divided into confusional state and paranoid thinking. In the former, the subject often struggles to "rationalize what is happening to him but visual imagery and ideas flood into his awareness at so high a speed that he cannot keep up with them. . . . He rapidly falls behind and loses the con-text." In the latter, the subject is so taken aback by the altered perceptions that he "mistrusts his own sense data and begins to question the validity and reality of everything he does and perceives."[29] The subject sometimes intuits the feelings of others present in the session to an unusual degree, leading him to "hide his incapacities and imperfections from those around him."[30] As important as the psychotomimetic aspect might have been in understanding madness, it had little or no therapeutic value. Only upon reaching the psychedelic levels of experience are the greatest therapeutic benefits realized, because at these levels the subject becomes stabilized and "is no longer concerned with escaping or explaining the drug effects but accepts them as an area of reality worthy of exploration."[31] At the fifth level of experience, the subject "accepts his apparently enhanced intellectual capacity and his ability to empathize with and to appreciate, accept and understand others." Finally, at the sixth level, the individual sees the experience as "offering a new and richer interpretation of all aspects of reality."[32]

Instead of interpreting these levels as distinct, Blewett and Chwelos viewed them as points ranging along a continuum through which the patient progressed gradually. Some patients reached the psychedelic stage considerably faster than others, managing to bypass the more uncomfortable levels with little or no difficulty. Other patients stagnated at the first levels, which usually meant either that the treatment failed or that further treatment was required to work through problem areas. Here again the levels that subjects attained became heavily dependent on the set and setting of the experiment. When considering the physical setting, Stanislav Grof noted that "it makes a great difference whether the session takes place in a busy laboratory milieu, in a comfortable homelike environment, in a sterile medical setting with white coats and syringes or in a place of great natural beauty."[33]

The Saskatchewan researchers ensured that the treatment environment was comfortable and that the mood in the room was one of complete openness and trust. Taking their cue from Hubbard, they used carefully selected props with certain goals in mind:

> The auditory stimuli consisted mainly of music supplied by a record player. Usually classical or semi-classical and relaxing music was played. The person was encouraged to lie down, relax and listen closely. Visual stimuli consisted of various pictures which the patient examined and concentrated on intently. Other visual stimuli...aided him in getting his mind off himself....For emotional stimuli, photographs of relatives were often used....The suggestion was made that he could be markedly aware of unhealthy attitudes toward people in the photographs....He was also asked to concentrate on a list of questions that he had previously compiled about his problems.[34]

The physical and psychological makeup of the patient also had an enormous influence on the outcome of the treatment procedure. Blewett, Hoffer, and Osmond diverged to some extent on this subject. Whereas

Hoffer and Osmond came to emphasize the biochemical factors of the psychedelic experience, Blewett saw it mostly in psychological terms: *"Self-unacceptability is the root of all functional psychopathology*; and if self-acceptance can be achieved through the self-understanding provided by psychedelics, the greatest of strides toward recovery will have been achieved."[35] These objectives—self-understanding and self-acceptance—were at the heart of the manual by Blewett and Chwelos. As they noted in its introductory pages, "on the basis of self-knowledge...the patient can, with the therapist's help, clearly see the inadequacies in the value system which has underlain his previous behaviour and can learn how to alter this in accordance with his altered understanding."[36] A critical variable was whether or not therapists provided the necessary support and guidance to subjects. If the subject was to reach his or her fullest potential with the drug, then it was largely up to the therapist to impart some organization to his or her "unhabitual perceptions." This job as psychedelic guide resembled, in many ways, that of the shaman or sage of ancient cultures.

The Saskatchewan therapists employing LSD and mescaline devised their own approaches to dealing with their patients. Nonetheless, all of them aimed to have their patients attain the psychedelic experience, and all of them accentuated the importance of having some personal experience with the drug before any therapy commenced. This practice sparked controversy because it was seen as antithetical to the objectivity of the experiment.[37] One innovation that Blewett and Chwelos introduced, and one that caused concern even among established LSD therapists, was that the therapist take the drug with the patient. They believed that there were a number of advantages to this method, including enhancement of the doctor-patient relationship and formation of a much stronger empathic bond. As Blewett phrased it, "the doctor-patient relationship where the doctor assumes a role of authority cannot persist if the patient is to achieve a high degree of self-acceptance for the relationship implies an inferiority in the patient."[38] By sharing the drug, doctor and patient were placed on an equal level. [39]

In addition to their alcoholism treatment project, Blewett and Chwelos attempted LSD therapy, on a much smaller scale, with inmates at the Regina penitentiary. Their goal was to effect changes in inmates' values in order to prevent recidivism. The results, however, were not nearly as positive as those obtained with the alcoholic group. Blewett and Chwelos contended that in the penitentiary "the subject, should he come to the realization of the possibility and advisability of altering his value system, goes back into a cell block in which day to day living is conducted on the basis of the mores of a community in which values are almost diametrically opposed to those he has encountered in the LSD experience."[40] This psychedelic approach to criminal psychology was later adopted by Timothy Leary when he and some of his Harvard University associates administered psilocybin, the synthetic version of the magic mushroom, to a small group of inmates at the Massachusetts Correctional Institute in the early 1960s. The short-term findings of this study revealed that only 25 percent of those who had the drug experience returned to jail, whereas the average rate (for those not having the experience) was 80 percent.[41]

Following the modifications to Saskatchewan's LSD therapy program in 1958, Chwelos, Blewett, Smith, and Hoffer released the findings from a second trial group of alcoholic patients. Even though this new group, sixteen, was smaller than Smith's original pilot study with twenty-four refractory patients and the follow-up time somewhat shorter (six–eight months compared with Smith's eighteen), the results of the second trial indicated a marked improvement in the patients. When the trial was finished, ten of the sixteen patients stayed completely sober, five showed a substantial reduction in their alcohol consumption, and only one remained unchanged.[42] The Saskatchewan researchers concluded that "self-surrender and self-acceptance are more easily achieved in the LSD experience...[and that] from the psychological point of view the resolution of the problem of the alcoholic lies in this surrender."[43]

The Saskatchewan research received wide acclaim, in the wake of this study, both at home and abroad. Within Saskatchewan, the Bureau of Alcoholism and AA hailed it as a startlingly effective innovation in treatment. The response was so enthusiastic, in fact, that similar therapeutic operations began to spring up. Under the watch of Osmond in Weyburn, Dr. Sven Jensen carried out one of the more intensive treatment programs to date in the province. With the assistance of AA and an expanded bureau, with its Counselling and Referral Centre, Jensen combined individual and group therapy, weekly AA meetings, and occupational and milieu therapies into a two-month treatment program that culminated with a guided LSD experience. Modelling his approach on the modified techniques of Chwelos and his colleagues, Jensen also obtained what to many were unbelievable results. Of the fifty-eight patients whom he treated and assessed during a six- to eighteenth-month follow-up, thirty-four remained totally abstinent, seven had improved, thirteen had not improved, and four had "broke[n] contact."[44] This provided further compelling evidence of the efficacy of LSD treatment in alcoholism.

In line with the other studies, Jensen assessed the dilemma of the alcoholic, noting that "the final breakdown and the final commitment to a new way of life come in the course of LSD therapy during which a transcendental or spiritual experience may further aid in establishing self-acceptance."[45] Jensen prepared prospective patients well ahead of time, letting them know that LSD would not "prevent them from drinking but...make them understand why they drink and what they can do about it."[46] He stressed that the LSD pretreatment phase would help them to develop the necessary motivation for the therapy to succeed.

Jensen's program proved to be one of the more convincing comparative studies. While the fifty-eight alcoholic patients were being put through the full program, two other groups received an alternative treatment. The results from a second group of thirty-five patients who did not receive LSD were four improved, nine unimproved, and eighteen lost to follow-up.[47] Jensen wrote that this group was "composed of patients who left the hospital early, who were considered unfit for LSD therapy for physical reasons, or who refused this treatment."[48] Another group consisting of forty-five controls was given conventional psychotherapy on an individual basis with other psychiatrists. Of that number, seven were abstinent, three improved, twelve unimproved, and twenty-three lost to follow-up.[49]

The next LSD treatment program to appear was at Moose Jaw's Union Hospital under the direction of head psychiatrist P. O. O'Reilly. Like the studies in Regina and Saskatoon, O'Reilly's therapeutic session was designed as a short-term procedure in which the patient was normally admitted to the hospital ward and prepared for the drug experience over two or three days. O'Reilly added to his program the position of "special nurse," who became a surrogate mother to the patient undergoing treatment. It was up to this nurse to "establish rapport with the patient [and]…explain and discuss with him all aspects of LSD treatment and…carry the patient through the LSD experience."[50] What was interesting about the study was that O'Reilly had initially debunked the whole idea of using LSD in therapy. Figuring that the drug would have little if any merit in the process, he agreed to test it for himself. His own positive findings forced him to rethink his misgivings. In one of the first reports to surface from the Moose Jaw tests, thirty-three patients (twenty-nine male and four female) had been treated and followed up over periods ranging from seven to eighty-eight weeks. Of the thirty-three individuals, seven were abstinent, seven improved, ten unchanged, and six unavailable for report.[51]

As a separate side project, O'Reilly decided to test the hypothesis of Blewett and Chwelos that the therapist ought to take LSD with the patient in order to increase the empathic bond. O'Reilly, however, was unable to validate this theory. According to his account of the experiment,

> Two psychiatrists (one who had two previous LSD experiences) were given 200 gamma of LSD together. Three observers consisting of two psychiatrists and one psychologist acted as therapists. The psychiatrist who had previously had LSD had a pleasurable experience and the other psychiatrist taking the drug for the first time had an unpleasant one. Throughout the six hours that both psychiatrists were under the influence of the drug neither had any idea of how the other felt. A complete block seemed to exist between them while under the drug, yet both were communicating their feelings to the observers. At one stage the psychiatrist having the bad experience became upset over his feelings but this did not register with the other subject.[52]

The limitations of the method were also cited in an independent follow-up report (two months to five years) filed in 1962 by the Bureau of Alcoholism on 150 patients whom it had referred to the four psychiatric institutions for LSD therapy. Although the method of Blewett and Chwelos benefited some patients, the recovery rate (22 percent) was much lower than anticipated.[53]

Despite the limited success, Hoffer and Osmond thought that this particular mode of therapy was "of some importance in determining the general nature of the LSD experience, not the least of which was that further studies should be made on the significance of the group LSD experiment."[54] The technique could "have value in the broadening or expanding of individual awareness and understanding of interpersonal relationships in groups of individuals who have gained self acceptance through previous experience."[55] And there might be some advantage in the therapist and the patient doing LSD together, but "the long-range effects of repeated dosages of that chemical upon brain and psyche remain to be determined.... Given the paucity of knowledge in this area, repeated exposure to LSD seems too risky and this risk should deter the guide from regular LSD self-exposure whatever the possible advantage in terms of therapeutic gains."[56]

The actions of Blewett and Chwelos provoked fear even among their closest colleagues. As Blewett recalled,

> After about 120 of these sessions we felt that we had proven our point. Unfortunately, our superiors had come to an opposite view and had become deeply concerned that we were addicted. We had to knock off taking acid for 6 months before the absence of withdrawal symptoms satisfied our compatriots. It was the end of our work together because, by the time the 6 months were up, the situation had changed, acid was being frowned upon by the psychiatric establishment and the Government Research with LSD ground to a halt.[57]

At the height of its popularity, the province's LSD treatment program produced rates of recovery of about 50 percent.[58] That outcome might not have impressed some, but it was unprecedented considering that conventional means of treating chronic alcoholism usually ranked in the lower percentiles (e.g., 15–20 percent with AA). Even more important, the favourable outcomes were achieved with patients who had been heavy drinkers for years and suffered from every conceivable health complication arising from severe alcohol abuse: delirium tremens, blackouts, cirrhosis of the liver, and so on. Many of these individuals came from extremely troubled backgrounds, such as prison terms, poor work histories, marital problems, physical abuse, and barbiturate addiction.

One of Smith's patients, a psychopathic alcoholic who fit into many of the above classifications, regarded his LSD session as transformative:

> The experience I have had as a result of the drug was unique and most certainly constructive. It manifested itself in many ways, the most important being my ability to recall certain events, some pleasant, some very unpleasant. It allowed me to see myself as I really am. . . .
>
> I recalled many of my war experiences. I even saw the crucifixion which would possibly indicate to me that my life is lacking in spiritualism. Many other outstanding incidents where I lied to people, lost jobs and affected the lives of many other people through my drinking passed through my mind. . . .
>
> I had just begun to recall things which happened in my immediate past, my domestic life, etc. when the drug wore off so if possible I am going to request permission from Doctor to go through the experience again and, if necessary, a third time.[59]

This patient did have another LSD experience and a short time thereafter attained sobriety and completed a university education. Smith was well aware that "one case, of course, proves nothing,"[60] but such reactions were common in many of the drug sessions.

The treatment programs had their share of failures, too: patients who either failed to respond to the therapy or relapsed shortly afterward. As Smith noted, many alcoholics had "literally nothing to return to after leaving hospital except their friends in Skid Row and the bottle. They were jobless or odd-job men and often little could be done to aid them in this respect. . . . With better selection of cases and more adequate rehabilitation facilities, the results could have been greatly improved."[61] It is also possible, as Hoffer and Osmond theorized, that individual biochemistry was a factor.

Questions arise about the validity of these research findings. Although questionnaires filled out by patients who had gone through the procedure were of some value in determining factors such as motivational change, such subjective reports could hardly be regarded as the only criterion for judging the results. Likewise for the therapists' own conclusions. Fortunately, with many of the Saskatchewan studies, reliability of the data was backed up by objective observations from patients' family members, friends, employers, as well as outside agencies such as AA and the Bureau of Alcoholism. A senior counsellor for the bureau, Angus Campbell, pointed out that in its follow-up study the officials "purposely were conservative in making our counts. In cases of doubt, we made it a rule to mark the results negatively. We therefore feel that a careful re-check might indicate some higher rates of recovery than our tables show."[62] Severe restraints on both time and money made it nearly impossible for therapists to carry out extensive follow-up studies.

Nonetheless, the pioneering efforts of the Saskatchewan research team earned them respect and praise from many of their peers who drew inspiration and ideas from them. Reaching its peak in 1961, Saskatchewan's psychedelic research program stood apart in using LSD to treat alcoholism. Although official research ended within approximately a year, the amazing claims regarding treatment (improvement and sobriety rates of between 50 percent and 80 percent) lived on in prominent scientific journals and the popular press.

Given its many successes, LSD treatment in the province had become official policy of the Department of Public Health. As noted in a Bureau of Alcoholism report, the single, large-dose LSD treatment for alcoholism was "no longer [seen] as experimental" but should "be used where indicated."[63] At the turn of the decade, LSD treatment programs operated out of four medical facilities in the province (Weyburn, Regina, Saskatoon, and Moose Jaw), all with comparably successful outcomes. Hoffer and Osmond reckoned that their program was "faster, cheaper, and twice as effective as any other program. By contrast, standard psychiatric treatments, including group therapy, psychotherapy, psychoanalysis and so on,...take longer, cost more and do not produce satisfactory results."[64]

Such bold statements stirred the ire of many in the field of research on alcoholism as well as in the larger medical community, but Saskatchewan's psychedelic research program and therapy continued to fascinate and inspire others, both locally and internationally. Hoffer later wrote about a visit with Florence Nichols, a psychiatrist who had experience working with LSD in England and subsequently picked up on the psychedelic technique and applied it in her own treatment of alcoholics at the Gordon Bell Clinic in Toronto:

> She approached the Ontario Alcoholism Foundation saying she knew how to use LSD and that it would be an excellent treatment for alcoholism. John Armstrong would have nothing to do with her. Later Al [Hubbard] went there and gave Bell LSD. Nichols treated approximately 100 alcoholics, all failures from other treatments including Bell's three weeks of lectures. She claims 90% success, sober or much improved.[65]

The Saskatchewan work also drew interest from outside the country. American researcher Myron Stolaroff wrote to Hoffer that he and his colleagues "were quite stimulated by the exciting news of the progress of your work in Saskatchewan and the report of your experience there has done a great deal to broaden our understanding of this whole field of mental illness."[66] This success led Hoffer to make the following observation to Osmond on December 27, 1961:

> It is curious how easily LSD as a treatment is spreading....Except for one noisy critic on the West Coast there has been no major opposition. Perhaps our suggestion [that] it must be combined with psychotherapy...helped—it proves that physicians will readily accept a treatment if it can be fitted into a well-known theoretical system (which may be true or false). LSD can be described in well-known terms such as abreaction, revivication of memory, psychotherapy, insight, empathy etc.—but with nicotinic acid for schizophrenia or for arthritis the response goes counter to accepted ideas and there is great resistance even though it is better established as a treatment.[67]

Within Saskatchewan, the LSD treatment elicited positive responses from the agencies and staffers tasked with referral and follow-up of cases. Local media outlets often provided "heartening" case histories of individuals who had entered the provincial program with seemingly intractable alcoholism and come out of it changed for the better, maintaining sobriety on discharge into the community and going on to achieve success in their personal and professional endeavours.[68] The exuberance over the results extended into the political realm, with local politicians waxing poetic on the virtues of LSD treatment. Speaking on radio for a 1960 alcoholism info week, CCF Minister of Health Walter Erb noted that

> some real advances had been made here in Saskatchewan and...I should...take this opportunity to pay tribute to the wonderful research that has been done in connection with the new drug LSD 25. It should be a matter of pride to Saskatchewan citizens that our own Psychiatric Services Branch has spear-headed the research in this area.[69]

For a time, the work on alcoholism continued unabated, treatment centres continued taking on new patients, and because of the long wait list another operation began at the Saskatchewan Hospital–North Battleford. There was even talk in the bureau of beginning a separate treatment outlet. But soon a host of factors combined to destroy the potential held out by advocates of LSD treatment.

CRITICISM OF SASKATCHEWAN'S LSD TREATMENT PROGRAM

The province's LSD treatment program had its critics as early as 1958. They consistently pointed to a lack of scientific evidence to back up what they considered to be spurious and preposterous claims. University of Virginia psychiatrist John Buckman summarized the mood in the psychiatric community:

> Some of the early reports were so unreservedly enthusiastic and so wild in their claims of success...that they succeeded in antagonizing much of the informed psychiatric opinion. Many therapists were outraged because of the threat to their own omnipotence. Many

were justifiably concerned about the irresponsible use of a powerful drug on unsuspecting patients or volunteers. As a reaction to the early reports that here, at last, was the answer to the problem of mental illness, there began to appear publications stressing mostly the dangers of suicide and psychosis, and accusing those who were using LSD of charlatanry and self-deception.[70]

Whenever treatment claims surfaced from the province, psychiatric and other medical professionals were quick to question the efficacy of LSD therapy and scrutinize perceived weaknesses in the research such as experimenter bias, the uncontrolled treatment environment (i.e., patients and therapists knowing the drug being administered and therapists taking the drug with patients), and the failure to use double-blind methodology in particular. And from 1960 on reports began to warn about the adverse effects of LSD.

An early and vociferous opponent of the findings was the head of the University of British Columbia's Department of Psychiatry, Dr. J. S. Tyhurst. When he discovered that the *Vancouver Sun* was reporting on Saskatchewan's success in the psychedelic treatment of alcoholics and similar outcomes achieved by Drs. Ross Maclean, D. C. McDonald, and their colleagues in British Columbia, Tyhurst issued his own belief that LSD precipitated psychoses and had negligible benefits. As far as he was concerned, the efficacy of the drug in treating alcoholism was far from conclusive. To further discredit the Saskatchewan research, he even circulated rumours that Hoffer was funnelling LSD to the underground and should therefore be fired from the University of Saskatchewan. Tyhurst went public in 1959 and insisted that LSD treatment had no therapeutic value and in no way satisfied the criteria for orthodox scientific methods; as he saw it, the LSD research was "more characteristic of a cult than of responsible scientific investigation."[71] Yet he had little understanding of LSD and its effects and no experience using the drug in treatment settings, so he was not in the best position to make authoritative statements. Researchers knowledgeable in the area paid scant attention to Tyhurst. As Blewett sarcastically remarked, Tyhurst knew "as much about it as I do the type of ballet danced in Persian harems."[72] Osmond and Hoffer also realized that much of the early skepticism was fuelled by an aggressive campaign of misinformation led by the likes of Tyhurst, who had no conception of the drug or its use in therapy. The campaign that he launched nevertheless fomented a scandal in British Columbia that eventually spilled over into the rest of the country.[73]

Perhaps Tyhurst's efforts would not have had much power or backing were it not for the fact that psychedelic drugs were beginning to spread into mainstream society at an alarming and uncontrollable rate. In what quickly became a trend, people from all walks of life, not just scientists but also artists, students, clergymen, businesspeople, and politicians, flocked to these experiments to see for themselves what the drugs would do. The research also began to be reported more regularly in the lay press. The mounting publicity on LSD worried governments, some of which initially had other purposes in mind for mind-altering substances, and the drug's main distributor, Sandoz Pharmaceuticals. Responding to the widely circulated press report (in which Hoffer was quoted) in the *Vancouver Sun* in 1959 regarding LSD as a cure for alcoholism, officials began to express concerns. Dr. John Grosheintz, the scientific director at a Sandoz subsidiary in Montreal, wrote to Hoffer on September 22, 1959:

> As you know, LSD has been under investigation for a great number of years and some 600 major publications on this drug have been made. In compliance with the Food and Drug Act, therefore, a decision as to the status of this drug will have to be made. In the meantime no further supplies of LSD will be available for research purposes. You are probably aware of the deplorable lay publicity which recently started in B.C. in connection with the treatment of alcoholics with LSD. I would be much obliged if you could state anything concerning the source of supply of LSD available to the staff at the Hollywood Hospital, and whether the work there is truly in the nature of a research project.[74]

Concerned with the growing pressure and unwarranted attacks on legitimate scientific research projects, Hoffer responded to Grosheintz days later:

I wonder if you would let me know whether the Food and Drug division, that is Dr. [C. A.] Morrell, has advised you to discontinue shipment [of] LSD. I am sure that when the Macy-Sandoz Conference is published there will be even more sensationalist reports and you ought to be prepared for them.[75] When someone discovers that a simple treatment allows a really sick person to recover, you do not expect them...to mildly report these findings.[76]

In this and subsequent correspondence, Hoffer tried to reassure the Sandoz representative that the Saskatchewan project was well designed and that the province was even prepared to pay for the LSD since it was being used as a routine treatment. Hoffer maintained that he did not know where the BC group was procuring its LSD.[77]

Hoffer reiterated that it was his research team's strict policy to have results published in the medical press before any reports to the lay press. As he viewed it, Tyhurst's attacks were chiefly to blame for the uproar in British Columbia, even though Tyhurst had no first-hand experience with LSD. Regardless of Hoffer's efforts to temper the controversy, though, the Food and Drug Directorate in Ottawa inevitably followed the course of its American counterpart and placed LSD under new regulations. Obtaining the drug for research purposes soon became a much more complicated process. Numerous forms had to be filled out prior to any shipment being cleared. This restricted the drug's availability to a few chosen investigators.[78] Also, the Sandoz patent on LSD expired in 1960, so after that year any drug producer could have made LSD legally. This left governments scrambling to do something to prevent widespread production of it. For Blewett, who no longer had access to the drug, the situation was entirely avoidable:

During the 17 years that LSD had been available, the government had never faced up to the problem of how LSD should be distrib-uted. The Food and Drug directorate did not wish to open the use of LSD to the medical profession. LSD was, therefore, declared an experimental drug and the distribution was left up to the chemical firm which produced it under patent. This control could no longer operate. Isolated places continued to use LSD for a short while but within a few months all the centres which had produced the early research were shut down.[79]

From the standpoint of Hoffer and Osmond, the new controls, though inconvenient and unfortunate because they barred fellow experienced professionals such as Jensen, Blewett, and Chwelos from using the drug in research and treatment, were not necessarily a bad thing. As they saw it, the new controls could prevent the growing indiscriminate uses and abuses cited by critics, hence rendering professional outbursts à la Tyhurst null and void. For Hoffer, the survival of psychedelic therapy depended a great deal on having trained professionals, and he laid out a possible solution to his partner in June 1961:

In Ottawa I met Dr. Murphy who is the physician in charge of Food and Drugs for the Department of Health and Welfare. I outlined to him the entire LSD problem in Canada....He suggested that he might set up a committee which would...examine the credentials of physicians who are interested in using LSD as a treatment either for research or for service. We would only insist upon these people having an MD and having had some experience in the use of LSD but we would not of course insist that they have themselves taken it. Any physician in Canada could then apply to this committee and would send along their qualifications. If they had no experience with LSD they would be required to go to some centre to get it which could either be Weyburn, Battleford, Saskatoon or Regina....[The] committee would notify Sandoz or [an]other company providing psychedelic drugs that they ought to supply these physicians. [This]...will then make it possible for people like D.C. McDonald and Nick Chwelos to get LSD to use in their private practice.[80]

The perceived dangers associated with LSD, however, continued to take precedence over any counterarguments about its merits.

CHAPTER 5

NEW FRONTIERS IN PSYCHEDELIC RESEARCH

I n 1957, the experimentation with psychedelic drugs in Saskatchewan focused on the treatment of schizophrenia and alcoholism. However, Abram Hoffer, Humphry Osmond, Duncan Blewett, and their group of investigators also prepared the way for studies in areas where few researchers had gone before. With increasing use of LSD and mescaline in psychotherapeutic processes in the late 1950s, medical interest shifted toward wider "psychedelic" implications and away from the narrow "madness-mimicking" interpretation upon which much of the initial research had been founded. Osmond played a critical part in this movement, but he never lost sight of the importance of hallucinogens as psychotomimetics, and he maintained that they could remain valuable in areas of research on schizophrenia other than his and Hoffer's biochemical-based work. Osmond saw no reason why the psychedelic and the psychotomimetic could not be systematically explored. This openness permitted him and others to chart new research paths with hallucinogens. Two of their undertakings focused on applications of hallucinogens in the design of mental hospitals and the field of parapsychology.

Using the psychotomimetic experience as their basis, Osmond and prominent Regina architect and friend Kiyoshi Izumi came up with a new concept in mental hospital ward design (the "sociopetal"), which they hoped would benefit long-term patients, particularly schizophrenics. This novel idea eventually became a source of inspiration for a host of intriguing perceptual experiments and the design of a radically different psychiatric facility. It also became a significant contribution to a relatively new field, environmental psychology. On another front, Hoffer, Osmond, and Blewett, further examining the psychedelic experience, found that hallucinogens' ability to bring about increased levels of sensitivity/empathy could result in parapsychological phenomena such as ESP (e.g., telepathy).

Of course, these announcements, like the other Saskatchewan findings with hallucinogens, received a lukewarm response from the psychiatric community and wider medical establishment. Looking back at this period of the research, Osmond recalled how it had run into "a very unusual situation...in which a telescoping has occurred and several phases have got mixed up....The normal more or less orderly process of hypothesis, experiment, confirmation, technological application and general wide use has been distorted so that confirmation and technological application are overlapping with experiment."[1] He understood the confusion that likely arose among outside observers, but he was troubled by wholesale rejection of the research among its opponents. As he commented to Hoffer in a February 1960 letter, "the curious thing is that they do not select any particular finding and subject it to great criticism but...pass global judgment upon our whole program."[2] Many who considered themselves informed thought that the research was muddled, lacked proper focus, and had no coherent methodology. Hoffer and Osmond countered that methodology "should remain the handmaiden of scientific discovery and should not attempt to become its taskmistress."[3]

Parapsychology provides another example of disagreement among scientists over what constitutes legitimate scientific inquiry. Years before psychedelic drugs became a factor in the investigation of such phenomena, parapsychology was disregarded by a large majority of scientists. In his analysis of altered states of consciousness and spiritual phenomena, Charles Tart asserted that "they are purely internal experiences. Thus, to a physicalistic philosophy they are epiphenomena, not very worthy of study unless they can be reduced to a physical basis."[4] Carl Rogers argued that

> the evidence for extra-sensory perception is better than, or certainly as good as, the evidence for many principles which psychologists believe. Yet, with very few exceptions, psychologists reject this evidence with vehemence....The psychologist falls back on his subjective knowing. The evidence does not fit with the pattern of knowledge as he expects to find it. Therefore he rejects it.[5]

That is the problem with revolutionary science. The majority of scientists rejected the research with hallucinogens simply because it went against long-held and cherished beliefs about how science should unfold. The new lines of inquiry of the Saskatchewan researchers were no exception.

PATIENT-CENTRED ARCHITECTURE AND THE SOCIOPETAL CONCEPT

With the gradual transition from custodial care to community mental health in Saskatchewan in the 1950s, there arose an interesting debate over the design and purpose of psychiatric facilities. The fundamental flaw with many of the mental hospitals constructed before 1950 was that the patients' stake in their design had been completely ignored. Administrators began to realize the impacts of ill-conceived designs on patients. According to sociologist Erving Goffman and others, the poor designs of mental hospitals actually contributes to, rather than amelio- rates, patients' illnesses. Goffman wrote that "the new inpatient finds himself cleanly stripped of many of his accustomed affirmations, satis- factions, and defenses, and is subjected to a rather full set of mortifying experiences: restriction of movement, communal living, diffuse authority of a whole echelon of people, and so on."[6]

When a special committee of the World Health Organization (WHO) examined this issue, it concluded in a 1953 report that, "if the hospital is to become a therapeutic community,...it must model its architecture and its plan on that of the community. If it is to support and recreate the sense of individuality in patients, it must not dwarf them by its size and by herding them together in thousands in giant monoblock buildings."[7] Shortly following this report, a movement took root in Saskatchewan dedicated to emphasizing the mental patient. As the embodiment of this movement, the "Saskatchewan Plan" made extensive proposals to alleviate the weaknesses of existing mental hospitals (e.g., huge size). Among the proposed changes was a return to the small, cottage-type dwellings advocated by people such as Drs. Thomas Kirkbride[8] and David Low.[9] As Dr. Stanley Rands, deputy director of the PSB, pointed out, the Saskatchewan Plan "envisaged dividing the province, except for the areas immediately surrounding the two existing mental hospitals, into regions with a population of about 75,000 each and building in each region a small mental hospital to provide all psychiatric services in the region. The hospitals were to consist of cottages each housing 30 patients, subdivided into groups of ten."[10]

The plan also prepared for in- and out-patient services, life skills, occupational therapy, and a host of follow-up services (e.g., travelling clinics, home visits, and consultations with family members) provided by tightly knit teams of psychiatrists, psychologists, nurses, and social workers.[11] The underlying purpose was to aid the patient after discharge. The proposed new mental hospitals, or "regional psychiatric centres" as they came to be called, would adopt a preventative approach to treatment. In other words, every possible effort would be made to keep the patient out of the hospital and in the community and home as a fully functioning member of society. Hospitalization would occur only as a last resort. For patients requiring long-term care, every conceivable effort would be made to accommodate them.

Given the increasing emphasis on deinstitutionalization and reduction of patient numbers in the hospitals in Weyburn and North Battleford, Osmond and Hoffer questioned the proposal to create smaller regional psychiatric hospitals. They believed that the Saskatchewan Plan was outdated even before it was implemented.[12] They believed that money would be better spent on a research hospital in Saskatoon that could attract mental health professionals to work on new ideas and methods in training, treatment, and care.

When it came to designing the new psychiatric facilities, many professionals speculated on how a hospital could be built to best reflect the needs of the patient. The province's research on psychotomimetic/psychedelic drugs led to some revolutionary design elements. Front and centre in this debate was Osmond, long interested in the effects of various physical environments on patients. Many of his initial thoughts on the subject flowed from his thorough readings of patient autobiographies as well as the writings of ethologist/zoologist Heini Hediger on animal behaviour in zoos. Also valuable to Osmond in his exploration of the world of the mental patient was the psychotomimetic experience. As early as 1953, he recommended to Hoffer that the use of psychotomimetics would give them "a very clear idea of the impingement of ward life and surroundings of the sick person."[13] He continued:

> We have often heard it said that the grievously mentally ill are not aware of or do not worry about their surroundings. All the evidence we have is exactly to the contrary and yet we repeatedly subject them to conditions which are aesthetically unsatisfactory for well and hale people.... The advantage of seeing psychiatric wards through psychotic eyes (and brain) lies in having a description by a known observer.[14]

Osmond then issued a challenge to the rest of the psychiatric profession "to present the needs of mentally ill people so that these needs can be understood by members of another profession—architecture."[15] The main premise behind his reasoning was that "structure will determine form, unless function determines structure."[16]

Because mental patients, particularly schizophrenic ones, perceive things in such a distorted way, the psychiatric hospital "requires surroundings which would allow them to make the most of their assets and which aggravate their disabilities as little as possible."[17] For years, psychiatrists had debated whether schizophrenics experienced abnormalities in perceptual apparatus or not. There was sufficient evidence to verify that in fact they did. As Teddy Weckowicz found from his own research with schizophrenics, "size constancy becomes disordered," and it leads to "disturbed perception of distances between objects."[18] This problem contributed to patients' inability to form a "subjective space," which, Weckowicz noted, can in turn lead to "a loss of feeling of reality and distortion of the bodily image, as, for instance, feeling small."[19] So long corridors, overconcentration of patients in small and large spaces, and lack of privacy would have to be avoided in the designs of future buildings. Although patients differed significantly in their symptoms, Osmond noted that all of them had experienced "a rupture in interpersonal relationships resulting in alienation from the community culminating in expulsion or flight. They are to a greater or lesser extent socially isolated. The psychiatric ward then has to be designed to care for people whose capacity to relate to others has been gravely impaired."[20]

The psychotomimetic experience continued to demonstrate its value as an educational tool for psychiatrists and nurses. Osmond believed that it could also aid the architect trying to design buildings for mental patients. Izumi took up the challenge. In 1954, he was called on by Dr. Griff McKerracher to draw up renovations for the Saskatchewan Hospital–Weyburn. Surveying the hospital in the 1950s, Izumi recalled that

> a room full of mentally defective children, about 60 of them in cribs butted together with narrow aisles between, who had all kinds of physical deformities, thin heads, large heads, short legs, long bodies, each in a crib, appeared as a visual embodiment of how different people can be. And yet it was apparent that their helplessness had certain similarities.
>
> I suddenly realized that if they could be so different physically, how different mentally they must be—how they perceive things differently.... The really bad ones were in the hospital basement, mixed up with the mentally ill. The physical aspects of the hospital wards were really depressing. Frankly I couldn't understand how it could have been left like that even with the war. It was one hell of a mess.[21]

After further discussions with Osmond, Hoffer, and Blewett, Izumi agreed that a psychotomimetic experience might give him a clearer idea of what his clients had to live with daily. His experiences with LSD, as he attested, were of tremendous benefit to the design process:

I began to comprehend many of the patients' remarks and concerns. For example, how a room "leaked" and a patient saw himself flowing away....To be "startled" by the monotony of one color, such as beige throughout the institution, may sound contradictory, but there was such a phenomenon, which could immobilize a patient. Similarly, the ubiquitous terrazzo floor, suspended ceiling, and similar "uniformity" added to the patient's confusion in relating himself to time and space.[22]

Izumi became acquainted with the lack of privacy and personal space for patients and what it was like for them to contend with extended corridors and eerie noises from fan belts and generators. He began to work closely with Osmond and his colleagues to make renovations in the old mental hospitals and to lay down principles of design for the new ones. As Osmond remembered, these principles were based upon "the assumption that persons with disturbed perception need a simple, familiar environment in which they can enjoy friendship within a small group and yet absent themselves, physically, when they wish, rather than relying on psychological withdrawal, which is much harder to control."[23]

The collaboration between Izumi and Osmond resulted in what was surely one of the most radical alternatives to traditional psychiatric ward design.[24] The "sociopetal" design, as Osmond termed it, was unlike anything seen before in the design of psychiatric buildings. The focus was clearly on patients and not on administrators' overriding concerns with security, control, and economy. When the two presented their design for the new hospital in a 1957 article entitled "Function as the Basis for Ward Design," it met with some enthusiastic and critical responses from both psychiatric and architectural practitioners. After reading the article and scanning the designs, many of these professionals concurred with suggestions such as the elimination of long corridors, the use of smaller, more intimate spaces, and individual retreats. There was also widespread agreement that the sociopetal concept overall was intriguing and deserving of further experimentation. Speaking for his Alabama architectural firm, Moreland Griffith Smith noted that Osmond's "presentation of the characteristics of the psychotic brings us closer than we have been able to come before to the kind of information needed in order to design intelligently for mental patients."[25]

Strong doubts remained, however, about the circular/semicircular plan for the buildings. For some, it was too radical a departure from the standard rectangular design and could be too costly and structurally unsound. Some critics even lambasted the design for contradicting Osmond's insistence on providing the mental patient with "familiar surroundings." As psychiatrist Zigmond Lebensohn put it,

> if *familiar surroundings* seem changed to the schizophrenic, it would seem appropriate to construct a hospital which would be as familiar and reassuring as possible. It is doubtful if Mr. Izumi's semi-circular solution, provocative as it is, will be either familiar or reassuring to most psychiatric patients....The...design of the semi-circular building with the multi-leveled ceilings may be difficult for the patient to grasp.[26]

Even though Osmond's theories remained to be tested, they provoked interest at home and abroad. In Saskatchewan, his ideas led to further studies in perception and some important sociological studies of mental patients at the Saskatchewan Hospital–Weyburn under Weckowicz, anthropologist Edward Hall, and research psychologist Robert Sommer. His ideas also inspired construction of new psychiatric centres promoted under the Saskatchewan Plan.

Between 1955 and 1958, Osmond and Izumi met regularly with McKerracher and his PSB successor, F. S. Lawson, to discuss the design of psychiatric hospitals. The emphasis tended to be on long-term patients, whose average stay was about seven years, rather than short-term patients, usually admitted and released within four months. Using information gathered from their psychotomimetic/psychedelic studies and the ongoing perceptual work of Weckowicz, Sommer, and Hall, Osmond and Izumi sought to establish better principles in architectural design to account for the suspected perceptual difficulties of patients, especially schizophrenic ones. As Osmond explained to Hoffer, "these designs start with the assumption that ill people have disturbances in perception similar to those found in LSD, adrenochrome etc. and these assumptions have been considerably supported by our perception experiments."[27]

From the outset of their research with these drugs, Hoffer and Osmond had put forward a solid case for these chemical substances to allow the normal person to probe the inner world of the schizophrenic. And, though it might have been easy for skeptics to say early on that they were imagining that their patients experienced the same things, there was enough evidence by the late 1950s that much of what Hoffer and Osmond had been saying was correct in many respects. One effective way to reconstruct the *"Umwelt* or life-space" of the schizophrenic was to consult the small number of patient autobiographies available at the time.[28] The literature proved that schizophrenic patients had the same kinds of perceptual anomalies that individuals encountered under the influence of psychedelic drugs such as LSD and mescaline.

Expanding on their perceptual studies[29] at the Saskatchewan Hospital–Weyburn, Sommer and Weckowicz established that "disturbances of ego, self-concept and body image have always been regarded as central in the psycho-pathology of schizophrenia."[30] They went on to assert that "the boundaries of the self become loose or blurred and the patient may feel, for example, that parts of his body do not belong to him or that he is part of the plants, animals, clouds, other people or of the whole world and they are part of him. He may feel at one with the whole of mankind."[31] This aspect of depersonalization was a common trait of the LSD experience. As Weckowicz stated, "hallucinogenic drugs produce very often symptoms similar to depersonalization that are sometimes indistinguishable from the depersonalization symptoms occurring in psychiatric patients and occasionally lasting several days."[32]

When Izumi took LSD in the setting of the mental hospital ward, it had a lasting impression on his design ethic:

> You begin to realize the significance of even the smallest detail and, most important, to realize that it is not your perception but how other people perceive that is important in making design decisions. As a result of my LSD experience, my ability to anticipate, to project, almost to pre-experience some of the difficulties that a

patient might experience through distorted perception...has been heightened."[33]

Sommer commented that something as simple as tile patterns could have major repercussions for the disturbed patient: "Corridors having repetitious or homogeneous tile patterns appeared infinite in length. In one ward, the reflection on the uniform tile from a row of incandescent lights gave the impression of a highway with a white stripe down the center."[34]

Building upon Osmond's sociopetal ideas and his premise that design should follow function, Sommer initiated several studies of his own looking at the psychological effect of space, or the lack thereof, on humans: "Whether they are put in closet-sized rooms or castles, we don't know what happens to them. If they become ill and die, it is blamed on a weak heart, poor constitution or inadequate diet. The last thing that is ever blamed is the amount of space available."[35] Consideration of personal space therefore had relevance for mental patients. Looking at the structure of a traditional mental hospital such as that in Weyburn, Sommer stressed the debilitating impact that it could have on its occupants: "Before entering the hospital, how many of our patients had slept in dormitories containing 60 beds, or eaten meals in fifteen minutes along with 150 other people? So many aspects of institutional care are different from the outside community that the long-stay patient must be completely reeducated if he hopes to return."[36]

By the end of the 1950s, the movement toward community psychiatry in Saskatchewan was well under way. Under the leadership of Lawson, who had unveiled the Saskatchewan Plan at an APA convention in Denver in 1957, the push for the first regional psychiatric centre rapidly developed into a protracted struggle. A central question in the ensuing debate was whether the new psychiatric centre should be part of the general hospital or an independent structure. Inevitably, pressure from professionals such as Lawson, Osmond, and Hoffer, organizations such as the CMHA, and hundreds of community members forced the CCF government to agree to the latter option.

Yorkton was chosen as the location for the first of the psychiatric buildings called for under the Saskatchewan Plan, and the job of designing it was given to Izumi and his architectural associates. The original sociopetal design, however, had to go through extensive remodelling before it received official approval. The initial circular design was scrapped because it was considered, as Izumi recalled, "too far out,"[37] and the Y-shaped model that followed did not qualify for federal funding because it allegedly went against existing building codes, fire regulations, and hospital construction standards. After continual reinterpretation, the final design for the Yorkton Psychiatric Centre (YPC) resulted in a complex of several small rectangular cottages.

Although the YPC (completed in 1964) adhered to many of the ideas put forward in the sociopetal proposal, it was far from what Osmond and Izumi had originally envisioned.[38] Nonetheless, they had managed to construct a unique psychiatric facility: long hallways were eliminated, patients were provided with their own rooms, and minute details were considered. The responses of the professional community and the hospital staff were positive. The APA noted that the YPC "must certainly rank among the most attractive and architecturally advanced buildings ever constructed for psychiatric services....As for the physical plant it was a pleasure to view a facility that was not merely new. Creativity and imagination were evident in scores of details. We felt the wards and day rooms combined efficiency with comfort and cheerfulness to a very exceptional degree."[39]

Had the province's psychedelic research made a distinctive contribution to the design of the YPC? Osmond affirmed that it had:

> It may be argued that Kyo Izumi, an unusually gifted person, would have done this just as well without the use of LSD 25. I do not know how this could be proved or disproved, but I do know that he and I believe that it played a crucial part in deepening our understanding of the problem and so enlarging the communication between us. The fruits of this collaboration are there to be judged by any who care to go and look at them....Here, indeed, are some of those concrete results of the psychotomimetic experience which critics have been so keen to discover.[40]

Osmond and Izumi had opened new lines of communication between psychiatry and architecture, and in recognition of this initiative they were appointed by the NIMH to develop guidelines for the design and construction of mental hospitals.

By 1964, there was a new government in power in Saskatchewan under Liberal leader Ross Thatcher, and the commitment to mental health disappeared. Institutions such as the YPC soon lacked appropriate staff to put the sociopetal principles into practice and use the hospital to its fullest potential, and everything possible was done to keep patients out of the mental hospitals. The frequent use of tranquilizers made this more easy. Nonetheless, many of the sociopetal ideals continued to influence areas outside mental health, for example in the work of Sommer and his research on personal space.

PSYCHEDELIC EXPLORATIONS IN PARAPSYCHOLOGY

Even though most scientists conducting hallucinogenic drug experiments had abandoned the psychotomimetic approach by the late 1950s, Hoffer, Osmond, and Blewett became increasingly attuned to the potential of the psychedelic experience. With the success of the LSD treatment program for alcoholism, the Saskatchewan researchers began looking at other possibilities. Osmond suggested that psychedelics could be of use among terminal cancer patients, aiding them in the process of dying.[41] Blewett and Nick Chwelos conducted preliminary studies in the penitentiary and used LSD in group environments. There were moments in the group experiments when the empathic bond became so intense that participants could "feel a unison so complete as to establish a communication verging upon the telepathic."[42]

Parapsychological phenomena were of equal interest to Osmond and Hoffer. Together with Aldous Huxley, fascinated by ESP and other psychic phenomena, American investigator J. B. Rhine, the father of modern parapsychology, and famous medium Eileen Garrett, they began to study the effects of psychedelics in this relatively untouched area. In a letter to Huxley in February 1958, Osmond wrote about his group experiments with Hubbard, Blewett, and others: "The possibility of exploring group relations is one of the most exciting. We are very ignorant. Many multiple experiments have not involved the group but only separate mescaline or LSD universes existing side by side. But much more can be done."[43]

Ironically, the main thing blocking the way of this kind of exploration was science, which has always had a difficult time accepting phenomena that cannot be verified. That parapsychological phenomena usually faltered under standard scientific tests and statistical methods led many scientists to dismiss them as unreal. The idea that LSD increased empathic awareness was thus deemed ridiculous by many. Writing to Rhine in late 1957, Huxley mentioned that

> Osmond...found that there seemed to be telepathic rapport between himself and another man, while both were under the influence of the drug. They didn't do any systematic tests, however. And the trouble here is that people under LSD or mescalin are generally in a state of intenser, more significant experience—a state in which they are apt to become extremely impatient with the learned foolery of statistics, repeated experiments, scientific precautions, questions by investigators etc.[44]

Hoffer, Osmond, and Blewett knew that convincing others of the reality of such phenomena was a formidable challenge. When the brightest minds in the field got together for a conference in New York in 1958 that centred on psychedelics and parapsychology, the Saskatchewan researchers provided their own interpretations of the hurdles to be overcome. Hoffer listed the stages of the scientific method: (1) description of the phenomenon, (2) creation of a hypothesis to account for it, (3) an experiment designed to test the hypothesis, and (4) the final conclusion.[45] The problem with parapsychology, as Hoffer saw it, was that it was "still struggling in the first stage."[46] This did not deter them from launching exploratory studies.

Like any other experiment with psychedelics, the set and setting demanded great consideration. In trying to demonstrate parapsychological phenomena, Hoffer, Osmond, and Blewett found that participants had to reach a sufficient level of the drug experience for such phenomena to appear. The failure of experiments such as that of P. O. O'Reilly seemed to contradict the suggestion that the empathic bond could be established under LSD, but this experiment was not carried out under optimum conditions. One participant had no prior experience with the drug and was paranoid and uncomfortable for most of the experience. Such failures seemed to convince skeptics of the implausibility of the empathic theory. Yet, regarding LSD and hypnosis, "if one subject can be found in whom certain changes are present, it is safe to assume that these changes will also occur in other individuals; however, one will not know how many people there are, i.e., what proportion of any normal population will react in a similar way."[47] If researchers were to have any success in demonstrating parapsychological phenomena, then a whole new set of methods and techniques, a "new mathematics" as Hoffer put it, would have to be devised.[48] The Saskatchewan researchers never had that opportunity because by 1960 changes both within and outside the province brought the research with psychedelic drugs to a standstill.

Even if Hoffer, Osmond, and Blewett had been able to continue their research in this area, and produce more substantial evidence that psychedelics can be effective in revealing parapsychological phenomena, most of the scientific community would have ignored their findings. This appears to be the case whenever a new theory is inconsistent with an established paradigm, for "the two sides do not accept the same 'facts' as facts...and still less the same 'evidence' as evidence....Indeed, one side may disregard some of the evidence altogether in the confident expectation that it will somehow turn out to be false."[49] It is important to keep this in mind when reassessing the work of Hoffer, Osmond, and Blewett in Saskatchewan with hallucinogenic drugs.

Because of the ban on psychedelic research in the 1960s, many questions in this field went unanswered. Nevertheless, Hoffer, Osmond, and Blewett can be considered trailblazers for raising the questions in the first place.

CLOSING TIME FOR HALLUCINOGENIC DRUG RESEARCH IN SASKATCHEWAN

After nearly a decade of groundbreaking research, the experiments in Saskatchewan with psychotomimetic/psychedelic drugs had more or less come to an end. Hoffer, Osmond, Blewett, and their team of approximately thirty researchers had added a great deal of knowledge to what could be accomplished with hallucinogens. Still, scientists had only begun to comprehend the effects, good and bad, of these drugs. As Blewett wrote, "psychedelic research is and will very likely remain, difficult. To date there are no established norms and only the crudest of guiding rules of thumb have been evolved."[50] Yet he was brimming with hope for the future, reiterating much of what Osmond had said for years: "The psychedelic state offers to become a focal point for uniting studies involving not only parapsychology and psychology but also drawing upon the fields of biochemistry, physiology, psychiatry, philosophy, theology and metaphysics."[51] As promising as it was, though, the research did not last. At the pinnacle of its success, changes within and outside the scientific realm ultimately sealed its demise.

With the expiry of funding sources and psychiatric research in the province headed in a different direction, the psychedelic drug research, in all of its facets, came to an end in 1961. Some of the researchers stayed on as employees of the PSB; others went their separate ways to pursue new opportunities outside the province. By this time, Saskatchewan was experiencing a brain drain in the mental health field as some of the psychiatric research program's most talented people left for other locales. The Saskatchewan Hospital–Weyburn was particularly hard hit, losing Sommer, its chief psychologist, who went on to positions in Alberta and California to pioneer his concept of personal space and the field of environmental/architectural psychology; psychiatrist Sven Jensen (Ontario) and his assistant psychologist Ronald Ramsay (Edmonton); Blewett's partner Chwelos (private practice in Calgary); and a host of others. The lack of competitive salaries in Saskatchewan was certainly a factor in this exodus, but increasing restrictions on scientific freedom no doubt played a role as well. Sommer wrote that

> I was the first PhD psychologist employed at this 1500 bed mental hospital and, when I arrived, few people could understand why I came....I have found this a very stimulating setting for both practice and research....To me it is...being free in one's thinking and not tied down to traditional concepts....One doesn't have to worry about Professor X's ideas or the customary approaches to particular problems....There is also a professional freedom for experimentation not found elsewhere.[52]

Chief among those who left the provincial psychiatric services were Osmond, Hoffer, and Blewett. Scientific freedom was key to their success. Saskatchewan's relative isolation and a receptive government had allowed for scientific freedom unparalleled in other parts of the world. But by 1961 things were different. Rather than continuing to chart a radical course in mental health research and treatment, Saskatchewan fell back and adopted values and approaches in keeping with the rest of the psychiatric community. As a result, the unique foothold of the province in mental health was largely lost. The views of Hoffer, Osmond, and Blewett placed them at the periphery of the field and cast them as outsiders.

When it was decided in 1960 that the Saskatchewan Committee on Schizophrenia Research would be dissolved and replaced by a larger committee that would direct future lines of research and scrutinize papers more closely for publication, it was too much for Osmond. He thought that the smaller committee had done an excellent job. It had been responsible for the prodigious output of roughly 200 scientific publications. More importantly, it had enabled researchers to pursue independent lines of inquiry and produce important discoveries. Osmond therefore worried about further limitations on his and Hoffer's project. In a March 2, 1960, letter to his partner, he wrote that

our work is individual and much of it novel. The beauty of working in Saskatchewan so far has been that you could get work done without all this bloody red tape....You are going to introduce a great top-heavy bureaucracy which will smother every idea that we have....Lots of our success arises from freedom of operation, combined with a clear understanding of our general function. In the sort of research that produces new ideas committees have no place.[53]

In many respects, and as Osmond himself had anticipated, Saskatchewan became more "organized" in its scientific research, with a much greater "emphasis on methodology, research design and planning by committee."[54] In his *The Organization Man*, Whyte comments that such direction is necessary but, taken too far, can exercise a devastating influence.[55] Increasingly dissatisfied with the shift in provincial mental health policy, the decision not to proceed with the planned psychiatric research centre, and restrictions on his scientific freedom (i.e., what he published), Osmond, the intellectual force that had driven the psychedelic research in Saskatchewan, decided in 1961 to return to England. As he confided to Hoffer,

> I've done what I can and the time is ripe for a move. The alternative would be to stay in Weyburn indefinitely trying to cope with its recurring and fairly simple difficulties. These have been difficult and irritable enough with [T. C. Douglas] but with his CCF successor or say Thatcher it would make life unbearable and work that much more difficult. The isolation which was once so useful isn't anymore.[56]

However, the options available to Osmond in the United Kingdom were not enough to entice him to stay, and within months he crossed back over the Atlantic and accepted an offer to head the Bureau of Research in Neurology and Psychiatry at the New Jersey Neuropsychiatric Institute in Princeton.

Hoffer persevered for several more years as the director of psychiatric research for the province and as a professor in the Department of Psychiatry at the University of Saskatchewan, but ultimately the same reasons that had driven his partner to leave the province led him to enter private practice in 1967.

Like his two former colleagues, Blewett felt disillusioned and limited in what he could do if he remained within the confines of government bureaucracy, so he took on a new role as the first chair of the Department of Psychology at the University of Saskatchewan–Regina Campus. He summed up his feelings at the time:

> In the seven years that I had worked with the government...there had never been a time when we could hire the psychologists we needed. We wanted to provide the best service possible and sought to employ 14 PhD's in the field. In those years, however, there were few to be found, particularly in Saskatchewan in the winter time. Sensing a chance we could train our own, I got a job at the university in 1961 and began moving toward that goal.[57]

Figure 1. Young Humphry Osmond at his home in the village of Milford in Surrey, England, circa 1930s. Courtesy of Fee Blackburn

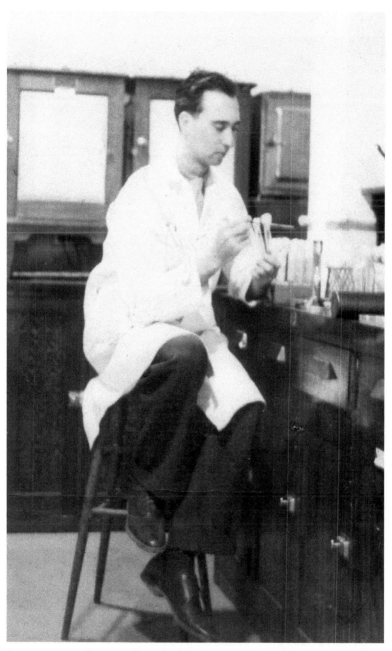

Figure 2. Young Abram Hoffer in the lab, circa 1930s. Courtesy of Miriam Hoffer

Figure 3. Young Duncan Blewett in the Canadian Army, 1st Battalion Essex and Kent Scottish Regiment, circa 1941–42. Courtesy of Mary Lowe

Figure 4. 1951 cartoon: Osmond arrives in Saskatchewan. Courtesy of Fee Blackburn

Figure 5. 1953 cartoon: Osmond begins reorganization of the Saskatchewan Hospital–Weyburn. Courtesy of Fee Blackburn

Figure 6. Aerial view of the Saskatchewan Hospital–Weyburn, circa early 1950s.
Provincial Archives of Saskatchewan, R-B9688

Figure 7. Aldous Huxley, British author, intellectual, and psychedelic forefather. Courtesy of Fee Blackburn

Figure 8. Weyburn research psychiatrist T.E. Weckowicz, Hoffer, and Blewett at a Native American Church ceremony outside of North Battleford, circa 1956. Photo by Gordon Skogland. Provincial Archives of Saskatchewan S-SP-B59839(1)

Figure 9. Kiyoshi Izumi, architect and co-creator of the "sociopetal" hospital design. Courtesy of Fee Blackburn

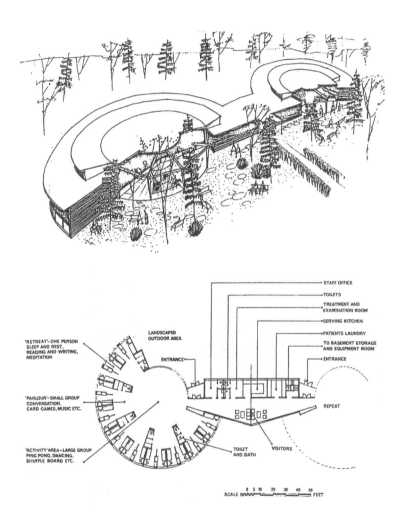

Figure 10. Izumi's architectural drawing and ground plan for the sociopetal hospital design. Courtesy of Doreen Chappell Arnott

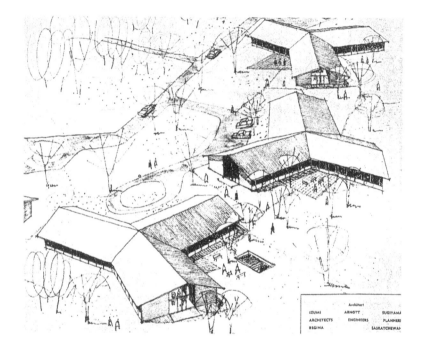

Figure 11. Proposed Izumi design for the Yorkton Psychiatric Centre. Courtesy of Doreen Chappell Arnott

Figure 12. Hoffer in his office at the Royal University Hospital, Saskatoon, circa 1960s. Courtesy of Miriam Hoffer

Figure 13. Blewett in his role as Chair of the Department of Psychology, University of Saskatchewan–Regina Campus, circa late 1960s. Courtesy of Mary Lowe

Figure 14. Osmond during his tenure as Director, Bureau of Research in Neurology and Psychiatry, New Jersey Neuropsychiatric Institute, Princeton, circa mid-1960s. Courtesy of Fee Blackburn

Figure 15. Osmond and the man who discovered LSD, Swiss chemist Albert Hofmann, in Le Piol, France, circa mid-1960s. Courtesy of Fee Blackburn

Figure 16. Hoffer and Alcoholics Anonymous co-founder Bill Wilson, circa 1969.
Courtesy of Miriam Hoffer

Figure 17. Osmond and Hoffer versus the American Psychiatric Association, Washington, D.C., circa 1973–74. Courtesy of Miriam Hoffer

Figure 18. Blewett in retirement at his home in Gabriola Island, British Columbia, circa 1990s. Courtesy of Mary Lowe

PART TWO

The Scientific Fallout:
Psychedelic Science on Trial
(1961–75)

Validity and verification in science primarily consist of a willing conspiracy among scientists not to challenge the authorities in the field and to take the sincerity of colleagues as insight.
—Vine Deloria Jr., *Red Earth, White Lies* (35)

At the beginning of the 1960s, a seismic shift began in the field of mental health. The predominance of psychoanalytical approaches to understanding, diagnosing, and treating mental illness began to subside. This waning can be attributed to various factors. Perhaps the most significant were the psychopharmacological revolution and the concomitant resurgence of professional interest in neurotransmitter systems and related biochemical theories of mental disorders. To adapt to the changes in the field and acquire more scientifically respectable reputations, mental health professionals had to adopt more objective clinical research methods.

By the early 1960s, psychedelic drug research in Saskatchewan was winding down. However, debate about the published claims intensified as numerous counterclaims appeared from 1963 on and raised serious doubts about the scientific validity of the research and the supposed efficacy of megavitamin treatment for schizophrenia and LSD treatment for alcoholism. As the 1960s progressed, support for the Saskatchewan research rapidly disintegrated, while the number of dissenting views multiplied, largely because of the supposed inability of the research to adhere to scientific standards of the day. Eventually, the research and its proponents came under the same unscientific light as that reserved for parapsychology, astrology, and psychoanalysis.

The fact that so many clinical studies were unable to confirm the Saskatchewan research leaves open to question whether it can be said that the Saskatchewan discoveries were truly revolutionary. Indeed, if one were to use Popper's rationale and his scientific criterion of falsifiability, then it appears that many of the Saskatchewan findings clearly failed empirical tests. Continuing along this Popperian line of inquiry, some have questioned whether the Saskatchewan research constituted a scientific controversy at all, given Hoffer, Osmond, and Blewett's determined efforts to rescue their theories whenever confronted with negative findings.[1] As Popper put it, the use of such "ad hoc strategems" to escape refutation of a theory, while "always possible," is done "only at the price of destroying, or at least lowering, its scientific status."[2] Yet the Saskatchewan researchers constantly maintained that their studies had been developed and carried out following a genuinely scientific approach. Or at least the research had the *appearance* of being scientific, a phenomenon that led Popper and other experts to further emphasize the need to distinguish between empirical and pseudo-empirical modes of investigation.

For Feyerabend, Popper's former student, "the idea of a method that contains firm, unchanging, and absolutely binding principles for conducting the business of science meets considerable difficulty when confronted with the results of historical research."[3] Breaking ranks with his master, he elaborated in *Against Method* that

> one of the most striking features of recent discussions in the history and philosophy of science is the realization that events and developments...occurred only because some thinkers either *decided* not to be bound by certain "obvious" methodological rules...or because they *unwittingly broke* them. This liberal practice...is not just a *fact* of the history of science. It is both reasonable and *absolutely necessary* for the growth of knowledge. More specifically, one can show the following: given any rule, however "fundamental" or "necessary" for science, there are always circumstances when it is advisable not only to ignore the rule...but to adopt its opposite. For example, there are circumstances when it is advisable to introduce, elaborate, and defend *ad hoc* hypotheses, or hypotheses which contradict well-established and generally accepted experimental results, or hypotheses whose content is smaller than the content of the existing and empirically adequate alternative, or self-inconsistent hypotheses and so on.[4]

Following Feyerabend's argument, then, all methodologies have their limitations, and the concept of a fixed scientific method, like Popper's contention about Freudian theory, is grounded more in myth than in fact. The important point that merits further consideration in trying to develop a more informed and balanced understanding of Saskatchewan's psychedelic research, and Hoffer, Osmond, and Blewett as individual scientists, is *how* the scientific communities to which the three men belonged framed scientific activity and *how* the Saskatchewan research came to be labelled as anti-scientific.

Commenting at length on the organizational aspects of scientific circles, Polanyi determined that

> any attempt to define the body of science more closely comes up against the fact that the knowledge comprised by science is not known to any single person. Indeed, nobody knows more than a tiny fragment of science well enough to judge its validity and value at first hand. For the rest he has to rely on views accepted at second hand on the authority of a community of people accredited as scientists. But this accrediting depends in its turn on a complex organization. For each member of the community can judge at first hand only a small number of others by whom he is recognized as such in return, and these relations form chains which transmit these mutual recognitions at second hand through the whole community. This is how each member becomes directly or indirectly accredited by all. The system extends into the past. Its members recognize the same set of persons as their masters and derive from this allegiance a common tradition, of which each carries on a particular strand....Suffice it here to say that anyone who speaks of science in the current sense and with the usual approval...accepts this organized consensus as determining what is "scientific" and what is "unscientific." Every great scientific controversy tends therefore to turn into a dispute between the established authorities and a pretender...who is as yet denied the status of a scientist, at least with respect to the work under discussion. These pretenders do not deny the authority of scientific opinion in general, but merely appeal against its authority in a particular detail and seek to modify its teachings in respect of that detail.[5]

Polanyi acknowledged tradition and its organized authority but stopped short of giving his commitment unconditionally; as he said, he accepted "existing scientific opinion as a *competent* authority, but not a *supreme* authority, for identifying the subject matter called 'science.'"[6]

Many sociologists of science, beginning in earnest with the contributions of Robert Merton, have also drawn attention to the social aspect of science. As Merton insisted, science is fundamentally a social process, and to really comprehend any scientific research endeavour necessitates viewing it in its complete historical and social contexts. Barry Barnes observed that it was Merton who

> identified the entire enterprise as a social unit functioning in a larger social system...and noted how its operation was necessarily affected, for good or ill, by the whole of which it is a part. He offered various accounts of the nature of this social unit, treating it as a set of peers, a moral community, and not least as a sub-culture with a distinctive normative order....And he addressed the question of how normative order in science is sustained...and thereby contributed to our understanding of the system of recognition and honorific reward in science and how it encourages competent and original research.[7]

In demarcating the boundaries of science, it becomes the responsibility of the scientific community to "police the existing boundaries of science, to avoid the intrusion of whatever may detract from its reputation, [and] to seek to expel anything potentially disreputable which arises within it."[8] The methods and criteria employed in all of these activities, rather than being set in stone, are in constant flux, changing from one historical context to another. In addition to their characteristic variability, demarcation criteria "must be regarded as conventional and their application in all cases as situated in human action."[9] The social groups that engage in the boundary-making process also do so to "protect and promote their cognitive authority, intellectual hegemony, professional integrity and whatever political and economic power they might be able to command by attaining these things."[10] According to Polanyi, a group of scientists typically succeed "through the control of university premises, academic appointments, research grants, scientific journals and the awarding of academic degrees which qualify their recipients as teachers, technical or medical practitioners."[11] "By controlling the advancement and dissemination of science, this same group of persons, the scientists, actually establish the current meaning of the term 'science,' determine what should be accepted as science, and establish also the current meaning of the term 'scientist' and decide that they themselves and those designated by themselves as their successors should be recognized as such."[12]

The important point, as Shapin once wrote, is that "there is as much society in the scientist's laboratory, and internal to the development of scientific knowledge, as there is outside."[13] This fact rarely comes through when reading the usual scientific journal articles because that social/human element is customarily absent. Merton regularly picked up on this in his own observations about the sociological nature of science. In his review of James Watson's autobiography and the highly personal recounting of the professional struggles that occurred in the lead-up to discovery of the DNA molecule, Merton noted that Watson's stories had aptly displayed just how human scientists are and helped "to dispel a popular mythology about the complex behavior of scientists."[14]

In much the same way that Watson's scientific tales shattered the myth of the scientist as calm, rational, and objective, Hoffer, Osmond, and Blewett and their opponents could be seen at times to display all-too-human behaviour in their investigations. In the historical records that the three men left behind, both the public and the not-so-public ones, we see scientists being passionate, ideological, highly competitive, spiritual, bitter, derogatory, arrogant, overenthusiastic, sarcastic, defiant, and even deceitful, quite unlike some of Hoffer, Osmond, and Blewett's opponents' portrayals of themselves and their research as firmly rooted in objectivity. Many would agree with the following assertion made by Tullock: "Pure detachment is a myth. Even the best investigator may have his judgment clouded and his inability to read instruments affected by strongly held opinions. The man who has discovered and promulgated a theory is more likely to find evidence supporting that theory in a given experimental design than is another man who has opposite personal biases and interests."[15] Nonetheless, "scientists are remarkably good at keeping [the] influence of these factors to a minimum in [their] work."[16]

Most scientists will admit a deeply personal connection in scientific practice, but the scientific community will allow it only to a certain point. There has always been debate about both the degree and the role of this factor on the road to scientific truth. For example, referring to Popper's description of scientific behaviour, Tullock emphasized that "the method by which we reach our hypothesis is less important than the question of whether the hypothesis is true, and this latter question can be answered only by testing it."[17] In a similar vein, Israel Scheffler insisted that

> scientific habits of mind are compatible with passionate advocacy, strong faith, intuitive conjecture, and imaginative speculation. What is central is the acknowledgement of general controls to which one's dearest beliefs are ultimately subject. These controls, embodied in and transmitted by the institutions of science, represent the fundamental rules of the game. To devise fair controls for new ranges of assertion, and to guarantee the fairness of existing controls in the old, constitute the rationale of these rules.[18]

Thus, there have always been expectations of individual scientists to adhere to established standards. In clinical psychiatry, randomized controlled testing eventually became the standard that, many believed and official protocol stated, provided the best way to protect against the interference of more personal factors such as experimenter bias in testing scientific theories. Whether this methodology has "guaranteed fairness" and completely objective results in every case is open to question.[19]

These were precisely the kinds of scientific battles in which Hoffer, Osmond, and Blewett were enmeshed. For their research findings and theories to be accepted, they would have had to gain general acceptance by the mainstream scientific community, but they were unable to do so, at either the peak of the research in 1960 or in the following years. Should one take their failure to achieve widespread acknowledgement and corroboration of their theories as an indication that the research was unscientific? No. A close reading of the historical record indicates that it was the transformation within the scientific community and its standards that accounts for the pariah status of the Saskatchewan researchers. Yet neither can we frame them in heroic terms, claiming that they were wronged at every turn, that all of their research was scientifically valid, and that their opponents were deceitful and unscientific (which might have been true occasionally).

Hoffer, Osmond, and Blewett had their own opinions about what it meant to be scientific and how scientific research ought to be conducted. Although they might not have necessarily subscribed to Feyerabend's anarchic "anything goes" model of science, they agreed that methodology had its uses and limitations. On other occasions, the subject was a source of conflict and disagreement. At various times, Hoffer and Osmond thought that Blewett's behaviour and evolving LSD research interests bordered on the unscientific and threatened the scientific credibility of their own research. In his own defence, Blewett maintained that there was scientific purpose behind what he was doing. At a much broader level and more intensely, the views and actions of all three men regularly clashed with the accepted standards of the scientific community. This had been the case throughout the 1951–61 period, when the Saskatchewan research project was active, but it became even more pronounced after its conclusion.

Certainly, the inevitable blacklisting of Hoffer, Osmond, and Blewett in the scientific community had much to do with the wider social, political, and economic changes taking place in the 1960s. When it came to the psychedelic drug issue, the scientific community and wider society crossed paths; in fact, the end of legitimate scientific experimentation with LSD in the 1960s can be attributed in many ways to what was happening outside the laboratory. The three Saskatchewan researchers' wading into the public hysteria that erupted over the recreational use of psychedelic drugs; their questioning of the perceived dangers of LSD; their persistence in their pro-psychedelic opinions; and their defence of the value of the original research caused their scientific colleagues, as well as some members of the public, to react in alarm. Perhaps, had LSD not become the countercultural phenomenon that it did in the 1960s, the research of Hoffer, Osmond, and Blewett—primarily their theories of schizophrenia and their use of LSD to help treat alcoholism—would still have met with much resistance from the scientific community. It is therefore important to keep the primary focus on society as it existed *inside* the scientific community because it was chiefly within this realm that professional judgment on the subject was shaped.

CHAPTER 6

THE GREAT SCHIZOPHRENIA CONTROVERSY

I n the 1960s, schizophrenia remained as enigmatic as ever. There was a host of competing interpretations and approaches. Leading psychoanalytical psychiatrist Silvano Arieti noted that "some authors consider schizophrenia an illness, others a syndrome, still others a mental mechanism or even a way of living. There is some truth in each of these views, and yet at closer analysis all of them prove to be unsatisfactory."[1]

Some facets of the research on schizophrenia had not changed much from the 1950s. North American thinking on the issue was still largely psychodynamic, heavily influenced by the theories of Eugen Bleuler, who perceived schizophrenia mostly in terms of underlying intrapsychic processes. Some thought that schizophrenia originated in the abnormal familial setting and the complicated interactions and communications between the developing child and his or her parents. There were also some radical, "antipsychiatric" newcomers to the scene, particularly Thomas Szasz and R. D. Laing, even more sociologically inclined in their interpretations of schizophrenia, seeing it as a social construct rather than a disease in the true medical sense. The opposite notion, that schizophrenia is indeed a mental disorder, whose manifestations can be traced to specific biological phenomena (e.g., prenatal development, hereditary genetics, and metabolic malfunction), continued to be the preferred route in other professional quarters. For those in the field who felt uncomfortable siding with any particular view, Adolph Meyer's holistic psychobiological approach (the precursor to what is now often referred to as the biopsychosocial model) seemed to be the most reasonable when dealing with schizophrenia.

Naturally, this professional muddle created much inconsistency in the diagnosis of schizophrenia, especially because it was considered not one illness but a group of illnesses with various subtypes and degrees. This was highlighted by Dr. Norman Brill at a UCLA conference on schizophrenia in 1969: "One of the things that has plagued research in schizophrenia is the fact that patients with this disease (or disorder) do not constitute a homogeneous group. Different kinds of persons develop the disease. There are catatonic, paranoid, simple, hebrephenic, undifferentiated, and mixed types. There are acute cases and chronic ones."[2] This differentiation made diagnosis extremely complicated.

American and European approaches to diagnosis were truly continents apart. Dr. Charles Peters attributed the divergent views to Emil Kraepelin, considered by many the father of modern scientific psychiatry, and Bleuler:

> Kraepelin and Bleuler spawned two relatively distinct approaches to the concept of schizophrenia. They differed in their definitions of the disorder and in their analyses of the essential elements of the condition. Kraepelin proposed a narrow and descriptive formulation of schizophrenia that emphasized the longitudinal unfolding of the illness, whereas Bleuler's concept of schizophrenia was theoretically grounded, less descriptive, and more broadly inclusive. European psychiatry embraced the Kraepelian concept; Bleuler's influence was more apparent in the United States.[3]

Thus, patients in the United States and Canada were diagnosed with schizophrenia who would never have met the criteria had they been seen by a psychiatrist in Europe. Edward Shorter has maintained that "such international differences were unscientific and embarrassing, for they suggested there was no science in psychiatry but simply weight of national tradition, making the discipline a branch of folklore rather than medicine."[4] Because of this lack of scientific reliability, the psychiatric establishment eventually broke the psychoanalytical hold on diagnostic procedures and moved to a methodological system (i.e., *DSM-III*) more empirical, atheoretical, and descriptive in approach.[5]

The same issues facing diagnosis arose in the treatment of schizophrenia; prior to 1960, treatment depended more on what was in vogue than on what was scientifically and empirically sound. Subtle changes started to occur in the field, and one can begin to chart a movement toward more biochemical and neurological treatments of schizophrenia.

THE PSYCHOPHARMACOLOGICAL REVOLUTION

A principal factor of psychiatry's newfound ability to demonstrate its scientific worth was the rapid evolution of antipsychotic, or neuroleptic (literally "nerve-seizing"), drugs. The first neuroleptic to gain international stature was chlorpromazine (CPZ), which can be traced back to France in the early 1950s and the efforts of naval surgeon Henri Laborit to find an anaesthetic cocktail that could induce an artificial hibernation and thereby eliminate the hazards associated with surgical shock.

When it was first synthesized by Rhône-Poulenc Laboratories in 1950, chlorpromazine did not automatically trigger a worldwide response from psychiatrists. It was only after its testing on mental hospital patients by the Paris-based psychiatric team of Jean Delay and Pierre Deniker in 1952–53 and, shortly after, by G. E. Hanrahan and Heinz Lehmann at Verdun Protestant Hospital in Montreal that others in the field began to take note. Referring to the 1953 work conducted on seventy-one patients, Lehmann considered chlorpromazine revolutionary, for it produced "sedation without significant clouding of consciousness or disinhibition of affect....The drug has shown its capacity to shorten the duration of acute psychotic episodes and also to prevent psychotic breakdowns if given in the prodomal stage."[6] Lehmann maintained that chlorpromazine was better than existing treatments, producing a comparatively high rate of remissions in early schizophrenia.[7] Further clinical testing and comparison by others were recommended. Chlorpromazine soon eclipsed all other antipsychotic treatments, providing an alternative to insulin shock therapy, barbiturates, and the even cruder and more invasive psychosurgeries (e.g., leucotomy).

The discovery of chlorpromazine and its impact on psychiatry, and to a larger extent medicine as a whole, should not be underestimated. In ten years, chlorpromazine spawned more than a dozen similar phenothiazine drugs and thus heralded the beginning of psychopharmacology as we know it today. Chlorpromazine signalled a renaissance for theories purporting schizophrenia to be a disease centred in the brain and resulting from faulty biochemical and metabolic processes. Psychopharmacologist Thomas Ban, who later teamed up with Lehmann to test Abram Hoffer and Humphry Osmond's niacin treatment theories in the late 1960s, wrote that "by the end of the 1950s, six neurotransmitters were identified in the central nervous system.... There were high expectations that CPZ combined with spectrophotofluorimetry would provide a royal road to the understanding of the pathophysiology of schizophrenia."[8]

One might think that more openness to and recognition of biochemical interpretations of schizophrenia would have been beneficial to the work of Hoffer and Osmond, but that was not the case. Negative reactions to their findings by those who upheld, say, a psychoanalytical approach to schizophrenia was predictable, but the most intense backlash came from researchers and clinicians who suspected a biochemical etiology for the disease. Again, this reaction can be understood if the adrenaline-based hypotheses of Hoffer and Osmond are viewed as competitors of the paradigm that would guide future research on and treatment of schizophrenia. The debate over their hypotheses returned to the question of whether they presented a more scientifically valid understanding of, and approach to, schizophrenia than the hypotheses put forward by other biological proponents at the time. Most professional accounts tell us that the theories of Osmond and Hoffer represented "the failure of such an extreme reductionist approach in biological psychiatry."[9] Yet, if one looks more closely at the matter, their ideas ran counter to and were in conflict with a number of powerful and vested interests, particularly those attached to the idea that neurotransmitters other than adrenaline were the doors through which one might understand the mysteries of schizophrenia and that neuroleptics were the keys to unlock those doors. With psychiatry in the 1960s headed in a direction more amenable to biological theories of schizophrenia, the situation between Hoffer and Osmond and their competitors quickly developed into a great controversy. Each camp attempted to outflank the other by meeting the scientific standards of the day. When it came to clinical research in psychiatry, the true mark of science was to have one's theories proven, corroborated by others through purely objective means, in a way that could control for the biases of enthusiastic investigators and the expectations of their patients.

RISE OF THE DOUBLE-BLIND METHOD

In an assessment of psychiatry in 1961, Seymour S. Kety remarked that "it has been said, usually by those outside the field, that psychiatry depends too much upon subjective observations ever to become a science. To this I take a strong exception. It is not subjectivity itself which keeps a field from being scientific, but a failure to recognize, minimize, or compensate for subjective bias."[10] For Kety and others, the solution was to adopt sound research designs that could effectively control for subjective factors. In psychiatry, the double-blind method, Kety insisted, was "relatively easy to do...in drug studies and in other biological observations and all the more lamentable when it is neglected."[11]

Randomized controls for testing the clinical efficacy of treatments in medicine could be traced back to the mid-eighteenth century when British naval physician James Lind performed the first comparative test of the efficacy of citrus juice versus standard treatments for scurvy-infected sailors. From his time on, controlled methods were used and evolved to the point where, in the estimation of some, by the twentieth century they had been refined to "a nearly flawless level of efficiency."[12] It had long been noted about many medical treatments, particularly psychiatric drug treatments, that positive (or negative) outcomes could be attributed to any number of variables, two of the most powerful being the bias of investigators conducting the trial and the susceptibility of patients to suggestion (e.g., "the placebo effect").

The sudden desire within the psychiatric community to become more scientifically respectable sparked the need to isolate the effect of the drug being tested to determine if in fact it accounted for the outcome observed. The double-blind method appeared to fill this need because it allowed for the testing of one or more drugs, or inert placebos, where neither patients nor investigators evaluating them knew which medication was being administered. Through elaborate coding, it was believed, secrecy would be maintained and objective findings would result. Hence, this method was recommended. When correctly carried out, it could distinguish between authentic science and quackery.

Psychiatric researchers were soon expected to hone their methods when testing new drugs, a challenging task in a field in which professional subjectivity was so predominant and homogeneity of the patient population so rare. Commenting on treatment specific to psychiatry, Michael Shepherd notes that the controlled trial "is likely to be most valuable when rational treatment is administered to a clearly defined disorder of known aetiology whose outcome can be established by unequivocal criteria."[13] Conducting trials for a pharmacological treatment for a common bacterial infection therefore poses far fewer difficulties than doing so for schizophrenia, with its multiple definitions, no known or agreed upon etiology, and occasional tendency to produce a spontaneous remission without any treatment.

Whether the double-blind method was the most scientific had both advocates and detractors. The former professed that it was the closest thing to a gold standard in clinical drug trials; the latter, while acknowledging its merits, drew attention to its limitations and potential for abuse.[14] Some questioned its ability to guarantee complete objectivity, as when "the nature of the drug or its dosage may be such that dramatic reactions are produced which destroy the 'blindness' of the procedure."[15] Others cited ethical dilemmas, such as giving a placebo for weeks on end, simply for the sake of comparison, and withholding a treatment that could benefit a seriously ill patient.

Hoffer and Osmond, ironically among the first in psychiatric history to carry out double-blind studies, recognized that the method had both advantages and disadvantages. Whatever the reservations of professionals regarding the method, it became standard protocol in clinical psychiatric research and, from the 1960s on, greatly influenced which research projects were approved and funded, which findings journals deemed publishable, and by which criteria research was judged and recognized within the scientific community.

HOFFER AND OSMOND VERSUS THE PSYCHIATRIC COMMUNITY

When it came to the research on schizophrenia in Saskatchewan, many in the Psychiatric Services Branch were becoming uncomfortable under the critical spotlight and sought to curb the influence and restrict the operational freedom of Osmond and Hoffer. Following his decision to leave Weyburn, Osmond remarked to Hoffer that "the implications of this ten years' work are formidable enough in psychiatry but I doubt whether they will long be confined to psychiatry."[16] In one sense, his prediction was accurate: more was to come for the two psychiatrists and their theories, though not necessarily what either one might have wished. Both could look back with pride at what they had accomplished, but their ideas no longer fit with the direction in which the province, and psychiatric services, were headed. For Hoffer and Osmond, their research with psychedelic drugs had always been a means to a much larger end, that of better understanding and treating schizophrenia. From this time on, they maintained a professional interest in psychedelics, closely following research in the area, participating in conferences and interviews, and writing journal articles and books. However, they focused on disseminating their theories on and treatment of schizophrenia, which soon developed into a protracted contest of elimination with their rivals.

To recap their work on schizophrenia, by the mid-1950s Osmond and Hoffer had produced several important but contentious discoveries, the highlight being the first specific biochemical theory of the disease, the adrenochrome hypothesis. They proved that adrenochrome, and a similar adrenaline derivative, adrenolutin, could be classified as psychotomimetics; synthesized the first pure samples of these substances in the lab; invented new diagnostic tests to distinguish schizophrenia from other types of functional psychoses (the mauve factor and HOD Test); and originated a new form of treatment (megavitamin) for mental illness. The model psychosis experiments also improved mental hospital design and led to a host of interdisciplinary studies examining the world of the mental patient. In assessing the claims of Osmond and Hoffer, many critics aimed their attacks at the veracity of the adrenochrome hypothesis.

A number of reviews assessing biological research on schizophrenia singled out their hypothesis as questionable and premature in its findings, immediately placing Hoffer and Osmond on the defensive. Although he welcomed the research, University of Colorado psychiatrist John Benjamin "fear[ed] that the field will again, as in the past, become somewhat discredited through poorly conceived and designed investigations and particularly through overenthusiastic and premature publications of unsubstantiated findings."[17] Turning to the mysterious oxidized adrenaline derivative compounds, he reiterated what most other critics were saying: although the chemicals had long been known, there was "no evidence whatsoever for in vivo occurrence of these substrates either in schizophrenics or in non-schizophrenics."[18] Benjamin applauded the brilliant research undertaken by the NIMH on the metabolism and effects of adrenaline in normals and schizophrenics. For him, the NIMH research promised "to supply the sort of critical and controlled research that is so badly needed in this as in other areas of biological research."[19]

Commenting on the rapid pace at which neurochemistry was developing, Theodore Sourkes, a biochemist in the Department of Psychiatry at McGill University and a senior researcher at the Allan Memorial Institute of Psychiatry in Montreal, singled out the research in Saskatchewan, claiming that all of its findings "have now been questioned."[20] In casting doubt on the purported psychotomimetic qualities of adrenochrome and adrenolutin, Sourkes referred to an article in the *Lancet* by Osmond's former partner, John Smythies, who claimed that some of the adrenochrome/adrenolutin experiments had been "unconvincing" because double-blind methods had not been used, that the results might well have been placebo effects, and that when Hoffer had conducted double-blind tests on adrenolutin the results had been negative.[21] Using Smythies as a base from which to question the credibility of Hoffer and Osmond, Sourkes proceeded to cite a lack of confirmation for basically every other finding produced by them.

Hoffer and Osmond soon replied to Sourkes.[22] Regarding the damaging statements made by their old comrade, they emphasized that they had distinguished between adrenolutin and placebo using double blinds but that the compound had deteriorated partway through the experiment. They also argued that Smythies had based his assumption on unpublished communications with detractors of the research, which for them was "a priori reasoning" and did not constitute scientific evidence.[23] Although they would use the double-blind research conducted by Stanislav Grof and colleagues in Czechoslovakia as confirmation of the psychotomimetic effects of adrenochrome in humans, Hoffer and Osmond remained disenchanted with the professional worship of double-blind procedures. They concluded their rebuttal to Sourkes by conceding that the adrenochrome hypothesis "may or may not be corroborated...[and] may or may not lead to the solution of schizophrenia....Only the efforts of scientists will settle that question."[24]

THE NIMH AND THE ADRENOCHROME HYPOTHESIS

This quibbling over the psychotomimetic properties of adrenochrome and adrenolutin continued for some time. It did not matter much for biological research on schizophrenia if the substances in question were artificial (in vitro) and not natural (in vivo). For nearly everyone who contested the adrenochrome hypothesis, the most damning scientific proof was supplied by Hoffer and Osmond's adversaries at the NIMH, led by none other than Kety, the reputed father of modern biological psychiatry. Most of his colleagues at NIMH, like many other mental health professionals in the United States, were steeped in psychoanalytical theory and practice. Freudian thought reigned, and undergoing psychoanalysis was a common right of passage for up-and-coming psychiatrists. In many ways, Kety heralded a changing of the guard in American psychiatry and a movement toward a more biologically inclined interpretation and study of mental illness.

Kety began recruiting some of the most talented scientific minds, drawing from a pool of disciplinary backgrounds and providing researchers with freedom to achieve their goals. Eventually, though, as chief of NIMH's Laboratory of Clinical Science from the mid-1950s on, he channelled most of his energies into the nascent research program on schizophrenia. A key researcher whom Kety brought on board in 1955 to work in his lab as its chief of pharmacology was a future Nobel laureate, chemist Julius Axelrod, regarded as one of the "greatest molecular pharmacologist[s] of the modern era of drug research."[25]

Axelrod was given free rein to work on "any problem that was potentially productive and important."[26] He began by looking more closely into the metabolism of LSD, but inevitably got pulled into the collaborative research program on schizophrenia and the controversy over the theories of Hoffer and Osmond. They would contribute indirectly to his predominant interest: catecholamine metabolism and related neurotransmitter research. In 1956, Kety presented to NIMH staff the publication of the two Saskatchewan psychiatrists and what was then touted by many as the infamous pink adrenaline story. Axelrod was curious and began trying to locate an enzyme that might transform adrenaline into the hypothesized toxin. He "spent four frustrating months trying to look for the enzyme that could convert adrenaline to adrenochrome and...could never find it. It was a lot of crap."[27]

For many outside observers, his adrenaline research was the first nail in the coffin of the adrenochrome hypothesis. Axelrod could find no proof that vindicated any of Hoffer and Osmond's adrenochrome claims; nevertheless, his findings marked the beginning of some pioneering research with other NIMH researchers, for which he was later awarded a Nobel Prize in Physiology and Medicine in 1970. His work also paved the way for the development of a long line of antidepressant treatments (e.g., Prozac, Zoloft, and Paxil).

Returning to Kety, his critiques of biochemical theories of schizophrenia, which appeared in the journal *Science* in 1959, are regarded by many as important hallmarks in the literature of biological psychiatry. He scrutinized what he labelled "sources of error" in all biochemical theories to date, among them heterogeneous patient populations, uncertainty in diagnosis, and investigator bias. Whether the theories centred on amino acid and carbohydrate metabolism, serotonin, or ceruloplasmin, Kety suggested that the "evidence supporting them was hardly compelling."[28] He drew on NIMH's own research team to judge adrenaline-based theories (e.g., the transmethylation and adrenochrome hypotheses). Even if some were convinced by the hallucinogenic properties of adrenochrome or adrenolutin, the possibility that the two substances had any relevance to the pathogenesis of schizophrenia was slim, if not completely unlikely, given the findings coming out of NIMH.[29]

Genetic research, like that undertaken by NIMH, held more promise.[30] However, Kety, in true psychobiological fashion, was cautious not to place too much emphasis on the evidence: "The genetic and the environmental approaches to the etiology of schizophrenia psychoses are...not mutually exclusive. Both are compatible with the hypothesis that this group of diseases results from the operation of socio-environmental factors on some hereditary predisposition, or from an interaction of the two, each being necessary but neither alone sufficient."[31] Kety then concluded his critiques with the following pronouncement: "The genetic factors in schizophrenia operate to determine inappropriate interconnections or interaction[s] between chemically normal components of the brain."[32]

For anyone believing his prognostications, it seemed improbable, even naive, to think that Osmond and Hoffer had discovered the elusive biochemical key to schizophrenia. Yet they were undeterred from continuing their work. As Hoffer said, Kety's review was "adequate but no more free from the sin of subjective bias than other reviews" and "the data used by Kety to disaffirm are not entirely free from error as listed by him."[33] According to Hoffer, Kety, in elaborating on the comparative studies that NIMH was performing on normal and schizophrenic individuals, failed to describe clearly the control group used. Furthermore, the schizophrenic group "consist[ed] of 14 patients selected from 14,000....Thus a sample of 0.1 per cent is taken as the ideal schizophrenic sample from which to extrapolate to all schizophrenics, including those in mental hospitals."[34]

From 1960 on, the competition between the Saskatchewan researchers and NIMH followed a predictable course. Whenever Hoffer and Osmond released a report summarizing their findings, it was greeted by a counterpaper or letter to a journal editor from NIMH disputing the claims, restating its stance on the Saskatchewan work, and noting the inability of its researchers to corroborate it. NIMH continued to publish research findings suggesting that any abnormality in the metabolism of adrenaline, in schizophrenic or normal people, was purely speculative and that the adrenaline hypothesis had been advanced "at a time when little was known about metabolism of this hormone and a crucial test of its validity unavailable."[35] Although Smythies had some doubts about the validity of the adrenochrome hypothesis, he was unwilling to relegate it to the dustbin of biochemical hypotheses.[36]

When Hoffer and Osmond, along with biologists Julian Huxley and Ernst Mayr, had a paper published in *Nature* in 1964 on the subject of schizophrenia as a genetic morphism, it was followed up by the commonly scripted reply by NIMH researchers that the views were "based on variable and uncertain data," that there was an absence of "any reliable data which indicates a specific biochemical abnormality in schizophrenia," and that "considerable work with little success has been directed to finding" one.[37] In line with previous detractors such as Kety and Axelrod, NIMH researchers Hans Weil-Malherbe and Stephen Szara also appeared to put to rest some of the claims of the Saskatchewan researchers. As they wrote in 1968, "despite the ease with which aminochromes [adrenochrome and adrenolutin] are formed in vitro, their formation in vivo has never been unequivocally demonstrated."[38] Similarly, they held the NIMH line when highlighting the tenuous nature of Hoffer and Osmond's mauve factor urine tests on schizophrenics, saying that "evidence for the identity of the chromogen with adrenolutin or adrenolutin-like substances...is less than convincing."[39]

The war of words between NIMH and Hoffer and Osmond over the adrenochrome hypothesis continued through the 1960s and beyond. In essence, anything associated with their research became suspect and a target for intense scrutiny and even ridicule. Much of the NIMH agenda was driven by Kety's political manoeuvrings and professional aspirations. It is obvious from the informal correspondence between Osmond and Hoffer that no love was lost between them and Kety, with the former personalizing the repeated attacks from the NIMH as the "Kety affair." To them, it was sheer arrogance for Kety to predict a last-place finish for biochemistry in the race to solve the greatest, and most misunderstood, mental illness.

Nonetheless, Hoffer and Osmond had to approach the NIMH with caution and patience because, by the 1960s, it had largely succeeded in planting the seeds of doubt in the psychiatric community about the legitimacy of their theories. The decisive battle on the subject, however, had not yet begun. Another fight was brewing for Osmond and Hoffer on the treatment side of their work on schizophrenia.

MEGAVITAMINS AND SCHIZOPHRENIA

To understand the premise behind Hoffer and Osmond's addition of vitamins and other nutrients—particularly the B₃ (niacin and its amide) and c (ascorbic acid) vitamins—to the treatment of schizophrenia, one has to go back to the foundational years of the Saskatchewan research program. Their decision to treat their patients with megadoses of niacin and ascorbic acid can be seen as a direct extension of their adrenaline-based hypotheses and their psychotomimetic work with drugs such as LSD and mescaline. The Saskatchewan researchers undertook various studies using niacin, ranging from its effects on LSD to its use as an effective anticholesterol agent. From the outset of their research program, Osmond and Hoffer conducted studies of the clinical applications of niacin to schizophrenia. If they were correct in their assumptions, then a metabolic derivative of adrenaline was to blame for schizophrenia. Substances inhibiting the production of either adrenaline (i.e., the methylation of noradrenaline to adrenaline) or its theorized poisonous offshoots might yield a preventative or therapeutic treatment for schizophrenia.

Hoffer ordered bulk supplies of the vitamins from suppliers to test their theories. Their first experiment using large doses of vitamins arose out of desperation:

> I had just received four fifty pound barrels containing the vitamins that we wanted to test. I took some of that precious niacin to Humphry in Weyburn. As we were visiting the head psychiatrist came in and told Humphry that Kenneth was dying. A few schizophrenic patients in a catatonic state died and at autopsy no reason was found. Kenneth had had insulin coma and ECT, which had not helped. I suggested to Humphry that we should give him the two vitamins I had brought with me, niacin and vitamin C. We rushed to the ward and found Kenneth in a coma. We promptly put in a

stomach tube and poured in 10 grams of niacin and 5 grams of vitamin C. The next day he sat up and drank the mixture and thirty days later he was so well his parent insisted on taking him home.[40]

Based upon this miraculous recovery, which they followed for several years afterward, Hoffer and Osmond gradually began to treat other schizophrenic patients. Following their initial positive findings, they launched a small, uncontrolled pilot trial using varying doses of niacin on seven additional patients in the Regina General Hospital and Saskatchewan Hospital–Weyburn. All test subjects responded well and recovered. Writing about one early success, Osmond remarked that the patient was

> much improved [and]...not nearly as paranoid and agitated. With her it took about ten days for the stuff to act....It appears that there must be a period of time over which the treatment is accumulated before anything happens, which strongly suggests that some toxic factor is being eliminated, or dealt with in some other way.[41]

After these preliminary tests, the two researchers conducted the first controlled studies of niacin with schizophrenia patients in September 1952. A thirty-patient, double-blind study compared the standard treatment of ECT, barbiturates, and psychotherapy with the standard treatment plus three grams of the vitamin per day for thirty-three days. Patients selected from a general hospital psychiatric ward were divided at random into three groups (one with niacin, one with its amide, and one with a placebo) and followed up for two years after discharge. Results revealed that patients who received niacin and niacinamide did two times better than patients who received the placebo, who were well for an average of only eleven of twenty-three months.[42] Hoffer and Osmond then planned a much larger analysis involving all schizophrenic patients treated in the province between 1952 and 1955, 171 in all, to be followed up for a four-year period (1955–59).

By 1955, Osmond and Hoffer were convinced that niacin was a viable contender in the fight against schizophrenia. That year, in a funding proposal to the Ford Foundation, they argued that "if an abnormal diversion of adrenaline is causative in the schizophrenias, any process that will decrease the production of adrenalin will therefore decrease the production of the toxic indole and thus will have therapeutic value especially for early cases of schizophrenia i.e. niacin or nicotinic acid—large doses with early and chronic schizophrenia as therapy."[43] They stated that they had treated 100 schizophrenic patients with niacin, and of them sixty had been discharged from the hospital for at least two years.[44] Other interesting information came to light as the experiments unfolded. In some cases, Hoffer and Osmond found that, if the vitamin was discontinued, there was an immediate relapse. Additionally, though the treatment was startlingly effective for acute schizophrenics, it failed to produce much noticeable change in most chronic schizophrenics, a phenomenon that the two psychiatrists speculated could have been "a result of dosage or [an] innate lack of response to treatment."[45]

Beginning in 1957, Hoffer and Osmond published findings of their open, single-, and double-blind series, all highlighting the treatment benefits of niacin compared with standard treatments. Despite its apparent promise, the treatment method went relatively unnoticed, flying under the radar of the broader psychiatric community until the mid-1960s. One wonders why it took nearly a decade before anyone noticed this new treatment for schizophrenia. Two likely factors were the tight grasp of psychoanalytic therapy on the psychiatric field and the appearance and eventual succession of antipsychotic medications such as CPZ. Another possible reason was psychiatry's own scientific coming of age. When word got out about the amazing outcomes of niacin, many psychiatrists probably greeted it with the same professional skepticism that usually follows the announcement of any new wonder treatment in medicine.[46]

It was not until CPZ and other neuroleptic drugs, and the biochemical theories inevitably tied to them, overtook the field in the 1960s and '70s that real controversy began over niacin use in the treatment of schizophrenia. In the psychopharmacological era, much more attention was focused on what came to be popularly referred to as "megavitamin" treatment, not because it might have some efficacy in alleviating the symptoms of the odd schizophrenic, but because it presented a direct challenge to the more popular antipsychotics over which was more effective in dealing with schizophrenia and other mental illnesses.

Hoffer and Osmond's use of niacin in treatment was not a medical first. Extensive research on the vitamin had been carried out decades before, notably in the prevention and treatment of pellagra, a disease that, like beri-beri and scurvy, was a direct result of vitamin deficiency. Known for its three d's (dermatitis, diarrhea, and dementia), pellagra was a potentially deadly disease that rose to epidemic proportions in the early twentieth century in both Europe and the United States. In many cases, the symptoms of severe pellagra were indistinguishable from those of schizophrenia. For many years, the etiology of pellagra perplexed medical professionals. The most popular theories attributed its symptoms to either infections or dietary deficiencies.

One of the most promising leads came with the research of American bacteriologist Joseph Goldberger and his 1914–15 experiments with the effects of poor dietary conditions on inmates. The results of his experimental reproductions of pellagra in healthy individuals were among the first to put to rest notions that the spread of pellagra could be attributed to unsanitary conditions or communicable infections. As Goldberger put it in an early paper, when one considered all the evidence, "it clearly and consistently points to diet as the controlling factor in causation as well as prevention of the disease."[47] Despite his groundbreaking findings, however, pellagra continued to devastate the United States, leaving the midwestern and southern parts of the country hard hit and overwhelming many state mental institutions. In a 1958 paper, one of his associates, Dr. V. P. Sydenstricker, determined that "reported deaths exceeded 7,000 in 1928, 1929, and 1930 and there were probably more than 200,000 pellagrins in this country. The disease had truly become everybody's business."[48] In the 1930s, there was a flurry of research on nutritional deficiencies and the identification, synthesis, and commercial availability of vitamins, including niacin. Numerous studies followed Goldberger's pioneering efforts, many of which examined the human requirements for niacin and its metabolic precursor, the amino acid tryptophan.[49] With the knowledge provided by these discoveries, the enrichment of flour with niacin proceeded in the 1940s, and the incidence of pellagra fell rapidly.

In terms of niacin treatment in psychiatric practice specifically, Osmond and Hoffer were not the first professionals to look at it either. Other researchers had found that niacin was helpful in the treatment of various mental disorders.[50] Even the man who became Hoffer and Osmond's chief adversary on this issue, Heinz Lehmann, had earlier flirted with niacin's therapeutic possibilities.[51] What distinguished the Saskatchewan researchers was that they were the first to explore niacin not only as a vitamin important in daily physical and mental health but also as a psychopharmacolic agent to treat schizophrenia. No one before them had theoretically connected the vitamin to schizophrenia, nor had anyone used it in enormous dosage levels, which, as treatment design evolved, typically ranged between three and ten grams per day.

As psychiatry underwent a significant paradigm shift in the 1960s, there was a simultaneous change in attitude toward how best to treat schizophrenia. With more practitioners leaning toward biological understandings and approaches (i.e., neuroleptics), it became harder to dismiss the fantastic claims about niacin. In 1962, Hoffer's book on niacin therapy was released, and the British journal *Lancet* published an article by Osmond and Hoffer summarizing nine years of studies that they had conducted using the vitamin.[52] The *Lancet* article recounted the pilot trials and the first double-blind experiment (mentioned above). It reported on the findings from the larger investigation of 171 schizophrenic patients, followed up for four years, indicating longer stays out of hospital for those who had received niacin (seventy-three) and higher rates of readmission for those who had not (ninety-eight). The results of a second double-blind trial comparing niacin with a placebo were also reported. Eighty-two patients with similar clinical characteristics (e.g., age, symptoms, previous treatment, days in hospital) were followed up for two to six years. Of the forty-three comparison patients, eighteen showed improvement, whereas thirty-one of the thirty-nine patients who had received niacin improved. The paper also referred to niacin's success in achieving "five-year cure rates": that is, "any patient who has been out of hospital continuously for five years and is functioning satisfactorily (i.e., 'well' or 'much improved')."[53]

No known examples outside Saskatchewan corroborated or even attempted to replicate the results. Osmond and Hoffer provided the most up-to-date data concerning the vitamin and its effects in treatment. They then laid out their method, indicating that the best results were with sub-acute and acute schizophrenic patients, sometimes in combination with ECT, and emphasizing the need for some patients to take the vitamin for months or even years so as to prevent relapses. Regarding why niacin had remained a "neglected remedy," they thought that it had much to do with the "extraordinary proliferation of phenothiazine derivatives since 1954."[54] They continued that, "unlike these, niacin is a simple well-known vitamin which can be bought cheaply in bulk and cannot be patented and there has been no campaign to persuade doctors of its usefulness....If others confirm our work, then many sufferers from schizophrenia will spend longer out of hospital, in better shape than they do now, at negligible cost."[55] In another article in 1964, Hoffer and Osmond went slightly further in calling attention to the fact that there were "comparatively few ten-year follow up studies of any treatment and none that we know of for the phenothiazines which just entered the ten year zone of use."[56]

None of this suggests that the Saskatchewan mental health system was any more immune to the lure of antipsychotic medications than other jurisdictions or that Hoffer and Osmond denied that drugs such as CPZ were useful. When they began using niacin in 1952, however, CPZ was virtually unknown in the North American market, which gave them the time and freedom to experiment with alternatives to more traditional therapies. Like many other psychiatrists of the time, Hoffer and Osmond thought that neuroleptic (and later antidepressant) drugs were effective in reducing patients' anxiety, tension, hostility, and agitation and sometimes used them in combination with vitamins in individual cases. But they did not believe neuroleptics to be effective long-term treatments; in fact, they thought that the drugs were more symptomatic than therapeutic. For them, niacin was the optimum choice for therapy.

Naturally, the suggestion that a simple vitamin such as niacin, which could be bought over the counter without a prescription, could be used to treat schizophrenia and might be preferable to traditional treatment with neuroleptics provoked controversy in mental health circles. Many initial criticisms of Hoffer and Osmond's niacin work attributed the phenomenal results to the bias of the two researchers, their idiosyncratic clinical approach, the placebo effect, or scientific fraud. Perhaps the patients treated were not actually schizophrenic, the two psychiatrists were too liberal with their diagnoses, or they chose patients whom they thought would respond well. Even when Hoffer and Osmond performed double-blind experiments, their methodology was called into question (e.g., small sample size). Osmond addressed some of these common accusations in an interview:

> And we have done all the tricks that scientists demand. That is, we've done these where the doctor in charge has not known what is being given. We've done them...where the doctor in charge has not believed in these particular substances working. And we've done the necessary thing of following them up for years and years and years. To put it briefly, I can say that the evidence is very strong that something is happening. What people are not agreed about is what is happening.[57]

In following years, opponents also raised the issue of toxicity and danger (e.g., liver damage) in taking anything greater than the daily recommended dose of niacin or any other vitamin (interesting given the well-publicized toxic reactions associated with many antipsychotics). Common side effects of neuroleptics included weight gain, lethargy, dry mouth, glucose intolerance, and even devastating neurological disorders (e.g., tardive dyskinesia). Having treated patients with the vitamin for over a decade, Osmond and Hoffer were more knowledgeable than the average professional about the topic and thought that the fears sparked about niacin's toxicity were grossly exaggerated and scientifically unfounded. Hoffer admitted the odd allergic reaction, but he challenged "any physician to provide evidence that niacin is as toxic as tranquilizers, antidepressants and any other chemotherapy used in psychiatry."[58]

PSYCHIATRY AND THE WAR ON VITAMINS

By the mid-1960s, news of niacin therapy, and that of a closely related compound known as NAD (nicotinamide adenine dinucleotide[59]), became increasingly public in both Canada and the United States, thanks in part to the 1966 release of Hoffer and Osmond's guidebook for patients and their families, *How to Live with Schizophrenia*. Eventually, psychiatrists and other professionals faced pressure from interested patients, their families, and the public. Because no independent study had corroborated the niacin work, many in psychiatric and medical communities accused Hoffer and Osmond of engaging in unethical advertising of an unproven remedy. Professionally orchestrated attacks against their niacin findings commenced in earnest in 1965 on several fronts. One was mounted by the national office of the Canadian Mental Health Association in Toronto and its general director, Dr. J. D. ("Jack") Griffin, a pioneer in Canadian mental health. As a founding member of the Canadian National Committee for Mental Hygiene, the precursor to the CMHA, Griffin took an active role in the promotion of community psychiatry, children's mental health, and the need to put psychiatry on an equal level with other medical disciplines.[60]

Griffin shared with his professional contemporaries a particular disregard for the claims about niacin treatment. What irked him most, however, was Hoffer's close connection with CMHA Saskatchewan and the fact that it was supportive of, and continued to publish, his and Osmond's findings. Mostly because of this uncomfortable relationship, Griffin staged a campaign to discredit Hoffer and force the CMHA to end its support for niacin therapy. Hoffer's relationship with the CMHA had not been much of an issue before the mid-1960s. Hoffer had worked closely with the organization since its inception in 1950 and was selected to serve as the chair of its Scientific Advisory Committee. He was also instrumental in assisting the CMHA to educate the public about mental illness and mental hospitals and to bring about much-needed improvements to the system. In 1966, he was far less involved with the organization than previously, but his views were deemed malevolent.

Griffin considered Irwin Kahan, executive director of CMHA Saskatchewan, as the source of the problem. As a research social worker and co-author of the original 1957 article, Kahan was responsible for following up and conducting blind evaluations of the patients of the first randomized controlled trials with niacin. When the results became available, he also became convinced of niacin's benefits in treating schizophrenia. He also happened to be Hoffer's brother-in-law. This was unacceptable to Griffin, who thought that Hoffer was manipulating Kahan and using the CMHA to attract attention to the niacin work. Perhaps there was some truth in this, but it is questionable whether Hoffer needed the organization or Kahan to draw attention to the niacin findings given his numerous publications on the subject and the positive outcomes in many of his patients. In any event, he alerted Osmond, noting that Griffin had written to Kahan that CMHA Saskatchewan was "pushing a cure concept" and that, if it was such a great treatment, CMHA Canada should test it.[61] Hoffer concurred, and within a year CMHA Canada, on the advice of its Scientific Planning Council, approved a $25,000 grant for a series of "replication" studies to test niacin as a treatment for schizophrenia, with psychopharmacological experts Heinz Lehmann and Thomas Ban as the lead investigators.

As the situation became more hostile, things became particularly uncomfortable for Hoffer and those working with him in Saskatchewan. He remained undeterred, continuing to publish and present his findings to both scientific and lay audiences. Soon, however, the government and academic colleagues pressured him to stop, which Hoffer thought was outright censorship. In April 1966, he received a letter from H. D. Dalgleish, acting registrar for the College of Physicians and Surgeons of Saskatchewan, regarding complaints about his publishing in the non-medical press, a departure from "traditional methods of publication."[62] This did not sit well with Hoffer:

> Your letter does raise a disturbing note: the restriction of scientific freedom as it applies to science generally....One of the reasons I have continued to work in Saskatchewan is that I have been given freedom to pursue my research—if there were any attempt to restrict this freedom (and I am certain this was not your intention) I would have to seriously question whether I could continue to put up with inadequate resources, monies and other inconveniences when other places have so much more to offer—it is only freedom of enquiry and publication which keeps me here and if this goes...so will I.[63]

As critics of the niacin work multiplied, however, many people who had stood behind Hoffer, such as McKerracher, began to rethink their association with the man who was fast becoming psychiatry's *enfant terrible*.

A similar confrontation was unfolding in the United States as the APA increasingly began to counteract the positive reception of niacin treatment by patients and their families, the public, and mainstream media. In an April 1967 issue of its newsletter *Psychiatric News*, the APA lambasted the American Schizophrenia Foundation (ASF), co-founded by Hoffer and Osmond, for its biochemical leanings and attempts to help sufferers of schizophrenia clearly dissatisfied with existing treatments. The article derided Hoffer-Osmond publications on schizophrenia and niacin treatment and cited a recent double-blind study that claimed to use Hoffer's method. The study, carried out by a research team headed by Nathan Kline, a well-known psychopharmacological pioneer and director of research at Rockland State Hospital in Orangeburg, New York, revealed no discernible difference between a placebo and NAD when used on a small group of chronic schizophrenics.[64] *Psychiatric News* closed on a negative note, invoking "the AMA Principles of Medical Ethics: 'When any sort of medical information is released to the public, the promise of radical cures or the boasting of cures...or of extraordinary skill or success is unethical.'"[65]

The *Psychiatric News* piece made a number of errors. It confused NAD with niacin (B_3) by saying that they were identical. Its use of Kline's NAD study as evidence of the ineffectiveness of niacin treatment for schizophrenia was a weak argument given that Kline and his associates did not follow Hoffer's original design, did not use niacin, and did not use the same NAD that Hoffer used (but some that Kline himself had prepared). He also used the treatment on patients with chronic schizophrenia, a subgroup that Hoffer had claimed all along did not respond favourably to the treatment.

Osmond noted these errors immediately and pointed them out to APA executive members in scathing letters. In an attempt to have *Psychiatric News* retract its errors, he insisted that the editor had failed to perceive that at no time had the ASF given support to the use of NAD (because it was not available). He emphasized that niacin treatment had been in use before tranquilizers and that several hundred doctors throughout the country had taken up the treatment and were reporting similar results. Regarding the accusation that the ASF (and Osmond and Hoffer) had shared information on niacin with the public too soon, he thought that they had actually been too slow in doing so, causing unnecessary suffering as a result. For Osmond, the accusation was the height of hypocrisy since many in the psychiatric establishment and pharmaceutical industry had been informing the public of the benefits of tranquilizers only a few months after they had appeared.[66] In another letter to APA's medical director, Walter Barton, Osmond criticized the APA's theoretical underpinnings and approaches to schizophrenia:

> You know as well as I do that patients and their relatives are being told that this grave illness derives from some faulty upbringing....The lives of ill people are daily being endangered by those psychiatrists...who...loudly advertise that schizophrenia is not an illness at all....I know, of course, that some, but by no means all, psychoanalysts and their hangers-on don't like any infringement on their vast and lucrative territory....Is character assassination a satisfactory way of trying to prevent the spread of ideas, which apparently don't commend themselves to some of our colleagues?[67]

With the exchange becoming progressively bitter, Osmond warned Hoffer of the impending battle: "We may have a heavy fight with APA but must use our many allies."[68] With APA's refusal to correct the errors in its newsletter, or at least provide space for Hoffer and Osmond to reply, contact between the opposing parties ended. It was another couple of years before they heard again from the APA regarding their practices.

Incontrovertible scientific evidence against the benefits of niacin therapy had yet to surface, but Hoffer and Osmond's credibility in the scientific community had been damaged. Although reputable professional journals had published their niacin articles for nearly a decade, many now refused to do so, proving detrimental to the dissemination and advancement of the niacin work. As Gordon Tullock notes, "the careful study which makes some minor, but respectable, improvement in knowledge in a given field poses few problems for the editor. An article proposing a drastic revision of existing theory or reporting dramatically unexpected research results is a much more difficult problem."[69] This might explain the blacklisting of articles by Osmond and Hoffer. By 1967, they and their niacin theories had become extremely unpopular in psychiatric and medical communities, and journal editors were loath to publish their work.

By that year also, the work environment in Saskatchewan had become too much for Hoffer to handle. Because of repeated efforts to silence him and curtail his research activities, he decided to resign from his positions with the government and university, opting to go into private practice. On May 2, 1967, he wrote to Osmond: "I do not yet know [the] government's reaction but I suspect it will be good riddance, and look how much money we can save.... In my new career, which may only last three years, I feel I will have time to expand very materially our joint research."[70] As much as it might have pained Hoffer to leave his long-tenured positions, he thought that there were significant gains in private practice: namely, "freedom of action [and] freedom from a rather stifling atmosphere at the Department of Psychiatry, College of Medicine."[71] He could better hold the government to task on deficiencies in the mental health system and psychiatric treatment. Being free from government and university restraints also meant that he could be much more aggressive in his promotion of niacin therapy.

MEGAVITAMIN THERAPY AND ITS ALLIES

It might seem that at this point Hoffer and Osmond had everything going against them, but they did have some staunch support that kept the spotlight on their theories and helped them in their fight for a fair assessment of niacin and megavitamin treatment. As controversial as it had become, megavitamin therapy continued to interest professionals and the public worldwide. Hoffer and Osmond continued to receive thousands of letters of inquiry regarding the treatment. More significantly, a handful of physicians and psychiatrists began to use the treatment in their own practices. Osmond and Hoffer took solace from the fact that their work was spreading and that professionals in Canada and the United States were producing similar positive treatment outcomes using their methods. Organizations that the two had helped to form, the American Schizophrenia Association and the Canadian Schizophrenia Foundation,[72] began to expand their influence, establishing a committee on therapy, publishing their own medical journal, and attracting hundreds of members. Hoffer and Osmond also received endorsements from powerful allies. One well-known figure who came to their aid on the megavitamin front was two-time Nobel Laureate Linus Pauling.

Pauling is perhaps best remembered as a forefather of structural chemistry and molecular biology. His pioneering explorations in the 1920s and '30s of the chemical bond and its implications secured him his first Nobel Prize in 1954. He also became known for his passionate antiwar activism. His efforts to highlight the dangerous levels of atmospheric radiation following nuclear tests and his successful antinuclear campaign led to an international testing ban and a Nobel Peace Prize in 1962. Pauling had a long-standing interest in medicine and the role of biochemistry in health and disease. From the 1950s on, he continued to study molecular diseases, examining phenylketonuria, a mental deficiency resulting from the inability to metabolize the amino acid phenylalanine. Because of his own illness in his early years, Pauling gradually focused on the role of diet in health. From the mid-1960s on, he began advocating the benefits of massive doses of vitamin C in treating the common cold and cancer, views for which he was castigated by the medical community.

It was also around this time that Pauling familiarized himself with the ideas of Hoffer and Osmond following his reading of their *How to Live with Schizophrenia*. This interest in using vitamins to treat mental illness led Pauling to write his famous 1968 paper in the journal *Science*, in which he put forward his ideas on "orthomolecular psychiatry" or "the treatment of mental disease by the provision of the optimum molecular environment for the mind, especially the optimum concentrations of substances normally present in the human body."[73] He provided an overview of existing research and laid out his theories on evolution, cerebral deficiencies, and the benefits of individualized amounts of megavitamins, essential amino acids, and other nutrients; he also cited the niacin work of Hoffer and Osmond as an outstanding example of research in the field.

Pauling's contribution to the debate invoked the ire of many medical professionals. NIMH clinician Donald Oken spoke for many when he wrote that the article illustrates elegantly the pitfalls which occur when an expert in one field enters another area. With his characteristic brilliance, Linus Pauling describes a biochemical mechanism which *could* be responsible for some forms of mental illness (or indeed for illness of many other types). Remote plausibility, however, no matter how intriguing and creative its nature, should not be confused with evidence. Unfortunately for Pauling's thesis, there is no adequate evidence to back up his view.[74]

Some were even less kind in their assessment of Pauling on this matter, suggesting that he had lost his scientific marbles and become senile in his old age. As one group of biographers put it, he had become more "partisan debater... [than] detached scientist," and his pronouncements on megavitamins were "scientifically as well as politically radical."[75]

Yet Pauling's scientific stamp of approval was a tremendous boon to Osmond and Hoffer. Within a few years, orthomolecular psychiatry acquired a number of adherents, who joined the three men to explore and develop therapeutic applications in schizophrenia and other psychiatric conditions.

Bill Wilson, co-founder of Alcoholics Anonymous, also came to the aid of Osmond and Hoffer. His awareness of, and interest in, their research arose from a constellation of common friendships. His first-hand experience with the effects of niacin came after they recommended it as a way to reduce his extreme anxiety and depression. Following a complete remission of his symptoms, Wilson spread the word about niacin to his friends and associates and launched a therapeutic movement in AA that sparked immense professional and popular curiosity. Over the years, he developed a close camaraderie with Hoffer and Osmond, championing the use of niacin and helping Hoffer to establish an anonymous support group for recovering schizophrenics modelled after that for recovering alcoholics. Again, having a spokesperson such as Wilson was enormously beneficial to Osmond and Hoffer. He had ties to physicians, senators, congressmen, and others in positions of power and could direct greater attention to niacin. In fact, largely because of his efforts, the NIMH agreed (if reluctantly) to sponsor Rutgers University in New Jersey to lead some trials of niacin for schizophrenics.

By the end of the 1960s, niacin therapy, and more generally the ortho-molecular school of thought, were mired in controversy. Many had weighed in on the niacin-schizophrenia issue, but it remained an unsettled issue in the psychiatric community. Both sides realized that the stalemate would continue indefinitely until independent, randomized controlled clinical trials repeating Hoffer's original protocol came out with their results. Expectations were understandably high for the replication studies, the most significant of which were launched by the CMHA.

When the studies received the go-ahead in December 1966, the CMHA selected as clinical leads McGill University psychiatrists Lehmann and Ban. Credited for the first research publications in North America on chlorpromazine and the antidepressant imipramine, Lehmann became a leader in the burgeoning field of psychopharmacology. Immigrating in 1937 shortly after obtaining an MD from the University of Berlin, he secured a temporary Canadian medical licence and a position at Verdun Protestant Hospital, a 1,500-bed mental hospital near Montreal, where he would eventually become clinical director in 1947 and research director in 1962. Lehmann also became extensively affiliated with the Department of Psychiatry at McGill, carrying out research and teaching duties and becoming its chairman in 1971. It was through his association with McGill that he developed a close working relationship with Ban, an up-and-coming psychopharmacological expert.

Ban, also an immigrant, arrived in Canada in 1957 following a medical education at Semmelweis University in Budapest. Upon his arrival, he took up a fellowship in neuro-anatomy at Montreal's Neurological Institute and a rotating internship, later enrolling in McGill's psychiatric training program and doing a residency under Lehmann. For his thesis, Ban focused on Pavlovian concepts (e.g., conditional reflex) and their applications in psychiatry, which he subsequently extended to studies of the effects of psychopharmacological agents on psychiatric patients. In just a few years, he became a leading authority in psychopharmacology, teaching and publishing extensively and writing the first comprehensive textbook on the subject. Along with Lehmann, Ban was the first to conduct clinical experiments with the newer groups of antipsychotics (e.g., butryophenones), antidepressants, and other psychotropic compounds.

Beginning in the 1960s, both men, operating out of the Verdun hospital, became intimately attached to the Early Clinical Drug Evaluation Unit (ECDEU) Network of NIMH, in which they, along with a handful of American-based research units,[76] received funding through U.S. Public Health Service grants to conduct extensive tests on the therapeutic efficacy and side effects of virtually every known psychotropic drug as well as new ones being synthesized by pharmaceutical companies. ECDEU investigators were expected to meet regularly to present their results and reach consensus on standardized designs and diagnostic and outcome instruments (e.g., rating scales) that could be used in future clinical testing. This work, which continued into the 1970s, ultimately formed the basis of the American Food and Drug Agency's (FDA) guidelines for clinical trials of psychopharmacological drugs. Based upon their early clinical experiences together, Lehmann and Ban developed an affinity for neuroleptics in treating schizophrenia and other psychiatric illnesses. Even before the CMHA studies began, Lehmann firmly believed that, despite its limitations, "no other single therapeutic procedure can compete with neuroleptic treatment in terms of rapid effectiveness, sustained action, general availability and ease of application."[77] Ban echoed this sentiment, insisting that "modern pharmacotherapy had by 1963 resulted in a remission rate of between 50 to 60 percent for patients who had been ill less than three years. Furthermore, a discharge rate of 75 to 80 percent was claimed for all acute hospitalized patients within the first year and more than 50 percent within six months."[78]

Hoffer thought the choice of the two McGill psychiatrists a good one. They were highly skilled clinicians and researchers, and in his estimation were fair and would do an honest job. The CMHA collaborative studies were to last for three years; they would consist of several separate, randomized controlled trials and focus on testing the short- and long-term efficacy of the megavitamin claims for schizophrenia, particularly the niacin ones, using both acute and chronic schizophrenia patients as research participants. Most importantly, there was an explicit agreement that the CMHA studies would adhere as closely as possible to Hoffer's original treatment protocol, which can be broken down into three progressive stages: Phase I, acute schizophrenics, ill less than a year, not requiring immediate hospitalization treated with B₃ (three to six grams per day) with or without the addition of vitamin C (one to five grams per day); Phase II, acute schizophrenics requiring hospitalization treated with a combination of B₃, C, and ECT (successful cases discharged and continued on niacin for one year); and Phase III, chronic schizophrenics or refractory Phase II patients treated with B₃, C, and ECT.[79] When Hoffer was presented with Ban's design, he agreed to it with a few minor variations, such as that neuroleptics would be added to B₃ as a supplementary treatment if required. The first trials were slated to begin in late 1968, with the main emphasis on niacin. As they unfolded, the enthusiasm that Hoffer initially displayed gradually disappeared.

By 1969, tensions had escalated in the megavitamin controversy, provoked in part by the continuing activities of Hoffer and the Saskatchewan Schizophrenia Foundation. As the fledgling organization expanded its influence in the province and beyond, through sponsored conferences and an aggressive public education platform (e.g., Schizophrenia Month), it caught the attention of professionals and the public. Many turned to the Saskatchewan government to determine where it stood on niacin treatment. Minister of Health Gordon Grant looked to the PSB for its input. Having met with Hoffer, Grant wondered if the province should invest in its own controlled studies to put the matter to rest. In response to the minister's request for more information, Colin Smith, now in the uncomfortable position of having to take a stance regarding his former boss, cautioned Grant about the inherent difficulty in evaluating the vitamin in the treatment of schizophrenia: "One matter which complicates such a study is that nicotinic acid is not used alone in the treatment of schizophrenia. It is combined with a variety of treatments. Separating out the effects of each treatment is a very complicated task."[80] Smith pointed out that no roadblocks had been put in the way of provincial psychiatrists to use niacin, and each had discretion to decide on the best treatment method. He also told Grant about the CMHA studies and said that he would follow up with Ban. On January 2, 1969, Ban provided Smith with an update:

> To make a long story short, the original protocol failed. With the exception of the research unit at St. Jean-de-Dieu and at our hospital we were unable to obtain sufficient participation (because of

the rigidity of the design), and we have decided to suffice with 30 cases (15 from each hospital) instead of 150. To compensate, we put forward a modified design.[81]

Responding to the favourable media attention garnered by Hoffer and his organization (soon to be the Canadian Schizophrenia Foundation) and, of course, niacin treatment, Griffin stepped up his counterattack, pressing George Peacock, registrar of the College of Physicians and Surgeons of Saskatchewan, to take measures:

> Hoffer himself in an interview implied that he could cure 80% of schizophrenics and went on to criticize Canadian doctors and particularly psychiatrists who apparently were not taking his work seriously or even willing to check his results "because they would all lose their jobs if schizophrenics were all cured" and "mental hospitals and psychiatric wards would all be closed down."...On behalf of the CMHA, I would like to protest most vigorously the statements made by Dr. Hoffer in this newscast....There is no general scientific agreement yet established in favour of the Hoffer hypothesis and efficacy of niacin in the treatment of this disease.[82]

About the same time that these events were transpiring in Canada, the APA made its next move in the controversy, accusing Osmond and Hoffer, still registered APA members, of breaching the APA Code of Ethics with their publications and "flamboyant promotion of a treatment method and the lack of substantial scientific validation."[83] Osmond, angered by the letter, was perplexed about which "hallowed custom" he and his partner had broken, and he sought clarification from Robert Garber, secretary of the APA, noting that "in matters of medical custom and science differences of opinion frequently occur."[84] Shortly after, Osmond grumbled to Hoffer that "just as we're getting a following of our colleagues the establishment behaves almost incredibly true to form.... [They] have never corrected the piece in APA news or published any of our letters—a savory piece for the history of niacin therapy when it comes to be written."[85] In response, Hoffer tried to reassure his partner:

My letter was a warning that if necessary I will force their action to be examined by a court of law. I doubt they would appreciate public scrutiny. If their Ethics Committee does call us for a hearing, I would want you and I and our colleagues using B-3 to appear.... We can bring out the facts, the silly report in APA news, their refusal to publish a correction or to publish our letters.[86]

The actions of the APA are intriguing. For years, many APA psychiatrists had vigorously promoted psychoanalytical approaches to the treatment of schizophrenia; Hoffer and Osmond's approach can therefore be viewed as competitive. According to orthomolecular physician David Hawkins,[87] much of the antagonism toward niacin therapy "stem[med] from psychiatrists who depend[ed] primarily on psychotherapy in treating schizophrenia, despite lack of evidence anywhere that psychotherapy is of any benefit in schizophrenia, and in the face of scientific studies which in fact prove that psychotherapy alone is totally useless."[88] The APA move against Hoffer and Osmond might have been logical, but there was more to account for its official message.

It is also important to consider the shift in direction in psychiatry that, by the end of the 1960s, had become increasingly apparent. Hoffer and Osmond had noticed a turn away from psychotherapy in 1968.[89] The metamorphosis of APA into an organization heavily influenced by biological psychiatry was another indication of this turn. Some within the psychiatric establishment continued to apply psychoanalytical techniques to schizophrenia, but generally they moved to the background, becoming secondary to biological techniques. Pauling's assessment of the transformation was blunt:

Psychoanalysis has failed, and psychiatry is now rapidly returning to the scientific approach, the recognition of the corporeal character of mental disease, with manifestations determined to some extent by environmental stress and past experience.... Recognition of the effectiveness of phenothiazines and other drugs (and the ineffectiveness of psychoanalysis) has accelerated the reacceptance of the concept that mental disease is a disease of the brain, and that the brain itself needs to be treated, by changing its molecular composition.[90]

Hoffer, Osmond, and other biological psychiatrists would likely have agreed; what separated them was what they considered to be scientifically legitimate. The major source of contention was the hypothesized biochemical etiology of schizophrenia (e.g., adrenaline or dopamine metabolism) and, by extension, the most effective course of treatment (e.g., pharmacotherapy with neuroleptics or megavitamins).

Another factor in the megavitamin controversy was the powerful role of pharmaceutical companies in psychiatric research. To secure approval for and market their products, companies came to depend on researchers to make scientific cases for their creations. Consequently, an almost symbiotic relationship developed between them, with companies pouring millions of dollars into the clinical testing, promotion, and prescription of drugs.[91] The CMHA studies themselves were financed in part by the NIMH ECDEU grants, in turn supplemented by drug company dollars.

The other significant change in the psychiatric research field was the reliance on randomized controlled methodologies (e.g., double blinds) for the scientific justification of clinical drug trials, a trend with which Hoffer and Osmond had become thoroughly disenchanted. Writing to Osmond, Hoffer said that

> because double blind methods have come in coincidentally with massive use of drugs in psychiatry, most psychiatrists fall into [the] gross error of thinking it was double blind which was responsible....There is every reason to believe there is no relationship whatsoever. Had the double blind never been introduced I doubt the history of drugs in psychiatry would have been altered one iota. Double blinds without tranquilizers would quickly have gone out of style.[92]

Such opinions further alienated Hoffer and Osmond from the psychiatric community, thoroughly indoctrinated with the supposed scientific advantages of the double-blind method in clinical drug testing.

Despite the many obstacles, the CMHA studies still held out the promise of true replication of the therapeutic approaches of Hoffer, Osmond, and their orthomolecular colleagues. It was to the great relief of many, then, when Ban released preliminary findings of the studies at the annual meeting of the Canadian Psychiatric Association in June 1969.

Here I will highlight their methods and findings, particularly in regard to niacin, because they had negative repercussions for the scientific reputations of Hoffer and Osmond and megavitamin treatment in general. The first two CMHA trials tested the usefulness of niacin therapy in schizophrenia as both the sole treatment and an adjunct to traditional treatment with phenothiazines. When Ban provided the Canadian psychiatric community with its earliest glimpse of the first niacin study, his findings appeared to corroborate some of the Saskatchewan work, revealing a statistically significant improvement after two weeks of treatment with niacin compared with a placebo. Although this result was promising, Hoffer indicated to Osmond that Ban was "well aware of the marked antagonism toward the use of nicotinic acid and he has written the paper in a rather interesting way in order to prevent direct attack on him."[93] Hoffer also expressed reservations about the ethics of another cross-over study of the Ban-Lehmann group that sought to examine niacin as a methyl acceptor, thereby testing Osmond and Hoffer's biochemical hypothesis of schizophrenia. This was done by giving methionine, a methyl donor known to exacerbate schizophrenia psychopathology, to a group of patients with chronic schizophrenia and then administering niacin, a methyl acceptor, to see if it would have a therapeutic effect (reducing the conversion of noradrenaline to adrenaline). The findings failed to confirm this effect.

By December 1969, findings from the second controlled study, which compared thirty "newly admitted" acute patients divided into three groups (one with niacin-phenothiazine, one with niacinamide-phenothiazine, and one with placebo-phenothiazine), were circulated, with the results again questioning niacin treatment. As Hoffer reported to Osmond,

> there was no sign of difference between any of the groups but response rate is measured by discharges and their scales show that all three groups were identical. I pointed out to Ban that with these kind[s] of data it would be impossible to show that any treatment is superior to another treatment. I suggested to him he should now follow up in the community very carefully because this is where the

significant difference will arise....This paper will probably...be used as an attack upon the whole vitamin concept, however, it will not be of much value in places where it counts, that is with people who are already using the treatment very successfully and are spreading its use.[94]

When the Ban-Lehmann group published their official results, their conclusion was that "no statistically significant difference was seen between the active treatment and the placebo groups; i.e., the addition of nicotinic acid or nicotinamide to the regular phenothiazine treatment regimen did not have any measurable therapeutic effect in this sample of patients."[95] Through the use of pre- and post-treatment Brief Psychiatric Rating Scales (BPRS), it was shown that niacin had no therapeutic effect on patients in the short term (six months), nor, as later pointed out, did it have any positive impact in the long term (after two years).[96] Ban also reported that niacin therapy was "not superior to the overall therapeutic efficacy of an inactive placebo [and that] the addition of nicotinic acid [niacin] to regular phenothiazine treatment prolongs the duration of hospital stay and increases the amount of neuroleptics needed in treatment."[97] With this information, Griffin felt comfortable in issuing the following warning in his introduction to a progress report on the CMHA studies:

It is important that this Progress Report be read with care by all physicians who are interested in the pharmacology of schizophrenia and particularly in the niacin treatment. There is little in this first report to support the public enthusiasm which has been engendered by the megavitamin therapy. In view of the potency of the substance and its potential for dangerous side effects the drug should be used with great care and caution....*It is definitely not the treatment of choice for every schizophrenic patient under all possible conditions.*[98]

Additional studies in the CMHA series focused on the effects of niacin in patients with chronic schizophrenia or tested the incorporation of additional vitamins (e.g., pyridoxine and ascorbic acid) presumed to enhance the effect of niacin, but they also failed to confirm Hoffer and Osmond's alternative treatment. The findings that inflicted the most damage, and resonated with the psychiatric community, were those that failed to verify the claims of niacin's therapeutic efficacy for patients with acute schizophrenia, which Hoffer and Osmond had always insisted yielded the best outcomes.

To many, the results of the CMHA studies called into serious question—some would even say destroyed—the scientific basis for niacin therapy. The official message, as enunciated in a 1973 publication prepared for the National Scientific and Planning Council, was that the

> Ban/Lehmann investigations concerning megavitamin treatment of schizophrenia must be considered to be a fair translation of the theories and treatment programmes of Dr. Abram Hoffer and the orthomolecular group of psychiatry into formulations which allow for testing....We support the conclusions arrived at: megavitamin therapy is not indicated as the treatment of choice in acute or chronic schizophrenia. Whether or not the therapy has any place in the treatment of the schizophrenias cannot...on the basis of current evidence be determined....Despite the pronounced biochemical bias of the investigators, negative findings nevertheless emerged.[99]

These and other condemnatory remarks about niacin treatment were reiterated in all subsequent CMHA publications. The Ban-Lehmann findings also sent reverberations throughout the psychiatric and medical communities. The message was clear: not only was high-dose niacin therapy ineffective in treating schizophrenia, but also its use was potentially toxic to the liver and kidneys. To practitioners not yet convinced of these findings, Ban issued some stern words: "If psychiatry is to become and remain scientific, it must meet the tests of scientific validity. Nicotinic acid does not do so at this time....If there is to be professional acceptance of a megavitamin or orthomolecular treatment program, it must be based upon adequately designed and carefully executed clinical experiments."[100] Many assumed that the CMHA studies came closest to fulfilling these criteria. In reviewing the Ban-Lehmann studies, John Mills wrote that "the authors displayed a complete grasp of the principles of experimental design, a good appreciation of the shortcomings of those principles in medical settings, and made praiseworthy efforts to modify the applications of the principles accordingly."[101] He even claimed that the "CMHA acted as the [adrenochrome] theory's mortician."[102]

There is no denying the lasting negative impact of the CMHA studies on megavitamin therapy and its proponents, but it is questionable whether the work of Ban and Lehmann was as thorough, flawless, and unbiased as many have claimed. In fairness to Hoffer and Osmond, some criticisms of their original studies can also be made of the CMHA studies. Even the ad hoc committee reporting to CMHA's National Scientific and Planning Council picked up on this similarity in its reference to some of the studies' "methodological problems...concern[ing] the size of the sub-groups under study, drop out rate, and the fact that statistics may be misleading when applied to relatively small groups."[103] In the committee's view, however, these concerns were minor compared with the overwhelming proof against the use of niacin in treating schizophrenia.

Nonetheless, the CMHA studies displayed some weaknesses. They were not true repetitions of Hoffer and Osmond's treatment protocol; use of the term "replications" to describe them is a misnomer because they were at best an approximation of the Hoffer-Osmond method. Ban conceded early on in the clinical trials that he had been forced to make modifications. For example, the studies never used ECT, which Hoffer and Osmond claimed potentiated the effect of niacin in Phase II patients. There was also some confusion over whether the subacute and acute subjects used for some CMHA niacin studies fit the diagnoses; although patients were referred to as having been "newly admitted," it was later found that some had actually been chronic schizophrenics transferred from other hospitals. If this was so, then the less than stellar results achieved with niacin therapy merely repeated, as other independent studies had,[104] what Osmond and Hoffer had already demonstrated in their earlier work: that niacin alone did not produce the best outcomes in chronic schizophrenics. One also has to wonder about the ethics of the Ban-Lehmann team in aggravating the symptoms of chronic schizophrenics through the use of methionine in order to disprove Hoffer and Osmond's biochemical hypothesis of schizophrenia (and the important role that niacin played in it). Besides, the study was set up in such a way that it guaranteed the poor performance of niacin. As Pauling and others emphasized, "the patients were given 20 g of methionine per day. Over 16 g of niacinamide per day would be required to accept the methyl groups donated by 20 g of methionine, but the patients were given only 3 g. It could have been predicted that the experiment would fail."[105]

Many of these and other criticisms were raised by Hoffer, Osmond, and their supporters in their responses to the negative conclusions of the CMHA reports. Interestingly, the proponents of megavitamin therapy did judge some of the data collected from the studies positively. As Pauling said, part of the resistance to megavitamin therapy stemmed from a "misunderstanding of the meaning of statistical significance."[106] According to him, "investigations described as attempts to replicate Hoffer and Osmond's results are reported to have failed to show a statistically significant difference between the subjects receiving niacin or niacinamide and those receiving placebo. This conclusion is then incorrectly interpreted as meaning that the investigations have shown niacin or niacinamide to have no greater value than placebo."[107] Hoffer's son John, a medical student at McGill University, waded into the debate. In a letter to the editor of the *Canadian Psychiatric Association Journal* in 1975, he questioned the CMHA studies, from their failure to repeat his father and Osmond's original design, to the dosages of vitamins administered, to test subjects, and to how the findings were evaluated. He even challenged the lead investigators' bias in conducting the trials:

> When the results of the five completed [studies] are viewed together, it becomes apparent that they are not nearly as negative as Ban and Lehmann have made them out to be in their report. Their interpretation of the data appears to be coloured, in that improvements observed in the treatment groups are either downplayed or ignored in two studies, while in other studies insignificant and meaningless differences are interpreted as demonstrating a significant difference in favour of the placebo group.[108]

Lehmann and Ban responded to the John Hoffer letter in the same issue of the journal, noting that "we started out our series of controlled megavitamin trials in schizophrenic patients nine years ago, with the sincere hope that we might obtain at least some positive results. Alas, so far the facts have turned out to be different."[109] They staunchly defended the studies and their interpretations of the data. This defence was countered by another letter by John Hoffer in which he wrote that the two psychiatrists themselves were "prone to an unscientific, hostile bias in their analysis of the therapy."[110] By this point, the fight had reached a draw. As W. T. Brown, editor of the journal, accurately observed, "a controversy that could not be resolved and contained in the research laboratory, [in] the hospital or in the halls of international science now moves ominously into the sociopolitical arena." His plea for "honest, humane dialogue between the opposing camps" was wishful thinking.[111]

A key point of contention in the wake of the CMHA trials was whether or not niacin had any value in treating some cases of schizophrenia. An important parallel study sponsored by the NIMH in the United States was carried out by Rutgers University psychologist and psychopharmacologist J. R. Wittenborn on a group of New Jersey State Hospital patients over a two-year period. It also tested whether niacin supplementation aided in the treatment of acute schizophrenia.[112] As with the CMHA findings, he initially failed to discern any advantages in adding three grams of niacin to traditional treatment. Subsequently, however, he discovered that, when he separated cases of "good premorbid personality," the niacin group performed better than the control group. As he noted, *"those patients with conditions diagnosed as schizophrenia who come to treatment with a history of strong interpersonal commitments will respond well to niacin supplemental therapy."*[113] Wittenborn concluded by calling for additional studies of his findings and later sent a reprint of this report to Hoffer with a written apology for his previous mistake.

Most critics of the megavitamin therapy ignored or minimized the significance of Wittenborn's secondary analysis. Ban did not deny that niacin might have some use in psychiatry, but he never deviated far from his original estimation that, "apart from pellagra and the encephalopathy of nicotinic acid deficiency, no other nicotinic acid responsive clinical psychopathology has been successfully and definitively identified."[114] With regard to the use of niacin, Ban thought that "the prescription of a therapeutic regimen which has not shown itself to be the optimal treatment for the average schizophrenic patient before the optimal treatment has been tried may be contrary to [a] physician's duty."[115]

Hoffer remained defiant in the face of the CMHA findings. Many psychiatrists came to see his unwillingness to admit defeat, his calls for new tests of the megavitamin work, and his way of explaining away the purported significance of the CMHA studies as pathetic attempts to rescue his theories. As one Canadian physician complained, "to obtain exact treatment schedules from Dr. Hoffer is difficult if not impossible because of his changes in criteria with each request. While Dr. Hoffer may feel that the [CMHA] is prejudiced against him, one must ask why anyone would be against any treatment that could help in such an agonizing disease."[116] Recommendations such as those provided by A. H. McFarlane, a professor of psychiatry at McMaster University, became the preferred response to Hoffer's pleas: "I would suggest that we stop chasing Dr. Hoffer down the many explanatory pathways he might take us."[117]

To attempt to bring some closure to the whole megavitamin debate, NIMH hosted a two-day workshop in April 1973 that brought together experts from the opposing camps to discuss the orthomolecular treatment of schizophrenia. Both sides made familiar accusations and methodological criticisms. When the unpublished data from Wittenborn's "retrospective" analysis came up, critics insisted that they were hypothesis-seeking and in need of replication and that the subgroup of those responding well to high-dose niacin was an anomaly. The subgroup was too small, the patients in question might not have had real schizophrenia, and they might have done just as well with other or no treatments.[118] Following Lehmann's review of the CMHA studies, and following criticisms by Hoffer, Lehmann admitted some weaknesses in some of the CMHA studies. For example, in one experiment to test induced psychopathology, more niacin would have been required to neutralize the large doses of methionine. Nevertheless, Lehmann stated that, "over and above the controlled clinical trials of efficacy, he had never had the clinical feeling or impression that niacin was an effective treatment for schizophrenia."[119] Lehmann and others thought that Hoffer's clinical skills, and maybe his inaccurate diagnosis of schizophrenia, had more to do with his good results than the treatment per se.

Concerning the more theoretical aspects of the debate, Pauling laid out the rationale for his nutritional hypothesis of mental illness and ortho-molecular treatment. Some researchers present thought that there was insufficient evidence to show that schizophrenia is a vitamin-responsive disease. They also raised concerns about toxic reactions associated with high-dose vitamin treatments. When Osmond and Hoffer's long-time foe Kety provided a closing summary to the workshop, he reiterated many of the criticisms of orthomolecular treatment and referred to the earlier NIMH work under him that discounted the original theoretical basis (e.g., the adrenochrome hypothesis) for using niacin in treating schizophrenia. As far as Kety was concerned, there was "no basis for a crash program of further controlled trials."[120] At the end of the workshop, the two sides had failed to reach a common understanding and were as far apart on the issue of orthomolecular treatment as before.

Perhaps the major factor in the demise of megavitamin therapy was the APA *Task Force Report 7* on the subject in 1973. Hoffer and Osmond had been reprimanded by the organization in the late 1960s over pub-lished articles, one by Osmond on the background of niacin therapy, one by Hoffer on five positive schizophrenia case studies from California. According to the APA, their writings had violated an ethical principle, an accusation to which Osmond had taken great exception. In his view, the ethical requirement that "all members of the APA shall be bound by the ethical code of the association" was "extremely vague and imprecise" and loosely interpreted to silence him and his partner.[121] In December 1970, the men had appeared, along with Pauling and other key allies, before the APA Ethics Committee in Washington, DC, but again the APA had gone away undecided on the matter. Behind the scenes, a team of experts had been put together to provide an in-depth and objective review of the exist-ing evidence on megavitamin and orthomolecular therapy.

The team was chaired by Morris Lipton, a professor of psychiatry in the School of Medicine at the University of North Carolina. He was a well-known and outspoken critic of the megavitamin approach, and this did not bode well for an objective review of the therapy. In fact, when one considers the list of members selected for the task force, the deck was stacked against orthomolecular therapy. It was akin to expecting a group of hardened psychoanalysts to objectively assess biological theories of schizophrenia. The only difference was that most team members were committed to the biological paradigm, albeit one centred on neuroleptics and biochemical explanations of their use. Other members of the task force also stood out, such as Ban, who had already made public statements against the therapy. Many of his earlier views in CMHA reports were replicated almost verbatim in parts of the APA report. And then there was Loren Mosher, chief of NIMH's schizophrenia studies, at the other end of the theoretical spectrum but equally against the use of megavitamins. He was an old-school psychodynamic theorist and no more open to megavitamins to treat schizophrenia than to antipsychotics.[122] In an earlier review, Mosher wrote at length on the dangerous side effects of high-dose niacin.[123] Wittenborn, not an official member, acted as a consultant to the team. He had just completed his preliminary findings from the NIMH studies, which found no discernible difference between a placebo and niacin in the treatment of schizophrenia (only after public release of the APA report were some of his initial assumptions retracted in his second publication). Hoffer, of course, had serious issues with the task force membership, particularly with Lipton as chair, with whom he had publicly and professionally butted heads on a number of occasions. Of more concern to Hoffer was the absence of an orthomolecular practitioner on the task force. To no avail, he took these concerns to Lipton and then directly to APA senior management, saying that they precluded a fair, unbiased review. It proceeded as planned.

In its fifty-four-page report, the APA task force presented its rendering of the history of megavitamin and orthomolecular treatments and the theories behind them. After a review of the literature for and against the therapy, the task force stated that most studies supporting megavitamin treatment were rife with "logical and methodological difficulties making objective evaluation and assessment of the validity of the treatment procedure extremely problematic."[124] Regarding the studies on the opposite side, particularly the not-yet-complete CMHA ones, they were to be commended for their carefully controlled nature and "thoroughgoing attempts at replication."[125] Having weighed all the evidence, the task force ascertained that "the credibility of the megavitamin proponents is low...[and]

further diminished by a consistent refusal over the past decade to perform controlled experiments and to report their new results in a scientifically acceptable fashion."[126] The concluding statement was that "under these circumstances this Task Force considers the massive publicity which they promulgate via radio, the lay press and popular books, using catchphrases which are really misnomers like 'megavitamin therapy' and 'orthomolecular treatment,' to be deplorable."[127]

Hoffer and Osmond had never pinned any hopes on the report and had expected as much from the APA. They knew that it was part of a concerted attempt by the organization to discredit them and their theories once and for all. Not surprisingly, they wrote a 128-page response to the APA report.[128] To them, the report contained many half truths and even some outright lies, and they painted it as a complete misrepresentation of their work. Even Pauling weighed in, calling the last sentence of the report a blatant example of bias: "This concluding sentence, like many others in the book, seems to me to have been written in order to exert an unjustifiably unfavourable influence on the readers of the report."[129]

And that it did. After publication of the report, Hoffer, Osmond, and their megavitamin colleagues were perceived as quack doctors, a lunatic fringe of the psychiatric establishment and wider medical community. Within a couple of decades, the pioneering efforts of Hoffer and Osmond in the treatment of schizophrenia had gone from the cutting edge in mental health research to the cutting room floor. In 1961, T. C. Douglas, a revolutionary in his own right, was confident that the work of the two psychiatrists would "be recognized as one of the greatest advances made by medical science in this generation."[130] Few would have made the same assessment in the 1970s. By the middle of that decade, enthusiasm about Hoffer and Osmond's schizophrenia work had waned, if not disappeared, and their work had become a target of ridicule in medical and scientific communities, some even noting that a "cult" had developed among followers of megavitamin therapy.[131]

For most observers, the APA report was the final nail in the coffin of Osmond and Hoffer's theories on schizophrenia and the death knell for megavitamin/orthomolecular treatments. Prospects were not good for either Hoffer and Osmond or the theories that they had spent so many years developing and testing. In 1976, the Canadian Psychiatric Association followed up the APA task force with an official disavowal of the treatments: "There is no good scientific evidence showing that they are therapeutically and prophylactically effective in psychiatric disorders [and] there is evidence that they are not entirely safe [and for these reasons] CPA recommends that physicians do not use these treatments except in controlled clinical investigations."[132] Battered and bruised, Hoffer had to admit that the task force report had dealt a heavy blow. It did not stop him or those already using the therapy, but, as he later wrote, "the momentum of the movement was destroyed."[133] It remained to be seen whether megavitamin therapy would fall by the wayside of scientifically viable psychiatric treatments, but the prognosis was gloomy. As reported by one physician in the *American Journal of Psychiatry*, "orthomolecular psychiatry has failed to sell in the marketplace of scientific investigators, and...we are faced with the unpleasant likelihood that this treatment will continue to be used without adequate scientific assessment."[134]

Although it became somewhat lost in the megavitamin/orthomolecular debate, psychedelic drug research in Saskatchewan played a crucial role in the evolution of Hoffer and Osmond's theories on schizophrenia. This research only complicated matters by leading to more confusion among, and fodder for, critics. As discussed in the next chapter, the groundbreaking use of LSD in the psychotherapeutic treatment of alcoholism, in particular, and the fantastic claims that resulted created another volatile scientific controversy, equal to the one surrounding Osmond and Hoffer's work on schizophrenia.

CHAPTER 7

LSD: A NEW HOPE FOR ALCOHOLISM?

The other major plank of the psychedelic drug research program in Saskatchewan was its use of LSD in the treatment of alcoholism. Between 1957 and 1962, spectacular claims about treatment came out of the province and were written about and widely debated in both professional and lay circles. At the peak in the early 1960s, the findings revealed astounding rates of recovery in treating alcoholism (50–70 percent), and LSD-assisted psychotherapy became a highly touted and officially sanctioned approach in the province.

What really set the research program in Saskatchewan apart from hundreds of other projects elsewhere employing LSD and other hallucinogens in psychiatric treatment was how it incorporated the drugs into treatment settings. Abram Hoffer, Humphry Osmond, and Duncan Blewett played pivotal roles in the development of an innovative approach to the treatment of alcoholism, better known as psychedelic therapy, in which the goal was to bring about a transcendental experience in the patient and through it lasting sobriety. It was largely because of their efforts that Saskatchewan became one of the few locations in the world in 1960 using this type of therapy.

Yet the popularity of their idiosyncratic approach to treating alcoholism was short-lived. Faced with a lack of sustainable funding and support, the reformation of provincial psychiatric services and research, and the departure of key researchers such as Osmond, the psychedelic drug research program in Saskatchewan gradually closed up shop. By 1962, it was all over. In the years that followed, the treatment claims that had come out of that program were laid to rest.

What happened exactly? Critics of the research pointed out that the claims did not hold up when subjected to more rigorous and scientifically respectable methods (e.g., double blinds). Other researchers failed to produce the same positive outcomes with psychedelic therapy. Regarding the favourable results achieved by the Saskatchewan researchers, critics noted that they stemmed from serious flaws in research methodology and some wishful thinking, if not outright deception, by the overzealous practitioners. The historical record thus reads that Saskatchewan's psychedelic treatment of alcoholism, much like Hoffer and Osmond's theories regarding schizophrenia, became a casualty of psychiatry's scientific maturation and the move toward a more scientific approach in all clinical research in mental health.

Although this has become the general perception of the Saskatchewan research within professional ranks, it seems to be a rather simplistic account of the sudden reversal of fortune for the treatment claims. Like the schizophrenia research of Hoffer and Osmond, the treatment of alcoholism in Saskatchewan with LSD has to be viewed in the context of massive changes within the field of psychiatry in general and the burgeoning field of research on alcoholism in particular. By the early 1960s, things had changed considerably in the scientific understanding of alcoholism and its treatment.

THE DISEASE CONCEPT OF ALCOHOLISM

By the mid-twentieth century, the study of alcoholism had developed into a scientific field in its own right. This was thanks in large part to overwhelming public and professional acceptance of "the disease concept of alcoholism," the premise being that alcoholism was a legitimate illness that deserved medical attention and treatment. This concept in North America has often been traced to the pioneering accomplishments of organizations such as Alcoholics Anonymous and scientific researchers such as E. M. Jellinek in the 1940s and 1950s. However, the roots of the concept can be found as far back as the late eighteenth century with the efforts of American physician and leading temperance advocate Benjamin Rush. As Harry Gene Levine insists, it is in the work of Rush, "taken as a whole, that we can find the first clearly developed modern conception of alcohol addiction."[1]

In contrast to the popular perception that alcoholism was the result of individual choice and wrongdoing, Rush drew attention to the toxic and addictive nature of alcoholic beverages. He was also the first to promote the notion that alcoholism was a progressive disease characterized by "loss of control" over drinking that could be cured only by complete abstinence. As Rush commented in his 1814 pamphlet *Inquiry into the Effects of Ardent Spirits upon the Human Body and Mind*, "it belongs to the history of drunkenness to remark. . .that its paroxysms occur, like the paroxysms of many diseases, at certain periods, and after longer or shorter intervals. They often begin with annual, and gradually increase in their frequency, until they appear in quarterly, monthly, weekly, and quotidian or daily periods."[2] In his plea for support, Rush called on "ministers of the gospel, of every denomination," to "save our fellow men from being destroyed by the great destroyer of their lives and souls."[3] His thinking eventually influenced other physicians and formed some of the major tenets of the temperance movement in the United States.

From the time of Rush on, medical opinions regarding alcoholism, like those of the larger society, often oscillated between competing views of it as a moral deficiency and an illness. Throughout the 1800s, many within the medical community were comfortable with the idea that alcoholism constituted a disease and were prepared to treat it as such, but this atmosphere did not last indefinitely. According to William White, with the onset of the twentieth century the stance of the medical community swung to the other end of the pendulum:

> While individual physicians continued to advocate various disease concepts of addiction, the overall definition of alcohol and drug

problems shifted away from a focus on a vulnerable minority of users to [a] focus on the inherent "badness" of the drugs and the persons and institutions profiting from their sale....Alcoholics and addicts, once "patients" worthy of sympathy, became "common drunkards" and "dope fiends" portrayed, at worst, as moral weaklings and criminals, and, at best, as dangerously insane.[4]

Public and professional indifference to the plight of the alcoholic and lack of tolerance for alcohol-related problems became the norm for much of the early 1900s, reaching their fullest realization with prohibition in the United States (1919–33). It was not until official repeal of prohibition that the disease concept of alcoholism began to make a resurgence, albeit in a slightly altered way. Many historians have come to attribute the return of the concept to the educational and research activities of groups such as the Yale Center of Alcohol Studies and the National Committee for Study on Alcoholism. The group that has received the most recognition for the remedicalization and modernization of the disease concept, however, is the famous self-help fellowship organization known as Alcoholics Anonymous (AA).

When its two founders, proctologist Dr. Robert Smith and stockbroker Bill Wilson, established AA in the late 1930s, alcohol addiction remained a target for scorn within mainstream America. Remnants of the disease concept were still around, but alcoholism was still poorly understood professionally and publicly. Alcoholics were regularly refused admission to hospitals and largely ignored by society and the medical community. Habitual drinking was deemed problematic, but it was typically framed in terms of the damage that it could cause to society rather than the individual. The response to the problem was often not a medical one but a correctional one, in which incarceration in either jails or psychiatric institutions was the norm. AA played a significant role in helping to turn the emphasis back in the other direction. As Levine has pointed out,

the heart of AA's reformulation of compulsive drinking was to shift the locus of alcohol addiction from the substance itself to the body of the individual addict....Wilson and Smith maintained that people who became alcoholics had a disease—they had something wrong with their bodies which eventually made them unable to control their drinking. For a number of years this was likened to an allergy, and often described as one....Other elements of this classic temperance position were brought back: alcoholism was a progressive disease; the chief symptom of it was compulsive drinking, usually defined as loss of control over drinking or the inability to abstain; and the only remedy for this was life-long total abstinence from alcohol.[5]

Questions of whether, and to what extent, AA was responsible for rediscovery of the disease concept or whether it incorporated disease concept language to further its cause have sparked furious debates over the years. AA historian Ernest Kurtz has been quick to note that "given the prejudices and issues involved, it is unlikely that the historical relationship between [AA] and the disease concept of alcoholism will ever be definitively resolved."[6] Yet he has asserted that "contrary to common opinion, [AA] neither originated nor promulgated [the disease concept]....Yet its members did have a large role in spreading and popularizing that understanding."[7] Undoubtedly, AA was partly responsible for changing public attitudes about alcoholism; it contributed to a broader, and more sympathetic, understanding of the phenomenon through its literature and its portrayal of alcoholism as a complex, "threefold" malady (physical, mental/emotional, and spiritual), but its primary focus was on serving its members in their quest for sobriety. As Kurtz and others have contended, AA's concentration on the spiritual element was a major contribution to the disease concept, but the organization did not aggressively promote alcoholism as a disease per se, and it rarely, if ever, got involved in the politics and semantics of the matter, leaving the definition of alcoholism to scientific and medical professionals.

THE SCIENTIFIC APPROACH TO ALCOHOLISM

One leading figure who came to be equally synonymous with the disease concept of alcoholism was Elvin Morton (E. M.) Jellinek. About the same time that AA was starting up, he began working at Yale University as

the director of its Center of Alcohol Studies. From the outset of his tenure there, his primary goal was to establish a scientific basis for, and an
approach to, alcoholism. Although the disease concept had been around
since the late eighteenth century, it was clear that many of the scientific
"facts" about alcoholism as a disease, or disease-like, lacked solid empirical
evidence. Jellinek sought to remedy this gap in knowledge through a more
scientific, multidisciplinary approach.

Like AA, Jellinek was convinced that alcoholism was a serious medical illness, and he directed his energies to impressing this conviction, as
well as the need for further research and treatment, on his scientific peers,
government, and society. Through his scientific writings and analyses, he
breathed new life into the disease concept. Jellinek was among the first
researchers to make a statistical connection between excessive alcohol
consumption and liver cirrhosis mortality, offering the first indicator of
alcoholism; his "estimation formula" was subsequently picked up by the
World Health Organization in 1951 as a means of determining alcoholism
rates in the United States and other countries. From the 1940s on, Jellinek
also attempted to define alcoholism, laying out its different stages, distinguishing between its physiological and psychological manifestations, and
providing ideas on various types of alcohol addiction. According to biogra-
pher Penny Booth Page, "for proponents of a disease model of alcoholism,
Jellinek's work gave them the symptomatology they needed, coated once
more with a mantle of scientific authority."[8]

By the mid-1940s, it appeared that the tide was turning once again
in favour of a more humane, and scientific, approach to the thorny issue
of alcoholism. As Page reported, "the scientific community took a leadership role in this shift of public attitudes....Scientists began to come
together in more formal groups to exchange ideas and concerns about
alcohol problems."[9] For the longest time, the medical community resisted
having anything to do with alcoholism; alcoholics experiencing delirium
tremens continued to be barred from general hospital admission. By
default, then, alcoholism fell to the large psychiatric institutions and their
health providers to handle. As with schizophrenia, alcoholism became a
fruitful area of research for the mental health profession, and scores of
researchers began to devote their time to the area. And like schizophrenia,

alcoholism provided a source of intense debate, with a host of competing, and at times controversial, theories (was it a disease, a syndrome, or a learned behaviour?) and treatment methods. Given the predominance of the psychodynamic school of thought at the time, most mental health professionals were convinced that alcoholism was a response to an underlying problem, and the first recourse to treatment was often psychoanalysis or some other form of psychotherapy. When various psychotherapeutic methods alone failed, psychiatry turned to its ever-expanding psycho-pharmacological armamentarium, conducting clinical tests of drugs such as barbiturates, antidepressants, antipsychotics, and even hallucinogens, to see if they had any efficacy as adjunctive treatments for alcoholism. One of the newer drugs to come along for use in chronic alcoholism was disulfiram, or antabuse, which produced an aversive effect (e.g., vomiting) in relapsing patients whenever they consumed alcohol.

Yet it was not psychiatric or psychological methods but AA that had the greatest impact in helping alcoholics to remain abstinent. As some have contended, the AA method stood as the "most successful single therapeutic approach in the field."[10] Naturally, the unparalleled success of its method intrigued many mental health and other medical professionals, and, from the 1950s on, psychiatrists and psychologists came to recognize the value of incorporating AA into a comprehensive treatment program. On the surface, the partnership between the two sides seems to make perfect sense, but on other levels it is puzzling. Perhaps most perplexing is that AA placed a heavy emphasis on spirituality in understanding and treating alcoholism, whereas psychiatry was in the process of adopting a decidedly more scientific approach. Most in the medical-scientific community avoided any religious or spiritual experience in scientific affairs. As history has shown, the two simply have not mixed well.

Reflecting on this conundrum, Kurtz highlights an episode in which Wilson was asked to speak at an APA conference in Montreal in 1949. His detailed description of the AA program received a positive reception from the psychiatrists, but apparently "'they were applauding the results more than the message'": "'You see, they know that they have not had much luck with alcoholics, and they are grateful to AA for getting alcoholics out of their hair.'"[11] As Kurtz notes, here was a disability...that had proven singularly unamenable to medical treatments. If physicians or psychiatrists could not cure or even treat it, how could it be a disease? One attempted answer, the major psychiatric one, was to suggest that alcoholism was not disease but *symptom*. Yet this hardly solved the problem if even as symptom the phenomenon resisted medical/psychiatric intervention but became amenable to the approach of [AA].[12]

Publication of Jellinek's *The Disease Concept of Alcoholism* in 1960 represented a watershed in the scientific community's acceptance of the concept. Yet there were some who questioned the validity of viewing alcoholism through a strictly medical model lens. When researchers began to present findings showing that patients diagnosed with alcoholism were able to return to "normal" drinking without relapsing into a pattern of addiction, they called into question notions such as loss of control.[13]

Even Wilson and Jellinek were careful to dissociate themselves from too narrow a medical interpretation of alcoholism. Speaking at a conference on the subject in 1960, Wilson stressed that AA

never called alcoholism a disease because, technically speaking, it is not a disease entity. For example, there is no such thing as heart disease. Instead there are many separate heart ailments, or combinations of them. It is something like that with alcoholism. Therefore we did not wish to get it wrong with the medical profession by pronouncing alcoholism a disease entity. Therefore we always called it an illness, or a malady—a far safer term for us to use.[14]

Jellinek, too, was cautious about putting forward the idea of alcoholism as a unitary disease. Page reveals that, "from his earliest publications, Jellinek had insisted there was more than one type of alcoholic (his preferred term was 'inebriate'). Some were physically addicted, some had physical or mental complications (disease resulting from excessive alcohol use), and some were psychologically addicted ('habitual symptomatic drinkers')."[15] His views on alcoholism, and the disease concept in particular, had advanced since he began his studies in earnest in the late 1930s. To provide clarity,

Jellinek charted a series of stages, or "phases," of alcoholism and a number of typologies (better known by the Greek letters that he used to designate them: alpha, beta, gamma, delta, and epsilon). He maintained that only two of his five typologies (gamma and delta) could properly be labelled as true disease entities because of the primarily physiological processes involved (e.g., increased tissue tolerance, "loss of control," and "inability to abstain").[16]

Many professionals expressed fears that the disease concept would be exaggerated and misused. Back in 1955, Henry Tiebout, one of the first and staunchest allies of AA and the disease concept within the psychiatric community, relayed his concern about too much emphasis on the concept. For him, "the idea that alcoholism [w]as a disease was reached empirically by pure inference. It had never been really proved."[17] Jellinek predicted that his theories would be misinterpreted rather than seen as hypotheses to be scientifically tested. Indeed, those who took his theories to extreme lengths made them, and the whole concept, easy targets for skeptics.

Yet the disease concept remained the leading model for understanding and treating alcoholism throughout the 1960s, though it was not the disease concept of old. By the 1960s, it and the field of alcoholism research of which it was a part had significantly and irrevocably changed. Finding a definitive definition of alcoholism proved to be elusive for many professionals. According to White, the debate over the disease concept was "plagued by too many definitions of alcoholism and too few definitions of disease."[18]

Finding consensus within the scientific community on alcoholism and its etiology, psychological and sociocultural effects, not to mention its diagnosis, treatment, and prevention, seemed to be impossible. Interestingly, the research fields of schizophrenia and alcoholism shared many parallels during this time. Both schizophrenia and alcoholism came to be seen for their heterogeneous nature (schizophrenias and alcoholisms). No longer could alcoholism be framed as a single disease entity; instead, it was seen as a collection of related illnesses, some more disease-like than others. There was also an attempt among the alcoholism research community to be more scientifically grounded in its endeavours (as evidenced by the efforts of Jellinek and others). Furthermore, the psychopharmacological revolution in research on schizophrenia had the same influence on the study of alcoholism. Soon chemotherapies came to the fore in the field of alcoholism, along with reputable methodologies (e.g., double blinds), to scientifically validate or invalidate them in clinical studies. The most controversial pharmacological aid being tested in the treatment of alcoholism was, of course, LSD.

Like the United States, Canada was active on the alcoholism research front and a leading voice in the scientific debate about the disease concept. Perhaps one of the most recognized research organizations in Canada in the 1960s was Ontario's Alcoholism (later Addictions) Research Foundation (ARF).[19] Established by the Ontario government in 1949 as an arm's-length agency, the ARF quickly became one of the world's centres of excellence for alcoholism research, education, and treatment. Jellinek's thinking had a pivotal influence on design of the ARF and its activities. Together with Jellinek, ARF founder H. David Archibald, a social worker who interned with the elder researcher at Yale University in the 1940s, worked diligently to map out the future of the proposed agency. As a result of Jellinek's advice, Archibald recalled, the ARF, "from the outset, developed a broad perspective on the problem and concluded that studies had to be taken in a number of areas."[20] The ARF came to symbolize the field's move in a more scientific direction, and many looked to it as a source of scientific innovation and expertise on alcoholism and other addictions.

ALCOHOLISM RESEARCH IN SASKATCHEWAN IN THE 1960s

What took place in Saskatchewan regarding the issue of alcoholism was not radically different from what went on in other communities across the United States and Canada. Although Saskatchewan had made much progress in the field and was among the first jurisdictions in both countries to officially recognize alcoholics as sick people who should be treated in general hospitals, the province was not ahead of other North American centres in 1960 in its thinking on, or approach to, alcoholism. It cannot be said that the disease concept had advanced at a faster pace there than in other provinces and states; to a large extent, the old moral vestiges of the issue lingered in Saskatchewan as they did elsewhere.

In terms of the formal treatment of, and research on, alcoholism, many of the same developments unfolded in Saskatchewan that occurred outside its borders. Collaborative relationships sprang up among government agencies (e.g., the Bureau of Alcoholism), health professionals, the clergy, and AA. Wilson was drawn to the experiments in Saskatchewan with LSD in treating alcoholics. Although it is unclear to what extent Wilson himself experimented with LSD, it is clear that he thought the drug, when used correctly, could be helpful in achieving sobriety. For Wilson, LSD therapy could evoke the spiritual transformation necessary to reduce ego barriers, remove obstacles, and provide incentives for alcoholics to quit drinking.[21]

As in other jurisdictions, AA performed a valued role in the Saskatchewan treatment program, with local AA members volunteering as "sitters" for patients going through withdrawal and assisting in follow-up services in the community. Many in Saskatchewan were similarly influenced by the disease concept of alcoholism, largely because of a core group of enlightened professionals working in the field. Saskatchewan was a prime example of multidisciplinary research on alcoholism, on the same level as the ARF in Ontario. The province became recognized as a centre for research, education, prevention, and treatment. Blewett, who worked closely with the Bureau of Alcoholism in an advisory, planning, and research capacity, strove to maintain the close connection between the bureau and the research program. Recalling his days with the bureau, Angus Campbell noted how Hoffer used to tell him to send their most difficult cases, those that "had come to be regarded as hopeless. What was surprising to many was that Dr. Hoffer and his colleagues were getting results."[22] Many of the intriguing results, of course, had to do with the one thing that set the Saskatchewan treatment program apart from many of its contemporaries: its pioneering use of psychedelic drug therapy.

UNDER THE SCIENTIFIC MICROSCOPE

Hoffer, Osmond, and Blewett, as the principal investigators in Saskatchewan's psychedelic treatment program, were firmly convinced of the efficacy of LSD-assisted therapy in treating alcoholism, but they approached alcoholism in different ways. Although they were strong proponents of various aspects of the disease concept and the need for more scientific exploration in the area, they often expressed contrasting viewpoints on how alcoholism should be understood and handled. Hoffer and Osmond once asked "why do people become alcoholics?...Suppose we asked it another way 'round—why aren't we all alcoholics? When you come to think of it, that may be the real mystery, and answering it might solve the problem of alcoholism."[23] In many respects, their interpretation of alcoholism was closely tied to their ongoing research on schizophrenia. Because they characteristically viewed most mental illnesses from within a medical model, they wondered what made alcoholics different chemically from non-alcoholics and thus more biologically susceptible to alcohol addiction.

Blewett took a radically different path from that of Osmond and Hoffer in alcoholism research and scientific research in general. By the end of the 1950s, Blewett operated from a more spiritual than scientific base, which became more pronounced from the 1960s on and helped to distance him even further from some of his professional colleagues. He did not distinguish between the two bases. Settling into his new position as the chair of psychology at the University of Saskatchewan—Regina Campus, Blewett soon began to direct his gaze to more parapsychological matters and psychic energy in particular, an area that he subsequently referred to as "psychotronics."[24] As he later described the turn, his psychological training had "led him to believe, until 1956, that we psychologists were justified in maintaining a taboo on the study of the spirit....I then changed my mind. I now believe psychologists should no longer hide from the responsibility of studying the spirit as the source point of all science."[25] LSD, of course, was instrumental in this personal and professional transformation.

This newfound perspective inserted itself into all that Blewett did, and he tackled the treatment of alcoholism from a much different angle than Hoffer and Osmond. Speaking at a North American Association of Alcoholism Programs conference in North Dakota in 1962, he told attendees that

> self unacceptability is the root of all functional psychopathology.
> Since self hatred and self alienation underlie all mental disturbance,
> if self acceptance can be achieved through the self understanding

provided by the psychedelics, the greatest of strides toward recovery will have been achieved.... Our self concept conditions our thinking. It makes us respond with trust, understanding, love, and affection or conversely with suspicion, prejudice, and hostility, toward ourselves and toward others. Because these feelings are the wellsprings of behavior, they color—indeed determine—all of our relationships.[26]

The technique that Blewett employed in psychedelic treatment, wherein he took the drug with his patients to increase the empathic bond, was an even greater source of disagreement between him and many of his associates. As Osmond wrote to Hoffer,

Dunc's arrangement is not only difficult to do but misses the essential nature of the relationship, which is not to get on well with the sick people but to concentrate their attention on their problems. They don't want to share the treating person's problems and why should they? Would one have special confidence in a surgeon because he decided to undergo the same operation?[27]

On further reflection, Osmond was more blunt: "Dunc's method of taking LSD with alcoholics is almost sure to fail....It simply makes the therapist another and not very competent alcoholic! These most skillful social manipulators will not be defeated by such simple tricks—serious and sustained attention must be given to them and their drinking."[28] Although both Hoffer and Osmond were generally open to scientific freedom, they did have their limits, and for them some of Blewett's activities threatened the scientific credibility of the entire Saskatchewan program and psychedelic research in general.

Given that Hoffer, Osmond, and Blewett had such divergent opinions on science, it is surprising that they were able to work as well together as long as they did. If one were to plot a continuum of those within the scientific community who advocated for psychedelic-assisted therapy, Hoffer would fall at the more conservative end, whereas Blewett would fall at the more liberal end. Osmond would occupy the middle ground. Because he was more spiritually and philosophically oriented than Hoffer, he could serve as the binding agent that held the three researchers and their mutual interests together. One result of this combination was a unique alcoholism treatment program that bridged the scientific/spiritual divide.

THE SCIENTIFIC CASE AGAINST LSD THERAPY

While UBC psychiatrist J. S. Tyhurst was rallying the psychiatric and medical communities against the use of LSD in therapy in Canada (see Chapter 4), a similar movement was building in the United States. In one editorial, R. R. Grinker, president of the American Medical Association and editor of the prestigious journal *Archives of General Psychiatry*, stated that those who had the most experience with LSD "became disqualified as competent investigators."[29] He went on to say that

> now the deleterious effects are becoming more obvious, [and] latent psychotics are disintegrating under the influence of single doses. Long-continued LSD experiences are subtly creating a psychopathology. Psychic addiction is being developed and the lay public is looking for psychiatrists who specialize in its administration. Here again is the story of evil results from ill-advised use of a potentially valuable drug due to unjustified claim, indiscrimina[te] and premature publicity and lack of professional controls.[30]

In a subsequent editorial, Grinker issued an even sterner warning: "There are increasing numbers of reports indicating that temporary or even permanent harm may be induced despite apparently careful pre-therapeutic screening of latent psychoses and careful precautions during the artificial psychoses."[31] This led him to surmise that drugs such as LSD "are indeed dangerous even when used under the best of precautions and conditions."[32]

That Grinker and Tyhurst lacked experience in treating patients with the drug did not matter; it was enough for many in the scientific community, government, and public that reputable psychiatrists and doctors were calling into question the safety of LSD in official treatment or any other setting. Concerns within the scientific community led to a more aggressive push by governments and their agencies to place stricter controls on the use of LSD. Making the situation more volatile was the increasing illegitimate use of LSD and other psychedelics in the wider society, a nascent black market trade in the drug (or versions of it), and the appearance of a counterculture that adopted psychedelic use as one of its platforms. By 1962, internal and external pressures forced the federal governments in both the United States and Canada to pass legislation that effectively terminated the use of LSD in therapy.[33] Most Canadian researchers received a letter from C. A. Morrell, director of the Food and Drug Directorate, outlining the new regulations.[34]

Blewett was dismayed at regulations on a drug that had shown such promise. He conceded, along with Osmond and Hoffer, that some controls might be required to ensure the continuation of psychedelic studies and therapy, but in his mind the regulations were too heavy-handed, largely because the government would now determine who was experienced enough with the drug to use it in clinical settings. No doubt it bothered Blewett that he did not meet the criteria and was effectively barred from an area of research in which he had made valuable contributions and devoted so much time and thought.

Al Hubbard was another individual negatively impacted by the regulations. Although reputedly he knew a lot about the appropriate application of LSD in treatment, his unorthodox ways and lack of formal medical training made him suspect among many professionals engaged in the debate over LSD therapy. By the time that the new regulations were in the works, he had cut his losses in British Columbia after the scandal instigated by Tyhurst and moved to Menlo Park, California, to assist in the work of another group of LSD researchers (including Charles Savage, James Fadiman, Myron Stolaroff, and others), known better as the International Federation of Advanced Study (IFAS). Here, too, it was not long before local medical experts, such as Kenneth Ditman, began to apply legal pressure on Hubbard. They were outraged by his bogus medical degree and wanted him withdrawn from the psychiatric research and treatment scene entirely.

Hoffer and Osmond tended to view the regulations through a more positive lens than others. Writing to Osmond on January 7, 1963, Hoffer noted that as far as we are concerned [they] will in no way interfere with our use. It [LSD] is available to institutions approved by the Minister for use by qualified investigators. There is no definition of institution and no definition of qualified investigator....The only beneficial result of all this ruckus may be that the use of LSD is regularized and critics like Tyhurst can no longer, in their usual underhanded way, continue to attack its use.[35]

Although there might have been some benefits to the regulations, placing LSD under the same rubric as thalidomide, for many closest to the psychedelic research, sent the wrong message. That more credence was given to people such as Tyhurst rather than those most knowledgeable about the drug also proved to be a lingering issue.

None of the Saskatchewan psychedelic drug researchers ever denied that LSD was dangerous or should be administered in anything other than a controlled environment. In answering queries about whether the drug was dangerous or not, Hoffer confirmed "of course it is! So are salt, sugar, water and even air. There is no chemical which is wholly safe nor any human activity which is completely free of risk. The degree of toxicity or danger associated with any activity depends on its use. Just as a scalpel may be used to cure, it may also kill."[36] Colin Smith responded in the same way to reports warning of the drug's dangers: "That the drug can be abused is clear, but this is true of a great many standard medications. Properly used it would appear to be reasonably safe. It is instructive to compare published reports of complications resulting from lysergide with those accruing from other psychiatric treatments such as insulin coma therapy, electroconvulsive therapy, continuous sleep therapy and tranquilizers."[37]

After doing interviews with forty-four therapists, American psychiatrist Sidney Cohen amassed information on nearly 5,000 individuals who had been administered either LSD or mescaline on more than 25,000 occasions.[38] From these data, he calculated that "no instance of serious, prolonged physical side effects was found either in the literature or in the answers to the questionnaire. When major untoward reactions occurred they were almost always due to psychological factors."[39] In his study, many of the myths about LSD were laid to rest (e.g., addiction), and Cohen merely reiterated what many of the Saskatchewan researchers already knew. Critics' reports of suicide being directly attributed to LSD therapy were grossly exaggerated. Hoffer noted that only 5 of the 5,000 had committed suicide and that even in those cases it would have been hard to pinpoint LSD as the culprit.[40] Smith weighed in on the subject of suicide as well:

> Where such individuals take these drugs and later break down the drug may be incorrectly blamed. In the final analysis it is the degree of risk which must be ascertained. No drug is completely without danger. Penicillin has on occasions killed, sodium amytal may lead to severe addiction problems, phenacetin may cause severe renal damage, and so on ad infinitum. In the field of alcoholism disulfiram and apomorphine are drugs obviously associated with definite dangers.[41]

Smith, Hoffer, Cohen, and others argued that in many instances suicide was probably averted given the right set and setting. Harsh complications arose almost always in cases of total negligence by the therapist and improper use of LSD.

Many initial claims against the therapeutic value of LSD came from individuals who had little if any knowledge of the drug and its effects or experience in its use in treatment. This resulted in a stalemate in the scientific community, with claims and counterclaims about LSD flowing back and forth in scientific journals and the lay press. No sustained effort—in a methodological fashion acceptable to most in the scientific community—had been made to test the scientific veracity of the psychedelic treatment claims coming out of Saskatchewan. The first peer reviews by scientific and psychiatric experts on LSD therapy assessed the overall history of psychedelics, their effects and use in treatment, and the situation in the early 1960s. Overall, the approach adopted within these reviews depended on which side of the debate the author fell. Some reviews were blatant attempts to destroy any therapeutic credibility of the drugs; others were more moderate in their analyses; all called for more intensive and objective investigation and appraisal of the treatment claims.

Lending scientific credence to Grinker's medical alerts over LSD and other psychedelic drugs, Jonathan Cole, the chief medical doctor with the psychopharmacology service centre of the NIMH, and Martin Katz, a research psychologist, wrote that "rather than being the subject of careful scientific inquiry, these agents have become invested with an aura of magic, offering creativity to the uninspired, 'kicks' to the jaded, emotional warmth to the cold and inhibited, and total personality reconstruction to the alcoholic or the psychotherapy-resistant chronic neurotic."[42] Like Grinker, Cole and Katz also suggested that the early investigators were to blame for the increasing illegitimate use of the drugs and the black market trade.[43] More alarming was the possibility that investigators who had "embarked on serious scientific work in the area may have been subject to the deleterious and seductive effects of these agents."[44] Cole and Katz suggested that investigators who obtained positive results were overzealous in evaluating their findings: "None of these claims [is] based on detailed, carefully controlled studies designed to be free from possible distortions due to bias or enthusiasm."[45] The use of spiritual terms (e.g., "transcendence") to explain the effects of the drugs and the supposed transformations in patients made the reviewers uncomfortable: "Such explanations may have a mystical or philosophical sound which appeals to the enthusiast, but they are likely to produce doubt or even violent disbelief and concern in physicians used to a more communicative language."[46]

Another early review came from research psychologist Sanford Unger of the NIMH.[47] He cited the successes of, but had reservations about, the alcoholism treatments used by Hoffer and Osmond and the more controlled experiments directed by Sven Jensen in the ten-bed alcoholism unit of the Saskatchewan Hospital–Weyburn between 1959 and 1961. Despite his concerns about the latter's methodology, Unger noted that "the difference in percentages of patients 'abstinent or improved' between the 'full program-LSD' group (41 out of 58, or 71 percent) and the 'individual psychotherapy' group (10 out of 45, or 22 percent) was highly statistically significant."[48] As Unger concluded, the results had to be "regarded with healthy skepticism. On the other hand, they are more than merely trifling."[49]

In Canada, the names of Reginald Smart and Thomas Storm became synonymous with scientific refutation of the Saskatchewan alcoholism treatment claims. When Smart, a clinical and social psychologist, began his tenure with the ARF in the late 1950s, the organization was quickly evolving into the country's premier institute for alcoholism and addiction research, education, and treatment. As he recalled, his arrival came at a time when the ARF was questioning its role in the incipient field of alcoholism and addictions research: "Most treatment facilities in Canada and the U.S. were dominated by AA philosophy.... The belief in [the] ARF was that alcoholism studies could be separated from AA concepts and be part of general scientific thinking and research on treatment effectiveness."[50] Under the mentorship of John Seely, a renowned sociologist and the director of research at the ARF, Smart steered his attention early on toward alcohol consumption (and the effect of alcohol pricing on consumption) as well as the problems associated with heavy drinking over sustained periods. In a few years, he came to be recognized as one of the country's leading experts in the field, particularly in understudied areas such as teenage use and abuse of alcohol and drugs. In the 1960s, when the ARF was expanding its mandate to address growing concerns over psychedelics and other drugs, Smart played a central role in the debate over the efficacy of LSD treatment. The claims of Hoffer and his colleagues caused immense controversy in the field at the time, and the scientific verification or refutation of these claims was an ideal opportunity for the up-and-coming ARF researcher to earn his stripes. Smart was joined by Storm and a couple of University of Toronto clinicians, and they became the first research team to publish scientific data debunking the Saskatchewan LSD treatment claims.

Smart and Storm offered their own commentary on the therapeutic efficacy of the drug. Like many critics of LSD therapy, they castigated the Saskatchewan therapists for failing to conduct "proper" (i.e., double-blind) experiments, and they emphasized that the weaknesses far outstripped the merits. In the opening paragraph of their review, they noted that a "general lack of sophistication is especially characteristic of recent efforts to examine the effectiveness of LSD-25 as an adjunct to the treatment of alcoholism."[51] In their assessment, reports like those coming out of Saskatchewan were "little more than the chronicling of clinical procedure."[52] According to Smart and Storm, there were mandatory requirements that any clinical research should meet to be acceptable, and the experiments in Saskatchewan were no exception.[53] If one were to examine the Saskatchewan research and findings with their requirements in mind, then it would "fare badly indeed."[54] They stressed that it "opens the door to all sorts of interpretations of the positive findings" and "makes it impossible to state whether the changes in drinking behavior were due to LSD or to a myriad of other variables, such as the greater staff interest taken in the patients during the study."[55] Other possibilities existed for the positive findings, such as spontaneous recovery. Overall, for Smart and Storm, the clinical use of LSD in Saskatchewan was premature because "its real effectiveness [had yet to be] properly assessed."[56] Although the trials there might have met some of the criteria (e.g., pre- and post-treatment procedures) outlined in their commentary, they were subject to criticism because of insufficient detail and "lack of objective or uncontaminated subjective information."[57] Smart and Storm insisted that the only hope of determining the therapeutic usefulness of LSD remained the double-blind, controlled trial.

Such reviews, though illuminating in a number of respects, had brought the psychiatric community no closer to a consensus on the scientific value of LSD-assisted treatment of alcoholism. What was needed to move beyond the polarization was clinical research specifically testing the psychedelic treatment methods. In June 1966, the APA issued its "Position Statement on LSD." Although it too was concerned about illicit use of the drug, it did not want it to be unavailable for legitimate research purposes. The APA emphasized that "neither laboratory nor clinical findings [had] yet adequately documented" the usefulness of the drug in treatment, but "they [had] elicited sufficient information to justify continuing research on its possible values."[58] Consequently, from the mid-1960s on, there was greater willingness among mental health agencies to inject funding and resources into controlled clinical trials to settle the debate once and for all.

Into this research void stepped the ARF. Following up on the suggestions made in the earlier review, Smart, Storm, Earle Baker, and Lionel Solursh undertook what they considered to be the first controlled study of the massive, single-dose LSD treatment for alcoholics. The researchers selected thirty volunteer ARF patients similarly matched (e.g., age, drinking history, previous failed attempts at therapy) and then randomly separated them into three groups of ten each: one that received an 800-microgram dose of LSD, one that got a sixty-milligram dose of ephedrine, and one that was not hospitalized but received the same treatments as the other groups minus the drugs. The researchers maintained that no one—therapists, patients, or post-treatment team—knew ahead of time which drug was being given to which patients (though many experimenters became acutely aware during the course of the trial of who had received the LSD). Following the one-day treatment, patients usually remained as in-patients for a week before being released into the community, where they were followed up six months later. Although results showed some improvements (e.g., drinking behaviour) across the board, there was no statistically significant difference between the LSD, ephedrine, and control groups "in percentage gain in total abstinence or in their longest period of abstinence."[59] The ARF group took its study to be a contradiction of the glowing reports coming from Saskatchewan and proof that "LSD failed as an effective adjunct to psychotherapy."[60]

In their article, Smart and his associates attempted to address criticisms that could be made of their study. For example, they conceded that it used an extremely small sample size (i.e., only ten people received LSD) but defended their experiment in saying that follow-up studies on larger groups of patients would have been too costly and time-consuming and that previous studies using larger groups (e.g., Jensen's) had failed to include non-LSD controls for purposes of comparison and therefore produced "faulty data."[61] As they phrased it, "a controlled study with a small number of carefully studied patients is preferable to one loosely controlled and less intensive."[62] The article emphasized that all treatment personnel had undergone personal LSD experiences (and were thus familiar with the effects of the drug); however, it also admitted that these personnel were "skeptical about its value" and that "no one was committed to a belief in its efficacy," which raised the possibility of experimenter bias.[63] The double blindness that Smart and his colleagues thought so critical in clinical drug studies was also blown, given that "in 19 out of 20 cases the therapist administering the drugs guessed correctly which drug the patient received."[64] And some patients might have known that they had received LSD (given descriptions about the effects of the drug in press reports). All of these points lent support to comments by Hoffer and others about the immense difficulties inherent in conducting double blinds with LSD.

Closer inspection of the study by Smart and his associates reveals more significant weaknesses. It had neglected the importance of set and setting in the overall therapeutic procedure, a key component of the Saskatchewan studies. Commenting on this point, Albert Kurland and his associates insisted that Smart and his colleagues had not properly utilized the technique employed in Saskatchewan; instead, they had used what could best be described as "psychedelic chemotherapy," which emphasized administration of the drug itself. The amount of psychotherapy used in pre- and post-treatment was minimal.[65] Other aspects of the study stand out and might account for why its use of LSD in treatment did not fare as well as it might have: patients were strapped to beds during the procedure; pretreated with a sedative known as dilantin (which would have interfered greatly with the natural effect of LSD and hindered the achievement of a psychedelic/peak experience); given a fixed dose of LSD (well beyond the dose customarily used); and subjected to a three-hour "interview" about their child-parent relationships, sexual habits, reasons for drinking, suicidal ideas, and other topics.[66]

In an effort to compensate for some of the perceived methodological weaknesses of the study by Smart and his colleagues, psychiatrist F. Gordon Johnson carried out a slightly modified study based out of the London, Ontario, ARF chapter using four treatment groups (ninety-five patients in total): (1) LSD with a therapist present; (2) LSD with no therapist present; (3) sodium amytal/methedrine with a therapist present; and (4) routine clinical care. Taking his cue from the trial by Smart and his colleagues, Johnson opted for a single-blind methodology "since many previous studies and our experience have emphasized that the LSD effect was so distinctive that the therapist could not fail to be aware of its use."[67] In contrast to the earlier ARF trial, the Johnson study also gave a nod to the importance of set in the LSD experience; for example, in the patient group receiving LSD and being supervised by a therapist, a patient's regular therapist was involved before, during, and after the treatment, with the thinking that this involvement might result in better rapport and transference between patient and therapist. However, as Johnson's results showed, the presence or absence of a therapist in the LSD experience made little difference in the long-term outcome.

To keep patient suggestibility to a minimum and not prejudice the response to LSD, patients were not told if they were receiving it.[68] The Johnson study also allowed for a flexible LSD dose "regulated to the patient's needs" rather than the high LSD dose employed by Smart and his researchers.[69]

Although the Johnson study represented some improvements over the Toronto ARF study, it still used similar restrictive measures, such as restraining patients to their hospital beds with a Posey belt, which no doubt would have negatively impacted the set and setting. Moreover, a sedative cocktail including chlorpromazine and other drugs was administered to patients six hours into the experience (cutting short an LSD experience, which typically lasts between ten and twelve hours). And the results of the Johnson study were comparable to those of the former study in that they indicated statistically significant improvements across all groups (e.g., drinking and employment indices) in a twelve-month follow-up by independent investigators, but "no significant difference between groups; i.e., overall improvement occurred, but use of drug therapy in hospital did not confer special advantages over routine clinical therapy."[70] For Johnson, the results lent "further weight to the rejection of LSD as a significant drug in the treatment of alcoholism."[71]

From the perspectives of other experienced LSD therapists, especially those partial to the psychedelic mode of treatment, the experiment of Smart and his colleagues (and in certain respects that of Johnson) were guaranteed not to show usefulness of the drug in therapeutic procedures. In agreement with Hoffer, Savage and his colleagues later concluded that the "'transcendental' or peak experience is an essential ingredient but...it is insufficient merely to give LSD. The experience must be embedded in a therapeutic matrix which includes careful preparation of the patient, a day-long LSD session conducted by nurse and therapist and follow-up treatment to work through and translate the new insights into practical everyday living."[72]

For Hoffer and Osmond, the work of Smart and Storm actually confirmed the earlier finding in Saskatchewan that the psychotomimetic experience was not valuable in treating alcoholics. In later correspondence with Osmond, Hoffer commented that "the Toronto study was sophomoric and [displayed] a profound ignorance of the psychedelic concept," and he expressed amusement at an earlier slip by Osmond in a letter in which he referred to the ARF researchers as Smart and "Smog": "Certainly they have thrown a lot of smog over the situation but their effort will be evanescent."[73] Referring to the procedure that Smart and his associates used, Hoffer later wrote that "the experience was not psychedelic, but was more in the nature of an inquisition."[74] Blewett had always expressed reservations about overly strict controls in psychedelic treatment: "A too hasty presentation of tests, or indeed, a cold, unsupportive, objective research set-up, in which the subject is cast in the role of guinea pig, is almost certain to alter the entire nature of his perceptions. Almost universally, results obtained from trials or tests under such circumstances will show decreased efficiency of one kind or another."[75] To Blewett, studies like those carried out by the ARF were characteristic of early trials of LSD treatment: they showed a gross lack of understanding of the importance of set and setting. Furthermore, the "need for restraint was often taken too far.... For example, in some settings individuals were heavily sedated before the session was over which thrust any unfinished business under the psychic rug and insured residual post-session anxiety."[76]

Yet not everyone saw the ARF studies in the same negative light. Their results made some researchers, who had earlier believed LSD to be a valuable adjunct to the treatment of alcoholism, question some of their initial assumptions. For others within the psychiatric and medical communities, the results achieved by the ARF trials provided ammunition to those who had doubted the efficacy of LSD treatment all along. Smart and his colleagues, who later wrote a book on LSD therapy in treating alcoholism, came to be regularly cited in some quarters. One reviewer claimed that the "authors *rightly* conclude that there is no adequate statistical evidence showing LSD to be of value in treatment of psychiatric conditions including alcoholism."[77]

Another major, and oft-cited, study that failed to corroborate the Saskatchewan LSD treatment claims for alcoholism was that of American psychologists Arnold Ludwig and Jerome Levine. Like many within the

psychiatric community in the late 1950s and early 1960s, Ludwig, director of education and research at Mendota State Hospital in Madison,
Wisconsin, and Levine,[78] chief of psychopharmacology research with the
NIMH, were captivated by LSD and began exploring the drug for its mysterious effects and potential uses in the mental health field. One aspect of
the drug that caught their attention was the immense variability in people's responses when it was administered in therapeutic settings. Ludwig
and Levine hoped to improve therapeutic results by better controlling how
patients reacted to the drug, so they developed a treatment technique
(referred to as "hypnodelic" therapy) that combined psychotherapy and
hypnosis with LSD. Working with a group of narcotic drug addicts, the
researchers first put them into a deep hypnotic state, gave them LSD, and
then actively encouraged them to explore and work through their issues.
After a few hours, the researchers brought the patients out of the hypnotic
state and planted post-hypnotic suggestions so that they could recall their
experiences and continue to work to resolve their problems.

Ludwig and Levine discovered that the therapist could "mold, structure, and direct the patient's experience into his emotional conflict areas,
rather than allowing patients to drift off into 'panoramic scenes' or 'the
beautiful world of colors'—experiences thought to be of lesser therapeutic
value."[79] Although cautious not to make any claim about treatment efficacy following their experiment, the research team nevertheless found the
results "highly encouraging" and called for additional controlled studies
and follow-up.[80] Ludwig and Levine soon came to be seen as specialists in
LSD treatment, conducting further testing of their hypnodelic versus other
(e.g., psychedelic) LSD treatment models in which they achieved some positive short-term results (e.g., attitude change). Their studies incorporating
LSD also led them down some other intriguing research avenues in which
they examined the role of therapists' attitudes in the outcomes of patients
who received various combinations of treatment.[81]

As the controversy over the use of LSD in treating alcoholism began
to heat up and criticism began to mount over the enthusiastic claims
coming from its proponents, Ludwig and Levine were drawn into testing
claims like those originating from Saskatchewan. In 1965, the researchers
received approval and funding from the NIMH to begin a controlled clinical

study of LSD treatment specific to alcoholism. Assembling a research team that included Mendota State Hospital researcher Louis Stark and George Washington University research scientist Robert Lazar, Ludwig and Levine began a three-year investigation of the "differential efficacy of three LSD treatment procedures and a 'no therapy,' or milieu treatment condition," with 176 male alcoholic patients.[82]

The researchers randomly assigned patients (all of whom would have received the routine treatment provided through the centre's thirty-day program prior to the experimental treatment) into one of four groups: (1) hypnodelic: hypnosis, psychotherapy, and LSD; (2) psychedelic: LSD and psychotherapy; (3) LSD alone; and (4) no therapy.[83] Patients receiving LSD were given doses of three micrograms per kilogram of body weight; to reduce bias, they were not told what type of treatment they were receiving, only that all treatments were equally effective and that the one provided to them was specifically suited to them based upon pre-treatment tests and interviews.[84] A group of thirteen psychiatrists served as therapists, each of whom was randomly assigned a treatment type and an even number of patients; the therapists did not know which treatment type they were providing until just prior to giving it. As noted in the report, none of these therapists had any previous experience with LSD or its use in treatment settings aside from a crash course provided on each treatment procedure prior to commencement of the study. To measure treatment efficacy, patients were evaluated (by a team independent of the therapists) using a battery of tests (e.g., Behavior Rating Scale, California Psychiatric Inventory, and Psychiatric Evaluation Profile) pre- and post-treatment, and they were followed up at three-, six-, nine-, and twelve-month intervals.

The results of the three-year study revealed that, though some improvements had occurred in all groups, no one treatment had shown itself to be superior. In the words of Ludwig and his colleagues, "it would be professionally gratifying to report significant results for the hypnodelic treatment technique as well as for the other LSD treatment conditions.... Unfortunately,...none of the LSD treatment procedures produces any greater therapeutic benefit than can be realized by the 'no therapy' condition in the context of the general ward milieu program."[85] As with the previous controlled studies conducted by the ARF, Ludwig and his co-workers were "forced to conclude that the dramatic claims made for the efficacy of LSD treatment in alcoholism were scientifically unjustified."[86]

In examining their 1969 report, Canadian medical doctor Seymour Kantor noted some inconsistencies between the conclusion and the data. Referring to the figures in the report, Kantor pointed to the one showing that "after one month less than 15 percent of the psychedelic condition group...had returned to drinking, compared to over 40 percent of the control group; although the difference diminished sharply over time, the psychedelic condition group yielded a lower percentage after three months."[87] In other words, the data did show that the psychedelic group had an initial therapeutic advantage over the control group. Ludwig disagreed: "Since presenting our original findings, we have used every conceivable statistical approach to tease significant findings from our data—all to no avail."[88]

Other critics thought that such controlled studies contained lessons on what *not* to do. As psychedelic researcher Walter Pahnke commented, "I think that they are very valuable because they demonstrate that if you give LSD with the method they used you will not get results any better than giving no treatment at all.... That's why I am even more convinced that it has to be done with another type [of] procedure."[89]

So were Ludwig, Levine, and company correct in their appraisal of psychedelic therapy in treating alcoholism? Were the positive findings obtained with LSD therapy merely a "mirage," as they later claimed, when subjected to the scientific method? Less scientific factors, however, might account for their about-face regarding the therapeutic usefulness of LSD, in particular the politics within the psychiatric community at the time. Charles Grob observed that

> from the mid 1960s onward, a split began to appear in the ranks of psychiatric hallucinogen researchers. For those who would maintain enthusiasm for the potentials of these singular substances [read Hoffer, Osmond, and Blewett], a path of professional marginalization would follow. For those who would take a stand against their perfidious threat, accolades and professional advancement would be forthcoming. For most, however, it was to be a process of quietly

disengaging, often from what had been a passionate interest, and re-directing their careers towards tamer and less disputable areas.[90]

Quoting Stanislav Grof's appraisal of Levine and Ludwig, Grob wrote that "at a time when LSD was popular, [they] had reported positive results. When LSD fell out of favor and the positive results became politically unwise, they obtained negative results. Unconsciously or consciously they built into their study a number of antitherapeutic elements that guaranteed a therapeutic failure."[91] Savage had this to say about Ludwig: "He initially published positive results but found them somewhat embarrassing and I think this shaped the direction of his thinking afterwards....Somewhere along the line he changed in his thinking and this seems to have been reflected in a switch from positive to negative findings."[92]

THE SPRING GROVE EXPERIMENTS

Although a number of controlled studies challenged the LSD treatment claims coming out of Saskatchewan,[93] some investigators remained convinced that psychedelic therapy for alcoholism, when carried out properly (i.e., with the appropriate set and setting), held scientific merit and compared well to standard therapeutic approaches. Perhaps the closest to replicating the psychedelic treatment approach in a more scientifically respectable way were the clinical trials in Spring Grove State Hospital in Baltimore.

Beginning in the mid-1950s under the leadership of Research Director Albert Kurland and Associate Director Charles Savage,[94] and later joined by other pioneers in the psychedelic field (including Sanford Unger, Walter Pahnke, Robert Soskin, Stanislav Grof, and others), the Spring Grove experiments became one of the longest and most systematic scientific explorations of the psychotherapeutic use of LSD.[95] Modelling their experiments on the earlier work of the Saskatchewan researchers, Kurland and his team tried psychedelic treatment in 1963 with a group of patients admitted to Spring Grove's Alcoholic Rehabilitation Unit. As Kurland and his associates described it, "from the very beginning, our approach to the use of this potent compound was marked by extreme respect. We started by implementing a treatment effort...which consisted of approximately three weeks of intensive psychotherapy incorporating one high-dose [450 micrograms], highly structured LSD session."[96]

Conducting a preliminary treatment with sixty-nine chronic hospitalized patients over a two-year period, their original objective was merely to increase their understanding of the LSD experience and assess its safety and therapeutic potential. They discovered that "in the 69 cases under discussion...both clinical and psychological data agree that no patient has been harmed..., [and] the extent of benefit in some cases has seemed considerable."[97] Encouraged by these initial findings, the researchers proceeded to conduct more rigorous controlled studies of the treatment. Securing an NIMH grant, Kurland and his team set out in 1965 to conduct a double-blind trial of psychedelic therapy with 135 male patients. Realizing the inherent difficulty in performing a double blind that compared an LSD group with a non-LSD group,[98] they opted to conduct a double blind that randomly assigned patients to either a high-dose (450 micrograms) experimental group or a low-dose (fifty micrograms) placebo control group, surmising that the higher dose of LSD would result in a greater number of psychedelic peak experiences and hence more lasting sobriety.

In contrast to previous controlled studies, the Spring Grove team paid special attention to preparing patients before undergoing the treatment:

> The preparation of the drug session involves an average of about twenty hours of intensive psychotherapy. During this period, the therapist aims at establishing close rapport with the patient and gaining intimate knowledge of the patient's developmental history, dynamics, defenses, and difficulties. In specific preparation for the session itself, the patient is acquainted with the basic effects of the drug and encouraged to trust the therapist, himself and the situation. This is a very important part of the preparation that enables the patient to utilize the session in the optimal way—to let go voluntarily of his usual ego controls and so be completely open to whatever experiences he encounters.[99]

Following the lead of the Saskatchewan studies, the Spring Grove treatment sessions were structured within a comfortable environment that incorporated all the familiar props (pictures, flowers, and music) and safety features (nurse and therapist support for the duration of the experience). Patients were independently evaluated by a social worker prior to treatment and at six-, twelve-, and eighteen-month periods following discharge from the hospital. Final data revealed that the high-dose group showed an advantage over the low-dose group in drinking behaviour and other psychometric and behavioural adjustment scales at six months; however, by the twelve- and eighteen-month follow-up, this advantage had lessened to the point where no statistically significant difference could be seen between the two groups. This finding aside, the Spring Grove researchers were impressed that the "overall level of improvement was considerably better for both groups than the usual improvement for alcoholics in the same setting without any form of LSD assisted psychotherapy."[100]

As the Spring Grove team was quick to emphasize on completion of the study, "psychedelic psychotherapy was successful in helping over half of the alcoholics in this program as opposed to a 12% improvement rate at 18 months follow-up for comparable alcoholics in this treatment facility at Spring Grove State Hospital."[101] For Kurland and his colleagues, the experiments had confirmed the value of psychedelic treatment and opened the way for further research.

Not surprisingly, their results drew attention in the United States and abroad. In a 1969 critique of controlled studies of LSD-assisted therapy, NIMH and Harvard University Medical School psychiatrist Carl Salzman remarked that the Spring Grove study stood out as the "most progressive study to date and was carefully designed to avoid many of the difficulties of previous research projects,"[102] though it, too, possessed some methodological weaknesses. As Salzman pointed out, the absence of a "concurrent, 'no treatment' or 'standard treatment' group, present in some of the previous studies, was not utilized in this program. Thus, it is impossible to ascertain whether the treatment and control groups improved significantly more than a similar group might if not given any LSD."[103] The Spring Grove team had acknowledged this all along, noting that it had been attempted at first but that the "tensions between the treatment and no-treatment groups proved so disruptive that it was necessary to abandon that phase of the study."[104] Another methodological problem associated with the Spring Grove experiment, and other LSD therapy research, was the ever-present issue of therapist bias. In the case of the Spring Grove research, Salzman called into question the role of therapist bias in rating levels of psychedelic reaction: "Although the therapist-rater may be blind to drug dosage, he most certainly is not blind towards his likes, dislikes, or expectations about the patient. Further, unconscious transference and counter-transference issues in the therapeutic relationship may be enhanced during the psychedelic session and may profoundly prejudice a therapist's rating."[105] Yet Salzman believed that a therapist's enthusiasm could be of value in treatment and should not be entirely excluded: "Suggestion and expectation may be gainfully incorporated into the research design, rather than scrupulously avoided. Positive therapeutic alliance within a psychedelic research setting seems to be considerably more rewarding than an indifferent, impersonal research relationship with the patient."[106]

Hoffer believed that the work at Spring Grove confirmed what he, Osmond, Blewett, and others had maintained all along. Writing to Osmond in November 1968, he remarked that Kurland was the "best [person] to bless our work because he is so obviously safe, solid, conservative, skeptical and cautious."[107] Although by the end of the 1960s the tide had turned against psychedelic therapy as a treatment for alcoholism, the results achieved by the NIMH-sponsored studies at Spring Grove gave a glimmer of hope that further controlled testing and refinement of the controversial therapy might still be possible.

THE ROAD AHEAD FOR PSYCHEDELIC THERAPY

The various clinical studies of LSD treatment for alcoholism in the 1950s and 1960s, controlled or uncontrolled, had methodological shortcomings of one sort or another, especially when judged by contemporary scientific research standards. It would be hard to make any definitive conclusion about the experiments except that they lacked consistency from one to the next, many operating within very different treatment frameworks. As Savage and his colleagues described the situation in 1969, "the main difficulty is that LSD has been used in widely differing ways for a diversity of purposes and yet all of these approaches have been loosely lumped under the mantle of LSD therapy or even psychedelic therapy."[108]

For many patients who underwent the therapeutic interventions that incorporated the single, overwhelming dose of LSD, a mystical conversion experience enabled them to maintain sobriety. But were such conversions lasting? Among individuals successfully treated with LSD, there were anecdotal reports of many who experienced positive changes to their lifestyles that lasted for many years after the initial therapy. Many professionals thus believed that psychedelic therapy offered "new possibilities for scientific investigation of transcendence."[109] According to Abraham Maslow,

> [psychedelics] give us some possibility of control in this realm of peak experiences. It looks as if these drugs often produce peak experiences in the right people under the right circumstances so that perhaps we needn't wait for them to occur by good fortune. Perhaps we can actually produce a private personal peak experience under observation and whenever we wish under religious or non-religious circumstances.[110]

One of the more interesting findings to come out of the experiments was that LSD therapy tended to produce better results with patients in early post-treatment follow-up (e.g., one to three months) than in longer follow-up periods (e.g., six to twelve months). This finding had been cited in studies negating the scientific value of LSD for alcoholism, such as the one by Ludwig and Levine, and it had been made in many of the studies supporting psychedelic therapy.[111] Although it was either ignored or downplayed by opponents of the therapy, the finding intrigued researchers who advocated use of the therapy, the next challenge being to "discover how to sustain and maximize the initial therapeutic benefits."[112] As Osmond and his new colleagues in New Jersey understood the issue,

> many observers have remarked upon the "LSD honeymoon," a variable period of one month or so following treatment in which the individual is said to preserve a sort of "rosy glow" of sobriety, beneficence and tolerance. However,...no careful and systematic work has

yet been done on the relationship between such marked attitudinal changes immediately following treatment and subsequent outcome[s], though studies including follow-ups as long as four years have been reported.[113]

Following his arrival at the New Jersey Neuropsychiatric Institute in Princeton, Osmond decided to take part in another round of testing the psychedelic treatment of alcoholism. The results of this study revealed greater improvement rates with psychedelic therapy at the three-month follow-up and a lessening of advantage over the comparison treatment at six and twelve months. Although opponents of LSD therapy took great pleasure in citing this as proof of the psychedelic master's inability to replicate the earlier claims in Saskatchewan in a more controlled setting, Osmond did not see the results as a refutation of the value of psychedelic therapy in treating alcoholism. Rather, they suggested new possibilities to be explored further:

> The challenge for the therapist would be to *maintain* the early gains which LSD can provide. Our results suggest...better follow up of the patient after discharge in terms of continued therapy, including possibly redosage at three month intervals for a variable period depending upon improvement....Indeed, it is at least possible that the superior results obtained in the Saskatchewan and some other studies are a direct consequence of making fuller use of this period of enhanced motivation to establish better habits in close relationship with AA.[114]

Based upon controlled studies of psychedelic treatment at the Veterans Administration Hospital in Topeka, Kansas, W. E. Bowen and his team concluded that, though a single conversion experience might occasionally result in "radical and enduring" changes to one's lifestyle and personality, it was extremely rare.[115] As they reported, most alcoholics had a "waning of the initial inspiration, euphoria and good intentions gleaned from the LSD experience" once the "former stresses and difficulties in their lives" returned.[116] Like the study in New Jersey in which Osmond was involved, the experiment in Topeka indicated impressive short-term results among patients who had been treated with psychedelic therapy, but a failure to carry this advantage into long-term follow-up. As for the Spring Grove investigators and Osmond and his colleagues in New Jersey, the question for Bowen and his researchers was not so much whether there was value in using LSD as an adjunct in psychotherapy but whether, and how, short-term changes could be maintained.[117] For J. H. Halpern, it was a matter of keeping the "afterglow" going: "What would occur if a series of treatments were spaced to provide a sustained afterglow? Would the patient have the motivation to deal more effectively with his addiction?"[118] The real challenge lay in "how to sustain and maximize the initial therapeutic benefits."[119]

In 1969, the value of LSD in the treatment of alcoholism remained a source of division within the psychiatric and broader treatment communities. Some, like Hoffer, Osmond, and Blewett, held fast to their conviction that it did have value, whereas others insisted that it did not, citing the list of negative reports in their defence. In the end, it was a treatment approach to which the psychiatric and wider medical communities had yet to give the official scientific stamp of approval. Nevertheless, psychedelic therapy held possibilities to be explored further with studies designed more to the liking of the scientific majority. As Savage and his co-authors observed at the time, the "controversy over LSD has reached a pass where it is difficult to predict what direction the future holds."[120] Yet, despite some optimism, the enthusiasm over a once promising treatment for alcoholism quickly withered, fuelled by the perception that it was not only unscientific and unethical, but also downright dangerous.

CHAPTER 8

PSYCHEDELIC DRUG RESEARCH, THE CIA, AND THE '60S COUNTERCULTURE

The far-flung LSD movement of the '60s counterculture was responsible for many of the misinterpretations of and myths about the Saskatchewan research of Abram Hoffer, Humphry Osmond, and Duncan Blewett. The spread of LSD into mainstream society greatly influenced debate on their research and professional reputations. Indeed, the '60s counterculture[1] and LSD movement complicate even further the story of Saskatchewan's psychedelic drug research.

As this research was coming to an end in the early 1960s, reports of the recreational use of LSD were increasing as the drug became available to those who wished to explore it for themselves. Concerned governments and their various enforcement agencies attempted to curb this development by placing further restrictions on the use of LSD in research. By classifying it as an experimental drug available only to a select number of investigators, they hoped that public access to it would be extremely limited if not curtailed altogether. Government agencies, however, were ill prepared for what followed. Although established scientific researchers who had carried out most of the early research dwindled in number as official sources of the drug dried up and funds disappeared, a black market trade soon flourished. Underground sources of the drug multiplied. To make matters worse, the counterculture began to adopt psychedelics as one of its platforms.

At the helm of this burgeoning counterculture were figures such as American psychologist Timothy Leary, whose psychedelic drug antics caused a storm of controversy. In what came to be known as the "Harvard Affair," Leary and his co-researcher Richard Alpert were dismissed from the university for championing the use of psychedelics among students and for engaging in various on- and off-campus escapades. By the mid-1960s, it was increasingly apparent that a revolutionary LSD movement had formed and threatened to undermine the social order. This development, along with the incredible media attention, further obscured and negatively affected debate over the scientific value of psychedelics. Years later it also became publicly known that the CIA and U.S. Army had covertly financed most of the scientific research on LSD and other psychedelics to determine their potential applications as military weapons. All of this, of course, would have profound and lasting implications for how Saskatchewan's psychedelic research, and Hoffer, Osmond, and Blewett, came to be portrayed by the scientific community and society at large.

The Saskatchewan researchers were enmeshed in the debate about the scientific value of psychedelics, particularly as they related to the study of mental illnesses such as schizophrenia and the treatment of alcoholism, and they had begun to venture into wider, non-medical uses of the drugs. For example, they were among the earliest investigators to explore psychedelics as tools in the creative process and advancing human understanding. From 1960 on, they were thrust into the public debate about whether the drugs had social value, at a time when many in the scientific community were disengaging from the area and when Western society was becoming caught up in mass LSD hysteria. This willingness to wade into the social debate over LSD, and their connection and identification with LSD pop culture icons such as Leary, made for headline news, especially since they continued to express their views on the benefits of LSD when it was both professionally and publicly unwise to do so. Although this apparent anti-establishment stance further infuriated critics, it won the admiration of those who believed psychedelics to hold promise. The Saskatchewan psychedelic research thus became subject to a host of myths. Two of the most popular were that the research was insepara-bly tied to—and in some way contributed to—the excesses of the '60s counterculture and that it was part of the notorious CIA program to use hallucinogens as mind-control weapons.

COUNTERCULTURE RISING

Toward the end of the 1950s, American society was in flux. The hum of postwar consumerism was still in the air, with Americans experiencing a level of affluence and demographic change (e.g., suburbanization) not seen before.[2] The decade had also been a time of fear, uniformity, and rigid social hierarchies. Cold War rhetoric remained feverish, racial and gender inequalities abounded, and individual freedom was restricted. As historian Terry Anderson wrote, "continual threats from beyond, and sup-posedly within, created a society in the early 1950s unusually concerned about security, a people bent on conformity and consensus."[3] With the approach of the 1960s, the fractures were there for all to see.[4] The United States was headed for a revolution that would drastically alter it.[5]

Much of what surfaced in the 1960s was a backlash against estab-lished '50s ideals. Rock 'n' roll, sexual liberation, civil rights, race riots, anti-Vietnam War protests, and radical groups pushed the country to the brink of collapse. As underground press reporter Abe Peck wrote of the time, "a volley of challenges w[as] aimed at the accepted order in the United States. War, racism, class, nationalism, the environment, sexuality, the nature of consciousness, culture, work, lifestyle—all were radically, substantially, sometimes explosively reconsidered. Fire power and flower power; rock festivals and rocks thrown at cops; body counts, body bags, body paint—all became part of the national landscape."[6] In the middle of all this was a burgeoning population of baby boomers ready to assert its power and influence in the world.

By the early 1960s, the seeds of social and political discontent sown in the '50s began to germinate in the minds of radical youth. They soon began to challenge authority, calling into question everything from societal values and mores to religion and government, and they sought to reform Western society. As many social commentators and theorists of the time saw it, the impetus behind this brazen "new spirit" had been staring America in the face.[7]

For the man who popularized the term "counterculture," sociologist Theodore Roszak, the movement arose simply because of the technocratic nature of contemporary American society, in which "specially trained experts" who relied on scientific forms of knowledge "assume[d] authoritarian influence over even the most seemingly personal aspects of life: sexual behavior, child-rearing, mental health, recreation etc. In the technocracy everything aspires to become purely technical, the subject of professional attention."[8] The ultimate aim of the technocracy was conformity, and it paralyzed American society, for it was "ideologically invisible," "a grand cultural imperative...beyond question, beyond discussion....Totalitarianism is perfected because its techniques become progressively more subliminal."[9] Such a society led to youth dissent. Given their number (50 percent of the American population was under twenty-five years of age) and sense of power,[10] many youth decided to confront the establishment head-on simply because no one else was prepared or willing to do so. The result was that "technocratic America produce[d] a potentially revolutionary element among its own youth."[11]

This dissent can be traced to the beatnik subculture of the 1950s in the writings of iconoclasts such as Jack Kerouac and Allen Ginsberg, the latter of whom would become a staunch proponent of LSD in the 1960s. The beats had formed their own countercommunities and railed against the dominant social ethos and its conformity. Not surprisingly, it was to the beats and other radicals of years past that many of America's troubled youth looked for inspiration and direction in the 1960s.

While scientists began their experiments in the 1950s to unlock the mysteries of hallucinogenic drugs, a small number of artists and intellectuals began to conduct their own explorations of the curious effects of these drugs on the mind.[12] By the early 1960s, the availability and use of the drugs had become much more pronounced among middle-class American youth. LSD in particular would take centre stage in the countercultural movement, serving as a mind-altering catalyst.[13] Within a few years, university campuses and streets became the new "social" labs for LSD experimentation and consciousness expansion. Why did so many youth experiment with psychedelic drugs, and how did LSD end up so entwined with the counterculture? Many of the psychedelic research projects were based in prestigious American universities such as Harvard and Stanford, and in many cases students, faculty members, and others became willing participants in the experiments and "turned on to" the possibilities of psychedelic drugs. By the mid- to late 1960s, however, LSD use had increased to the point where it ripped away at the fabric of mainstream society.

Speculating on the LSD phenomenon years later in his memoir, Albert Hofmann, who had discovered the drug, expressed mixed feelings on its release into mainstream society: "The joy of having fathered LSD...was tarnished after more than ten years of uninterrupted scientific research and medicinal use when LSD was swept up in the huge wave of inebriant mania that began to spread over the Western world, above all the United States."[14] Hofmann blamed the escalation in drug use on "deep-seated sociological causes: materialism, alienation from nature through industrialization and increasing urbanization...ennui and purposelessness in a wealthy, saturated society, and a lack of a religious, nurturing and meaningful foundation of life."[15]

THE PSYCHEDELIC IN-CROWD (PART I)

Many people interested in what psychedelics had to offer turned to the growing body of literature on the subject. One source that became an immediate cult classic on psychedelia was Aldous Huxley's *Doors of Perception*. The quest for knowledge naturally led many to the good doctor (Osmond) who had administered Huxley his first taste of mescaline, and from him to the groundbreaking research that had unfolded in Saskatchewan in the 1950s. By 1960, Osmond had already been immortalized for coining the term "psychedelic," subsequently used widely to describe the mysterious hallucinogenic drugs. Hoffer and Blewett were also recognized by many as pioneers in the area. Because they were among the most knowledgeable people on psychedelics and their uses and effects, the three men were sought out and would come to be seen as figureheads in the history of LSD, which proved to be to their benefit as well as their disadvantage.

From the time of his 1953 experience on, Huxley maintained a close friendship with Osmond. They continued to seek otherworldly possibilities inherent in the psychedelic experience. They were joined in this pursuit by a tightly knit band of followers that included the likes of Al Hubbard, biologist Julian Huxley, philosophers Gerald Heard and Alan Watts, AA co-founder Bill Wilson, and spiritual medium Eileen Garrett. In what came to be like French psychedelic salons, these early initiates and invited guests came together in small, private sessions to explore, and debrief on, the hidden potential of psychedelics and their relevance for humankind. As the years progressed, Huxley's interest in psychedelics became linked to his fascination with mysticism and the visionary experience, as evident in much of his later writings and lectures. Although many religious thinkers came to see Huxley and like-minded individuals as hucksters of "instant mysticism," they exerted a powerful influence on many other professionals, academics, and laypersons curious about the spiritual possibilities of the psychedelic experience.[16]

Another psychedelic-related interest of Huxley and Osmond was how the drugs might be used in problem solving and the development, measurement, and interpretation of creativity. They analyzed the effects of the drugs on society's "best and brightest." As the Saskatchewan research was getting under way in 1953, Huxley suggested to Osmond, Hoffer, and Smythies that they pitch the idea to the Ford Foundation in the United States, but the proposal never saw the light of day. Osmond carried on as originally planned, looking for a number of select individuals interested in guided experiences. An early initiate was his old friend and schoolmate Lord Christopher Mayhew, then a well-known member of the British Parliament and BBC commentator. In what was supposed to be a special broadcast, Mayhew was administered mescaline in 1955 under the watchful eye of Osmond. Although the BBC never aired the special because of fear of scandal, Mayhew recalled that the experience had been worthwhile: "I still look back on this mescaline experiment, professionally supervised and filmed, as the most interesting experience of my life....I thought then, and still think now,...that I had had an experience that had taken place outside time, that I had visited, by a short cut, the timeless world known to mystics and to some schizophrenics."[17]

As awareness of the Saskatchewan experiments grew, more individuals became eager to have the drug experience. By 1960, Hoffer, Osmond, and Blewett were fielding hundreds of requests from scientists, businesspeople, musicians, painters, engineers, writers, politicians, and students for more information, speaking engagements, and guided excursions into the psychedelic realm. But one man in particular developed a special interest in what the Saskatchewan researchers were doing, a relatively unknown middle-aged psychologist by the name of Timothy Leary.[18]

In 1960, he was establishing a research project at Harvard University to study the effects of another psychedelic drug, psilocybin, the synthetic version of the famed "magic mushroom." Having been introduced to the drug while in Mexico with some colleagues, he declared his first psychedelic drug experience to be transformational. "It was above all and without question the deepest religious experience of my life," he later noted.[19] The experience shifted the course of his mission to focus on the drugs and their uses in consciousness expansion. As he remarked, "since my illumination of August 1960 I have devoted my energies to try to understand the revelatory potentialities of the human nervous system and to make these insights available to others."[20]

Leary looked to Huxley for guidance in pursuing this newfound path and familiarized himself with his literary contributions on the drugs and visionary experiences. Leary arranged for the first of several meetings with the intellectual heavyweight to seek his input on the Harvard project when he found out that Huxley was in town as a guest lecturer at the Massachusetts Institute of Technology. Following a couple of months of careful planning between Huxley and Leary and his associates, the Harvard Psilocybin Project was officially launched in the fall of 1960. Over the next couple of years, the effects of Sandoz-supplied psilocybin would be tested on approximately 400 volunteers.

It was also through those first meetings with Huxley that Leary came to learn more of Osmond and the Saskatchewan research. Eventually, Osmond was drawn into the fold when Huxley introduced him to the Harvard professor. Their first encounter is documented in Lee and Shlain's *Acid Dreams:*

When [Osmond] passed through Boston, Huxley took him to meet Leary. It was the night of the Kennedy election. "We rode out to his place," Osmond remembered, "and Timothy was wearing his gray-flannel suit and his crew cut. And we had a very interesting discussion with him. That evening after we left, Huxley said, 'what a nice fellow he is!' and then he remarked how wonderful it was to think that this was where it was going to be done—at Harvard. He felt the psychedelics would be good for the Academy. Whereupon I replied, 'I think he's a nice fellow, too. But don't you think he's just a little bit square?' Aldous replied, 'You may well be right. Isn't that, after all, what we want?'"[21]

Leary later wrote to Osmond, in January 1961, commenting on how wonderful he thought the Saskatchewan research was. He also gave a ringing endorsement of one piece that caught his attention: "John Spiegel[22] has loaned me a copy of the Handbook prepared by your colleagues Blewett and Chwelos. I think it is a splendid manuscript. Practical. Vitally useful for anyone working in this area. I wish we had a copy before we began our work. Although it is doubly impressive to see that experience...independently support your conclusions."[23] Leary explained to Osmond how he and his team had completed the exploratory phase of their research and invited the elder psychedelic statesman to participate in an American Psychological Association symposium on empirical mysticism and the future of psychedelic activity.

Largely through Osmond and Huxley, then, Leary came into contact with some of the leading psychedelic intellectuals, artists, and other luminaries (e.g., Heard, Hubbard, and Ginsberg), many of whom subsequently became involved in the Harvard experiments. No one, however, not even Leary himself, could have anticipated what was to come.

THE HARVARD AFFAIR

At the start of the Harvard Psilocybin Project, Leary was a shining example of "the organization man," a well-respected, widely published, tweed-clad professor of behavioural psychology. It did not take long, however, for him to tire of playing the game of conformity. During its run, the project yielded some interesting, albeit controversial, results, stemming mainly from the correctional study with inmates at the Concord State Prison, where the psilocybin-assisted psychotherapy was used to reduce rates of recidivism, and a side project connected to the relationship between drug-induced and naturally occurring religious experiences. In the latter study, led by Walter Pahnke, then a PhD candidate at Harvard, the research group carried out a double-blind study in which psilocybin and a placebo (i.e., niacin) were administered to twenty Andover-Newton theology students in a religious setting. The study, which became known as the Good Friday Experiment or "Miracle at Marsh Chapel," concluded that those who had taken the drug had experiences "indistinguishable from if not identical with" classic mystical experiences.[24]

Any chance of the project being legitimate scientific research was lost, however, amid increasing reports of Leary and his associates taking the drug with research subjects and of raucous, psychedelic-fuelled house parties and other on- and off-campus hijinks. One story worth quoting is the famous visit paid to Leary by Ginsberg and his poet-partner, Peter Orlovsky:[25]

> Leary made them comfortable in an upstairs bedroom and then retired to the study to talk with [Frank] Barron. Both expected a replay of the quiet, contemplative sessions that had become the norm. So they were unprepared when the study door banged open, and two naked poets danced in. "I'm the messiah," Ginsberg announced to the startled professors. "I've come down to preach love to the world. We're going to walk through the streets and tell people to stop hating." Barron went and drew the shades. Persuaded that this was not the best moment to march naked through the streets of Newton preaching love, Ginsberg decided to telephone Kennedy and Khruschev and "settle all this about the Bomb once and for all." But he was unable to get the two most powerful men in the world on the telephone, and had to settle for Jack Kerouac.[26]

Such episodes soon became the norm, leading many to question the scientific nature of the Harvard experiments. These developments did not go unnoticed by others in the Harvard Center for Research in Personality and by officials, especially after reports of the wild transactions of Leary and his group (which now included Ralph Metzner and Richard Alpert) in the *Harvard Crimson*.

Leary's metamorphosis from distinguished behavioural psychologist to psychedelic "high priest" was under way by 1961, and his conscious-ness-shattering introduction to LSD later that year only hastened his transformation. With the addition of LSD to his research repertoire and mounting stories of psychedelic barn-burners on campus, the Harvard experiments were shut down, followed shortly thereafter by the widely publicized terminations Alpert and Leary from the university in the spring of 1962. By the time of his firing, Leary was beginning to question whether the profound experiences brought on by psychedelics ought to be reserved for certain settings and the elite few, as Huxley and Osmond had recommended. In further discussing the subject with Ginsberg and others in his group, Leary decided to broaden his agenda.

He aspired to attain more radical heights. Aided by Alpert, Metzner, and others, Leary began to espouse the virtues of psychedelic drug use with an almost messianic zeal, pushing for democratization of the drugs in wider society. In 1962, he and Alpert founded the International Foundation for Internal Freedom (IFIF) to "encourage, support and protect research on psychedelic substances" and ultimately "to increase the individual's control over his own mind, thereby enlarging his inter-nal freedom."[27] The IFIF (later replaced by the Castalia Foundation) was used to sponsor and publish an academic journal, the *Psychedelic Review*, to assist in the realization of this goal. In a short time, Leary became a psychedelic guru as one follower after another joined his ranks. Upon his failure to establish an *Island*-like psychedelic colony in Zihuatanejo, Mexico, he returned to the United States in late 1963 to set up his psy-chedelic headquarters in upstate New York in a posh mansion known as the Millbrook estate, provided to him by wealthy benefactor and new-found devotee William Mellon Hitchcock. For the next few years, Leary and his associates used Millbrook as a base from which to continue their psychedelic experiments, drawing hundreds of followers as well as hun-gry media.

All of this no doubt brought further public scandal to the psychedelic research community. According to William Braden, the earliest press reports on psychedelics concentrated on the scientific findings, but from 1963 on the mainstream press could not seem to dissociate Leary and psychedelics. Almost nothing positive was written about LSD (e.g., its benefits in psychotherapy and the treatment of alcoholism) following Leary's dismissal, and press coverage tended to be "of the cops-and-robbers variety, concentrating on police raids, drug-control bills, suicides and fatal plunges."[28] In Braden's estimation, this presented a grossly inaccurate picture of psychedelics, with newspaper reports of LSD that were "superficial at best and violently distorted at worst."[29]

Osmond and Hoffer feared that what Leary was doing was not only dangerous but also jeopardized the whole study of psychedelics. Writing to Hoffer in March 1963, Osmond told him about a serious discussion with Huxley and Watts regarding the Leary situation. Huxley seemed to be "genuinely concerned and afraid that Timothy's excess of enthusiasm may do much harm."[30] Huxley and Osmond had never intended to recruit people en masse to the use of psychedelics. Rather, they preferred to keep the non-scientific experimentation with the drugs small-scale and carefully controlled. Although initially sympathetic to this course of action, Leary came to see it as too cautious and conservative.

Osmond and Hoffer thought it necessary to distance themselves from Leary to protect their reputations, their research, and legitimate psychedelic studies going forward. Osmond had agreed to be on the editorial board of *Psychedelic Review* without realizing its direct connection to the IFIF, and he wrote to Leary's collaborator Metzner that

> looking back at the *Psychedelic Review* I find that it is sponsored by the IFIF and it may therefore be considered as its official organ— since I disagree strongly and emphatically with the methods being used by IFIF, although sympathetic to many of its goals, it would be illogical for me to remain a consultant on the board of a journal sponsored by it. . . . [It] seems pointless to continue as an advisor when one's advice stands no chance of being heeded.[31]

Hoffer and Osmond also had to contend with growing concerns about their former colleague Blewett, who of the three men came the closest to espousing the ideas of Leary and his followers.

DUNCAN BLEWETT: THE TIMOTHY LEARY OF CANADA?

Blewett went much further than either of his former associates in advocating the extramedical uses of psychedelics. Some of his psychedelic explorations, both during and after the Saskatchewan research, were not far from some of Leary's own exploits. For these reasons, Blewett came to be seen in some eyes as the "Timothy Leary of Canada."

Leary and Blewett did have many similarities, so the idea that Blewett took his cues from Leary merits exploration. Both had brief stints in the army. Both abandoned the military in favour of a promising career in behavioural psychology. And both, profoundly changed by psychedelics, parted ways with strict behavioural practice to pursue more humanistic studies on the mystical and spiritual applications of the psychedelic experience. Like Leary, Blewett was tremendously influenced by the thinking of Huxley and Osmond, buying fully into the latter's maxim that to truly understand the psychedelic experience one must begin with oneself. And, like Leary, Blewett impressed on others the spiritual significance of psychedelics and their potential value for self-awareness and self-understanding. Blewett also seems to have shared with Leary a rebellious streak and a trickster element. His quip to some that LSD should be put into the water system was comparable to some of Leary's provocative statements about the drug.[32]

When Blewett left his job with the Saskatchewan Psychiatric Services Branch in 1961 for the University of Saskatchewan—Regina Campus, he continued to expand on his psychedelic theories and pushed for the creation of a psychedelic research centre. No sooner had he arrived on campus than stories began to emanate out of the university that he was importing peyote and extolling the virtues of psychedelics to students, much like Leary was doing at Harvard University.

The unease of Hoffer and Osmond over Blewett's more liberal views about psychedelics and his sometimes unpredictable behaviour actually predated Leary's appearance on the scene. Their concern could be seen with Blewett's attempts in the late 1950s to test the widely held theory that LSD was physically addictive (using himself as a guinea pig) or his idea that taking LSD with his clients resulted in increased empathy and therapeutic effectiveness. Hoffer and Osmond were also aware of episodes involving Blewett that, had they come to light, could have jeopardized the Saskatchewan studies.[33] Before Leary's activities became disconcerting, and before comparisons were made between Blewett and Leary, Hoffer and Osmond monitored Blewett's growing identification with another man, "Captain Trips" Al Hubbard, and his mission to spread the benefits of LSD to the larger society. As Osmond wrote to Hoffer, "Dunc needs to be discouraged from such groups because they drift off into hedonism of a goalless sort. He needs to work with very different people.... We should not let our main work be diverted an inch for Dunc and Al's convenience."[34]

The concerns of Hoffer and Osmond about Blewett increased in the 1960s following his move to his university position. When word of some of his on-campus activities reached Hoffer, he attempted to persuade those closest to Blewett to convince him to exercise caution and, if possible, end his psychedelic drug adventures. Writing to Hubbard in December 1962, Hoffer indicated that

> I know that Dunc is using peyote illegally and getting it by shipping in the cactus plant. Some of the students at this university are also using it. I strongly disprove of this misuse of these drugs. There have been several close squeaks with LSD in the past, and only by the grace of God have our Regina friends escaped serious danger.[35]

Hoffer wrote to Osmond several months later, noting that he'd had

> a five hour discussion with Jake Calder[36] the other night. He is doing his best to keep Dunc in check.... Thank God he has no ready access to LSD. He has been importing peyote legally under the Department of Agriculture permit as a cactus. What he has done is legal but the uses of the plant may well be illegal. I wonder whether I should alert

Ottawa so they can alter the regulations. I don't like to do so for no law is being broken, [and] peyote is rather harmless.[37]

Hoffer and Osmond concluded that what Blewett was doing, like what Leary was doing, threatened overall psychedelic research.

Blewett claimed years later that he had been misunderstood and that his actions had been blown out of proportion:

> My conviction that LSD was a most important discovery led me to discuss with my classes, as part of the curriculum, the importance that psychedelics held for psychological theory and research. I sincerely tried to be fair-minded in discussing LSD. I believed that it was dangerous and I said so, warning students never to take it except under "controlled" circumstances. Because of this I became known as a proponent of the use of LSD.[38]

When rumours of peyote use by students began to swirl about the university administration and public, many automatically correlated it with Blewett. As he wrote, "because I had openly expressed my views on the importance of LSD suspicion was directed at me as the person who was probably responsible in some way for this worrisome behavior on the part of young people in the city."[39]

Regarding the contention that Blewett was the Canadian counterpart to Leary, much in the correspondence between Osmond and Hoffer suggests that even they sometimes had a hard time distinguishing between the two psychologists. Osmond pointed out that Blewett provided the template upon which Leary based much of his research. Writing to Hoffer, Osmond noted that

> Dunc is full of reverence for Leary the pioneer....What I want to know is what has Leary pioneered? Many of his ideas came straight from Aldous who was writing *Island* in 1960 when Leary was starting up. But most of his actual research ideas came from Dunc B's manual—the prison project and the idea of the therapist taking it with patient came straight out of the Blewett manual![40]

Truth be told, Leary took much from Blewett without ever crediting him. Although the psychedelic guidebook that Leary and his associates subsequently developed resembled the *Tibetan Book of the Dead*, it probably owed as much to the *Handbook for the Therapeutic Use of Lysergic Acid Diethylamide-25*. In some respects, then, perhaps Leary was the American incarnation of Blewett.

Yet Blewett rarely, if ever, fully sanctioned Leary's approach; he understood it but did not necessarily promote it. And Leary ultimately dismissed the wisdom of the medical authorities when it came to psychedelics, deciding instead to radicalize and "turn on" the youth of the world. Even though Blewett sympathized with some of his motives, in the end he, along with Hoffer and Osmond, realized that progress in the field could occur only through scientific routes and not a religious crusade. Although Blewett became a local legend among the student population, he never quite reached the epic countercultural status that Leary did, nor did he aspire to reach that status. If anything, Leary was a magnified version of Blewett in the same way that Blewett was a radicalized version of Hoffer and Osmond. They shared some views on psychedelics, but not others. By 1963, questions and concerns about what Leary, Blewett, and other like-minded individuals were saying about, and doing with, psychedelics shifted more to where they were getting their supply and where, in turn, it was going. This situation was particularly worrisome with the sudden proliferation of underground sources of LSD.

BLACK MARKET PANIC AND "GREEN LSD"

With the tightening of restrictions on LSD in late 1962, reports began to surface about the appearance of black market samples of the drug or facsimiles thereof. One substance in particular that caught the attention of Osmond was a mysterious liquid green LSD that had a reputation more for its psychotomimetic (madness-mimicking) than for its psychedelic properties. On hearing that a green "dud" had turned up on the West Coast and that it might have been to blame for a number of unfortunate incidents, one being the tragic suicide of a Los Angeles doctor who had obtained samples of the drug, Osmond immediately wanted to get to the bottom of the issue.

He was all too familiar with the effects of various psychotomimetics (e.g., adrenochrome and adrenolutin), and the thought of an easily obtainable LSD-like drug, which tended to cause prolonged psychotic episodes, was unnerving. That this green LSD might be available on the street or fall into the wrong hands was reason enough for Osmond to go on the offensive. The first challenge was to discover where this green LSD originated and, if possible, to obtain samples of it (and any other black market samples of the drug) for detailed analysis by recognized experts. Osmond also thought it necessary to begin issuing warnings to his associates and the authorities to be on high alert for the drug, which could inadvertently be used under the assumption that it was LSD-25. He was somewhat perplexed, though; if the drug was not coming from Sandoz, where was it coming from, and how, and by whom, was it being made?

Writing to Hubbard in California in May 1963, Osmond commented that "Sandoz says that few people can make LSD and they probably know—the military have certainly been enquiring into LSD—[and are] interested in a stable LSD....I do feel that you have a strong responsibility to warn Leary and any others who are using these dubious substances ('green lsd')."[41] Referring to the claim by Leary that he could get all the LSD he wanted, and the likelihood that black market versions of the drug had different properties than authentic LSD-25, Osmond insisted that Hubbard "warn Leary by phone perhaps but follow by regular mail and a copy to the federal people and Sandoz. We may be able to reduce or prevent damage."[42] Osmond began to speculate whether the drug might have been made by someone connected with the U.S. Army Chemical Warfare or someone who might have once worked there. He was aware of the interest in LSD as a weapon of war, but he also knew that, to be used as such, the drug would have to be stabilized[43] and could thus have very different effects. As he framed the issue for Hubbard, "if makers of black market LSD found it unstable they would probably stabilize it without giving much further thought to the consequences."[44]

Always at the forefront in Osmond's mind was who might come into contact with black market LSD. Corresponding with Hoffer, he emphasized that "such stuff getting into the hands of those who did not know it was very different could be very dangerous....Apart from you and I on adrenolutin only a few of our colleagues know how disconcerting and disorganizing even mild prolonged psychotomimetic experiences can be."[45] Osmond then alerted Huxley, noting

> that some outfit took to selling a "stable LSD" which does in fact have some of the properties of LSD. Albert Hofmann has analysed some samples of it....He finds that this substance contains LSD but only 60%. 40% consists of other adrenaline-like substances. He does not yet know what these are or why they should be present....I hope that not much green LSD got around, [and] much of the blame must go to Sandoz and the FDA for their fantastic lack of a coherent and sensible policy....Perhaps no one in LA has received samples of the green [LSD,] but if they do they should *on no account* use them. The new LSD sounds much more like a "model schizophrenia."[46]

Leary was the focus of Osmond's fears, but there was the possibility that Blewett, given his many connections and since he no longer had access to Sandoz-distributed LSD, would acquire samples of green LSD: "Dunc may obtain far more objectionable substances than peyote—especially some of the green LSDs that are around—if he did this without letting us know grave harm might come to Dunc, his family, his students and of course us being his associates."[47]

Hoffer and Osmond wondered how they could work with the American and Canadian governments to develop a plan to curtail the black market threat and return LSD research to its rightful place. "I think we can help them devise a policy which will repair much of the damage that has been done," Osmond said, "and lead to new and valuable developments."[48] The two psychiatrists wanted to prevent a widespread LSD movement, which, in Osmond's estimation, "would have many dangers and many possibilities."[49] Hoffer responded that "[I] am quite happy to help Ottawa set up a control system that can work....The most sensible thing might be to release LSD to everyone with proper warnings and proper legal safeguards....This would immediately bring good LSD back and remove the thrill of the youthful rebel."[50]

Clearly, Hoffer and Osmond were frightened by the prospect of Leary, or others like him, having access to a limitless supply of LSD, especially as wider public experimentation with it was becoming a reality. By 1963, Osmond had abandoned any hope that Leary would heed his advice, so he sought to enlist others in the cause, such as Watts, Hubbard, and Huxley. Perhaps more than anyone else, Huxley had the best chance of convincing Leary to rethink the destructive course on which he had embarked. Unfortunately, Huxley passed away from cancer on November 22, 1963. His death was a major blow to the psychedelic community because he was one of the most elegant and respectable spokespersons for responsible use of the drugs. For Hoffer and Osmond, his absence was felt both personally and professionally. As Hoffer summed up the loss to his partner, "Aldous was perhaps the one person who was most influential in shaping our international image and work."[51]

With Huxley's sage voice of reason now silenced, Leary and his psychedelic campaign continued to be a thorn in the side of many within the psychedelic research community. Writing to Al Hubbard, Myron Stolaroff, and Bill Wilson, Osmond complained that Leary was "plagiarizing discoveries made by you, me, Abe and others while making our work far harder."[52] In his assessment, Leary had transformed himself by 1964 into a sort of psychedelic Billy Graham. Although it still remained a mystery to most where the black market LSD came from, Leary and his psychedelic-inspired antics were possibly the main reason for governments' increasing crackdowns on legitimate LSD research. But was Leary alone responsible for such unwanted attention? As Lee and Shlain argued, "many LSD researchers were quick to point an accusing finger at Leary for bringing [the] government's wrath down on everybody. But is it plausible that one wayward individual was single-handedly responsible for provoking a 180-degree shift in government policy with respect to psychedelic research? Was the FDA simply overreacting to Leary's flamboyant style or were there other forces at work?"[53]

LSD: A WEAPON AS POWERFUL AS THE ATOM BOMB?

No story about the history of LSD would be complete without some discussion of the hidden role played by the CIA and other American military and intelligence agencies. Thanks largely to the release of thousands of declassified U.S. government documents obtained through the Freedom of Information Act in the 1970s, it is now general knowledge that as far back as the 1950s the CIA, the U.S. Army, and a host of other government agencies had become enthusiastic about the use of drugs such as LSD as weapons in the Cold War, a phenomenon thoroughly recounted in John Marks's *The Search for the "Manchurian Candidate."*[54] Regarding the research with LSD, Marks wrote that "intelligence agencies, particularly the CIA, would subsidize and shape the form of much of this work to learn how the drug could be used to break the will of enemy agents, unlock secrets in the minds of trained spies, and otherwise manipulate human behaviour."[55]

Through top-secret programs known by names such as ARTICHOKE, BLUEBIRD, and MK-ULTRA, the CIA set about to discover the myriad espionage possibilities of LSD and other hallucinogenic drugs. Initially, the hope was that LSD might provide an ideal truth serum, but having limited success with that route the agency shifted gears to look into the drug's potential as an interrogative and brainwashing agent as well as an instrument for political subversion. The possibilities seemed endless. As John Buckman explained the rationale behind the growing preoccupation of the CIA and other government agencies with the drugs,

> in the mid-1950s serious consideration was given by numerous governments on both sides of the Iron Curtain to the feasibility of using LSD and other hallucinogens as a form of chemical warfare. One idea was that the substance could be put into [the] enemy water supply, rendering the enemy temporarily psychotic....A later idea was to use LSD in a sort of aerial drop so that the bomb filled with LSD would explode at ground level or some feet above the ground and the local population or troops would through inhalation...become intoxicated, psychotic and ineffective.[56]

Fearing that LSD might fall into the hands of its enemies (if it hadn't already), the CIA devoted its energies to unlocking the secrets of the drug.

To better understand the effects of LSD on humans, the CIA conducted extensive field experiments, testing the drug on society's most vulnerable (e.g., prisoners and mental patients) and dispatching agents to drug and spy on unsuspecting citizens. Sometimes the CIA even resorted to in-house testing, some of which provided comical moments:

> Turn your back in the morning and some wise acre would slip a few micrograms into your coffee. It was a game with the most exalted of weapons, the mind, and sometimes embarrassing things happened. Case hardened spooks would break down crying or go all gooey about the "brotherhood of man." Once or twice things went really awry, with paranoid agents escaping into the bustle of downtown Washington, their anxious colleagues in hot pursuit.[57]

At other times, the results were not so hilarious, as when a scientist from the U.S. Army Chemical Corps, Frank Olson, dosed with LSD by a CIA chemist, weeks later jumped to his death from a New York hotel. Silly pranks and criminal misfortunes aside, the CIA knew that it had something powerful in its hands, and it continued to test LSD and other hallucinogens to their fullest potential.

As the CIA began its covert drug programs in the early 1950s, it developed a tight relationship with Sandoz, whereby the CIA secured supplies of LSD whenever required and the Swiss pharmaceutical company agreed to share any information that it had on the drug, from how much was produced to where it was shipped and to whom. As Lee and Shlain explained,

> by 1953 Sandoz had decided to deal directly with the US [FDA] which assumed a supervisory role in distributing LSD to American investigators. It was a superb arrangement as far as the CIA was concerned, for the FDA went out of its way to assist the secret drug program. With the FDA as its junior partner, the CIA not only had ready access to supplies of LSD...but also was able to keep a close eye on independent researchers in the US.[58]

Contrary to popular belief, Sandoz was not the only organization in the 1950s that could manufacture and distribute LSD. It held the official patent to the drug, but that did not stop others from trying to make it. The

CIA eventually turned to a domestic company to see if it could produce the drug, asking

> the Eli Lilly Company in Indianapolis to try to synthesize a batch of all-American acid. By mid-1954 Lilly had succeeded in breaking the secret formula held by Sandoz. "This is a closely guarded secret," a CIA document declared, "and should not be mentioned generally." Scientists at Lilly assured the CIA that "in a matter of months LSD would be available in tonnage quantities."[59]

Other agencies fell victim to the CIA agenda. According to Marks, "both the military and the [National Institute of Health] allowed themselves to be co-opted by the CIA—as funding conduits and intelligence sources. The [FDA] also supplied the Agency with confidential information on drug testing. Of the Western world's two LSD manufacturers, one—Eli Lilly— gave its entire supply to the CIA and the military."[60]

Revelations of the CIA research on LSD shone a spotlight on less ethical aspects of the scientific community's early efforts to study the psychoto-mimetic properties of the drug:

> When the CIA first became interested in LSD, only a handful of sci-entists in the United States were engaged in hallucinogenic drug research. At the time there was little private or public support for this relatively new field of experimental psychiatry, and no one had undertaken a systematic investigation of LSD. The CIA's mind con-trol specialists sensed a golden opportunity in the making. With a sizable treasure chest at their disposal they were in a position to boost the careers of scientists whose skill and expertise would be of maximum benefit to the CIA. Almost overnight a whole new market for grants in LSD research sprang into existence.[61]

The CIA and military soon extended their reach across the United States, directly or indirectly financing LSD research projects at some of the most prestigious medical schools and universities (e.g., Johns Hopkins, UCLA, Stanford, Harvard, and Columbia) and drafting into service some of the most respected psychologists and psychiatrists to assist with their mind-control experiments.[62] CIA funds even found their way into Canada and the hands of eminent psychiatrists such as Ewan Cameron, who experimented with LSD on a number of unwitting mental patients at McGill University (Allan Memorial Institute) in Montreal.[63] Some have speculated that the Saskatchewan experiments with psychedelics were part of this wider network. Did the CIA follow what was transpiring in Saskatchewan? Was the CIA behind the Saskatchewan research the whole time?

The CIA was indeed aware of the Saskatchewan research. Pressed on this subject many years later during a CBC Regina interview, Hoffer downplayed claims that the CIA had kept tabs on the Saskatchewan work: "Why would they want to come to a hick place like Saskatchewan to look for what we might be doing? I have the idea that they didn't even know we existed."[64] When Blewett was asked whether the CIA had collected records of the experiments, he was more forthright, responding with his trademark trickster humour: "Oh yes...yes, of course. I'm sure they got everything. And I must say I'm not ashamed of any of it. I think it's damn good."[65] Given that Saskatchewan was on the cutting edge of studies of LSD, the CIA would have been curious about the findings, particularly those pertaining to the use of LSD as a model psychosis, and because of its special relationship with Sandoz the intelligence agency would have had knowledge of any research on LSD, be it in Saskatchewan or elsewhere. Yet perhaps Hoffer, Osmond, Blewett, and the other Saskatchewan researchers did not know that they were being watched or that there were CIA operatives in their midst. Lee and Shlain observed that

> the CIA, ever intent on knowing the latest facts as early as possible, quickly sent informants to find out what was happening at Weyburn. Unbeknownst to Osmond and his cohorts, throughout the next decade they were contacted on repeated occasions by Agency personnel. Indeed, it was impossible for an LSD researcher not to rub shoulders with the espionage establishment, for the CIA was monitoring the entire scene.[66]

Whether the CIA had a role in funding and/or directing the Saskatchewan research is much harder to prove, and this is where any grand conspiracy theory falls apart. Hoffer and Osmond were surely not blind to the military possibilities of LSD when they began their research. They even commented at length in a 1952 memo on its use in military psychiatry.[67] Despite their modest interest in the field, one should not read too much into it or lose sight of the real focus of their research: battling mental illness, not communism.

Nonetheless, Hoffer and Osmond received funding to attend a couple of academic conferences in the United States dealing with LSD, clandestinely organized through CIA conduits (the Josiah Macy Jr. Foundation). Could *this* mean that the two psychiatrists had been co-opted into the CIA program? Indeed, Hoffer and Osmond, and even Blewett, had meetings or discussions at various points throughout their careers with scientists and others later found to be psychedelic drug consultants to the CIA. For example, Hoffer was sometimes in contact with New York physician and Columbia University professor Harold Abramson (one of the main LSD researchers for hire by the CIA and U.S. Army Chemical Corps) to share information on LSD and its effects.[68] Hoffer also expressed interest in Abramson's article on the effects of LSD on Siamese fighting fish. And it was Abramson who arranged for Hoffer and Osmond to present on the Saskatchewan work at the two LSD conferences sponsored by the Macy Foundation in 1959 and 1965. Osmond, too, had ties to CIA-backed researchers. Following his move to the New Jersey Neuropsychiatric Institute, he sometimes worked alongside researchers such as Bernard Aaronson and Carl Pfeiffer, both of whom had projects financed by the CIA.[69] There is little to suggest, however, that these CIA contacts had any influence on the Saskatchewan research or that the CIA indirectly funded it.

What about Hubbard? He was instrumental in the evolution of LSD in therapy in Saskatchewan and was a close confidant of Osmond, Hoffer, and Blewett; he also happened to have a shady past as an operative of the U.S. Office of Strategic Services (OSS), the precursor to the CIA. Should one thus infer that the invisible hand of the CIA was steering the Saskatchewan research all along? Hubbard had a powerful influence on the course of LSD experimentation (including in Saskatchewan), and he was held in esteem by many within the psychedelic research community, among them Huxley, Leary, and Hoffer, Osmond, and Blewett. Yet it is doubtful that Hubbard was carrying out CIA directives in Saskatchewan or anywhere else. According to Lee and Shlain, he was not kindly disposed toward the CIA and its shoddy LSD experiments: "'The CIA work stinks,' he said. 'They were misusing it. I tried to tell [them] how to use it, but even when they were killing people...you couldn't tell them a goddamned thing.'"[70] Like most others in the psychedelic research community, he would have had contact with agency people at one time or another, but that did not mean he did their bidding. Lee and Shlain noted that

> his particular area of expertise—hallucinogenic drugs—brought him into close contact with elements of the espionage community. The CIA must have known what he was up to, since Sandoz and the FDA kept the Agency informed whenever anyone received shipments of LSD. The Captain, of course, was one of their best customers, having purchased large amounts of the drug on different occasions.[71]

In the end, a careful reading of the records left by Hoffer, Osmond, and Blewett reveals that the CIA played almost no role, directly or indirectly, in the psychedelic drug research in Saskatchewan. In fact, the Saskatchewan research was headed in a different direction from that of the CIA and most others in the scientific LSD pack. The CIA had no interest in the therapeutic potential of LSD; it was not into self-actualization, self-understanding, or consciousness expansion. It cared less about the psychedelic applications of LSD and more about the psychotomimetic ones (and for highly unethical reasons).

The CIA and its partners were knee-deep in official—and not so official—LSD research, and the suggestion that they were partly to blame for the explosion of black market LSD in mainstream society in the late 1950s and 1960s cannot be discounted. Iconoclastic musician Frank Zappa was reported to have speculated once that "the whole 'mind expansion' hype had been a clever ploy by the CIA on behalf of the entire Western establishment to undermine the potential threat of the emerging youth culture."[72] Was such an Orwellian vision possible, or was this just another conspiracy theory? In many ways, the CIA had its prints all over Pandora's box when it was thrust open and LSD was released among an unsuspecting public, so the idea that the escape of LSD from the lab was not purely accidental might not be far-fetched.[73] As the street use of LSD increased, it resulted in a significant rift between young politicized radicals who preached revolution and hippies who preferred to turn on, tune in, and drop out. If indeed this was what the CIA had intended, it did not go exactly as planned. By 1965, the LSD movement had erupted across North America and parts of Western Europe, and it was beyond the control of the CIA or anyone else. A psychedelic revolution was at hand.

THE LSD MOVEMENT

By 1965, the street use of LSD was a major source of controversy in America and much of Western society. As word of the powerful mind-altering drug spread, so did its initiates, many of them disaffected middle-class youth and young professionals. Faced with an impending drug epidemic, government authorities responded by applying additional legislative pressure to prohibit the manufacture, sale, and use of the drug to halt its spread into wider society. In 1965, Congress passed the Drug Abuse Control Amendment, making the illicit manufacture and sale of LSD a misdemeanour, but the move had little effect. With black market sources of the drug proliferating throughout America and elsewhere to meet the growing demand, it was a difficult, if not impossible, threat to contain. As new sources multiplied, the recreational use of LSD skyrocketed and became inseparably linked to the larger countercultural movement unfolding.

Why did LSD attract people in the numbers that it did? As Todd Gitlin, a former underground newsman and president of Students for a Democratic Society, observed of the time, "disgruntled with affluence and the 'disenchantment of the world' characteristic of modern society, a critical mass of the young began to look to drugs as spiritual conveyances."[74] Given the regularly touted mystical and transcendental qualities of LSD, it seemed to fit the bill perfectly. In Gitlin's assessment, LSD and other psychedelics thus became "part of the whole complex of meanings, symbols, and practices that came to be known as the hippie world or, later, the counterculture. The drug was the ideological centerpiece of this revolt against authority and materialism—against the values of consumer society itself."[75]

Whatever the reasons behind the meteoric rise in popularity of illicit LSD use, it stirred up public hysteria and prompted tough questions about whether the drug was a boon or a menace to humankind. In 1965, the United States was deeply divided over LSD, and Canada, the United Kingdom, and other Western countries would soon confront the same issues and debates. On the one side was a loud and increasing majority, professionals and laypersons alike, who viewed LSD as a social catastrophe, citing corrupted minds, horrible side effects, and destroyed lives as proofs. On the other side was a growing number of people who, convinced of the untold promise and potential of LSD, championed its use. As Baumeister and Placidi noted of the LSD phenomenon, "the vehemence with which contradictory beliefs were asserted is perhaps its most salient feature. LSD destroys your mind; LSD expands your mind. LSD increases creativity; LSD does not increase creativity. LSD cures insanity; LSD leads to insanity; LSD experience is a model of insanity."[76]

Many blamed the astonishing use of LSD on the provocative messaging and proselytizing of psychedelic leaders such as Leary, and no doubt he played a part. By the mid-1960s, he had become determined to take the drug to radically new heights, promoting it as a religious means to a psychedelic new world order. In 1966, Leary turned his campaign up a notch, publicly declaring the "fifth freedom," the right of an individual to control his or her own consciousness.[77] In September that year, at a press conference in New York, he announced a new religion, the League for Spiritual Discovery, as part of his attempt to maintain legal status for LSD and other psychedelics, a move disdained in many quarters but greeted by the counterculture. It was at this press engagement that Leary introduced the mantra that would become an enduring catchphrase of the turbulent 1960s. He was quoted in the *New York Times* as saying that "like every great religion of the past...we seek to find the divinity within and to express this revelation in a life of glorification and worship of God. These ancient goals we define in the metaphor of the present—'turn-on, tune-in and drop-out.'"[78]

Leary's recommendation to take up LSD and reject social values, patterns, and obligations attracted many followers but alarmed others and caused moral panic. His message also frustrated authorities' efforts to curtail widespread use of the drug. For these reasons, attempts were made to arrest and imprison Leary on whatever grounds possible. But from 1965 on, the U.S. government would have more to worry about than just the psychedelic high priest. While Leary and his cadre were busy fine-tuning and readying to unveil their psychedelic-inspired platform on the East Coast, another LSD submovement was brewing on the West Coast in the San Francisco Bay Area.

PSYCHEDELIC PRANKSTERS

One of the new recruits to the LSD movement, and soon to become one of its acknowledged leaders, was Ken Kesey, "the sort of maverick in all American boy clothes that the corporate personality testers were paid to spot and eliminate."[79] Part scholar, part comic book superhero/trickster, part bohemian, part wrestler, he hailed from the American Midwest and was recognized early on for his athletic and academic prowess. Influenced by the works of Kerouac, Ginsberg, and other beat writers, Kesey eventually chose the literary path, graduating from the University of Oregon and then entering a creative writing program at Stanford University.

Ironically, his first major breakthrough in the literary world and his entry into the LSD frenzy came courtesy of the CIA. In 1959, Kesey volunteered for the agency-funded drug experiments based in the Veterans Administration Hospital in nearby Menlo Park, headed by noted psychiatrist and pharmacologist Leo Hollister. During this time, both as a volunteer and then as an attendant, Kesey came into contact with psychedelics such as LSD and psilocybin, and his experiences on the psychiatric ward ultimately provided the basis for his 1962 best-selling novel *One Flew over the Cuckoo's Nest*.

Following the immense success of his novel, Kesey moved from the bohemian Perry Lane neighbourhood in Palo Alto to La Honda, California. His new base of operations quickly turned into a psychedelic testing centre, drawing long lines of visitors and curiosity seekers to its doors. It was there that his core group of psychedelic adventurers was formed. Known as the Merry Pranksters, the group mounted in 1964 the now famous road trip across the United States in the ramshackle, technicolour bus Further, driven by no less than beatnik captain himself, Neal Cassady,[80] to spread the good word of LSD to the nation.[81]

Like Leary and his group, the Merry Pranksters soon garnered the attention of American authorities, and perhaps it was completely warranted, for over "several impetuous months...they managed to introduce more people to LSD than all the researchers, the CIA, Sandoz and Tim Leary combined."[82] La Honda, however, gave off a different vibe than Millbrook. Kesey was much less interested in a religious crusade; unlike Leary in 1964–65, he was not concerned with drawing an elaborate map to the other world. Rather, he wished to live in the here and now and "freak freely." He was more into exploding one's consciousness than methodically and contemplatively exploring it. A chief means of doing so was by combining LSD, live rock music (e.g., the Grateful Dead), theatrics, and the latest in technological gadgets (e.g., strobe lights) in what came to be referred to as trips festivals and the aptly named Electric Kool-Aid Acid Test.

Trying to counter, undermine, or even halt the activities and messages of LSD martyrs such as Leary and Kesey was hard enough; trying to contain the underground production, flow, and distribution of the drug was much harder. Especially difficult for American authorities was locating precisely where the leaders of the LSD movement were obtaining LSD: with increasing control of official supplies of the drug, a host of amateur and semi-professional chemists tried to satisfy the hunger for it.

Perhaps the most legendary underground producer of LSD was Augustus Owsley Stanley III, or Owsley as he is popularly referred to, who would acquire the title "LSD King" and become the first LSD millionaire. The son of a wealthy Kentucky family and a Berkeley student, he was given 400 micrograms of Sandoz LSD in 1964, following which he decided to try making the drug. Working alongside a female chemistry undergraduate from Berkeley, Owsley managed to manufacture LSD in large quantities in 1965, producing several thousand capsules of the drug, known thereafter simply as Owsley acid, and later as "White Lightning" and "Purple Haze." Within a year, his acid was renowned for its purity and potency, and he went on to become the official supplier to San Francisco's Haight-Ashbury acidhead denizens, Kesey and his Merry Pranksters at La Honda, and numerous acid tests and be-ins, not to mention a number of rock 'n' roll luminaries.[83] Even the eventual bust of Owsley's operation by authorities in 1967 was not enough to stem the tide of illicit LSD; competitors merely stepped in to fill the void.

SCIENTIFIC LSD: THE BEGINNING OF THE END

When the 1965 amendments to the Drug Abuse Act came into effect in April 1966, Sandoz quickly withdrew from clinical and experimental research outfits and recalled all of its LSD supplies. The abrupt move caught many off guard, and was a major blow to the community of legitimate scientific investigators of LSD, even though they were aware of the social problems arising from, but not limited to, the black market trade, the misuse of and excessive statements made by some about it, and an overall lack of understanding of it.

There was by no means a uniform stance on LSD at the time within the scientific community or among its investigators. Indeed, from the early 1960s on, there was a widening schism over LSD and other psychedelics. San Francisco media professor Richard Marsh observed in 1965 that

> among serious students of the drugs, a war rages....One group claims big things. Thanks to the drugs and their capacity for expanding human consciousness, they say, psychiatry is at a turning point, and Utopia itself may be upon us before we know it. Another group is darkly dubious. The new drugs will lead us, they fear, not into Utopia but into psychosis...and in a pinch could be used by a future dictator as an insidious adjunct to thought control.[84]

The division, as Marsh characterized it, was between "psychotomimeticists" and "psychedelicists."[85]

Even in the psychedelic camp there were different opinions on how LSD should be understood and used. In the 1966 book *LSD*, *Life* photojournalist Lawrence Schiller teamed up with LSD stalwarts Sidney Cohen and Richard Alpert to provide a commentary on the "raging national controversy."[86] Schiller used pictures of people at various stages of their LSD experiences to encourage dialogue with the two psychedelic experts and to obtain their insights on the pictures and their responses to a list of popular questions about the drug. Cohen, like many of his psychedelic counterparts, considered the acts of Leary and Alpert to have contributed negatively to the field of study, leading him to declare that "some of the best friends of LSD are its worst enemies."[87] As he put it, "they have managed to shock the citizenry to the point that all hope of safely, cautiously and gradually introducing psychedelics into our culture is lost. This was the hope held by those who understood the nature of cultural change, like Osmond, Heard, Bateson, and others, years before Leary and Alpert knew of the existence of LSD."[88]

Despite attempts by those like Cohen to better educate the medical community and wider public on the pros and cons of LSD, the NIMH ended up reducing the number of LSD research projects in the United States from a hundred or so to a mere six. Why did the drug have scientific merit one moment, many LSD researchers wondered, and in the next became the scourge of society? Even some prominent U.S. politicians expressed confusion over the sudden change in LSD policy when the subject came up during a congressional probe headed by Senator Robert Kennedy in the spring of 1966.[89] Suggestions by Kennedy and others for sober thought on the LSD issue were to no avail. As the LSD movement intensified, the U.S. government became more intent on pressing ahead with stricter regulations. In October 1966, LSD was made illegal in the country; not only were possession and recreational use criminalized, but also all legal scientific research programs on the drug were eventually shut down.

The media did little to add a voice of calm and reason to the LSD debate. They frequently issued reports about LSD, a good number of them overblown or fabricated (though some journalists did try to communicate what was happening in the field and cast a more accurate light on the world of the LSD user). As Baumeister and Placidi wrote, "public information accused LSD of causing gangrene, leukemia, psychosis, homosexuality, blindness, birth defects, weakening of character and loss of motivation, suicidal and homicidal behaviors, and brain damage. . . . Unfounded warnings of physical and mental dangers were used to frighten people away from doing something that society wanted to prevent them from doing."[90] Paired with the increasing number of medical reports citing the dangers of LSD use, media reports made it virtually impossible for those in the scientific community unfamiliar with the drug, as well as the wider public, to distinguish between fact and fiction and make informed judgments, making it even harder for LSD investigators to uphold the scientific status of LSD.

Easy as it might be to blame the Learys, Alperts, and Keseys of the world, or the media and certain individuals within the medical community, for exacerbating the LSD situation, it would be hard to excuse the U.S. government's role. Contrary to some thinking, the government did not have its hands tied, nor was it merely responding to circumstances in society not of its own making. Much of the momentum in the LSD movement and the black market for the drug had been generated thanks to the actions, or inactions, of U.S. government agencies, in particular the CIA, always at work behind the scenes. As Lee and Shlain have argued,

> if the Agency had wanted aboveground LSD studies to proliferate, they would have. But this type of research was no longer essential as far as the CIA was concerned. The spymasters viewed LSD as a strategic substance, as well as a threat to national security, by virtue of its psychotomimetic properties, which had been fully explored in the 1950s. Creative or therapeutic considerations were not part of the covert game plan.[91]

Many researchers who still clung to findings showing LSD to have scientific and medical value questioned whether the heavy-handed regulations imposed by the government were truly warranted. Was wholesale prohibition the best means to handle the social crisis? If the government wanted to prevent people from seeking LSD experiences, then it seemed to be going about it in the wrong way. While the strategy prevented recognized researchers from continuing their studies and clinicians from using the drug in treatment procedures, it did virtually nothing to halt the booming trade in illicit LSD.

THE PSYCHEDELIC IN-CROWD (PART 2)

What did Hoffer, Osmond, and Blewett think about the LSD movement? What were their public and not-so-public thoughts on it? On the growing black market in, and the massive illicit use of, LSD and other psychedelics? On the LSD manoeuvrings of revolutionaries such as Leary and the policies of government agencies? Like many of their colleagues closest to the debate, all three men were troubled, and at times bewildered and bemused, by what they saw happening with psychedelic drugs in society. But they did not sit back and idly watch events unfold; they believed that they had a responsibility, as professionals, to speak openly and honestly about the subject; to educate the public on the dangers, merits, and legal regulations of LSD; and to offer what they saw as potential solutions to the escalating social crisis caused by the drug.

In late 1964, *Maclean's* journalist Sidney Katz was preparing a follow-up piece to his 1954 article on LSD to update readers on the status of the drug and opinions of some of its main supporters and detractors. Katz shared a draft of his article with Blewett to get his impression on it and what was then happening with LSD. Blewett stated frankly in a letter to Katz in mid-February 1965 that

> I fear that the whole structure of what has been accomplished by the Hubbards, Osmonds, Hoffers, Savages, Van Deusens etc., and by men like yourself, who had the guts to step out of the ruts and risk the unknown, will be lost. I believe that this area of investigation is the most important one man has ever broached....LSD offers the best and clearest path ever discovered to self-acceptance and self-understanding.[92]

On the issue of increasing recreational use of LSD, Hoffer and Osmond adopted a novel approach. Writing to Osmond in November 1965, Hoffer admitted that

> we cannot prevent young people from using hallucinogens. There are too many around. The only logical method is not to make it illegal

but to publicize widely all the dangers inherent in their use, to provide places where anyone who wishes can take them singularly, or in
groups, in controlled settings.... That would quickly remove every
element of rebelliousness from the young people.[93]

Blewett endorsed the idea of a psychedelic centre where the drug could be
administered in a safe and controlled setting. As he later noted,

> the funds and energies that will be expended in seeking to enforce
> prohibition should be applied instead to education and to the estab-
> lishment of legitimate outlets in the form of government-operated
> centers at which the drug would be made available, in controlled set-
> tings, to people seeking treatment or volunteering to act as research
> subjects. In such centers the dangers attendant upon taking LSD
> would be eliminated.[94]

Such an approach, the three men believed, could provide a viable counter,
and possible solution, to the growing black market problem. Hoffer and
Osmond also gave tips to those who were going to take the drug anyway
to lessen the dangers of its use. As Osmond wrote, "if young people take
LSD 25 which I don't advise but we can't prevent them then the least we
can do is to encourage them to take niacin and ascorbic acid for 2–3 days
after.... Having this available would itself I suspect prevent some prolonged reactions."[95]

Hindering efforts to persuade government authorities to back down
from their aggressive legal course were, of course, the activities of Leary.
Osmond and Hoffer were all too aware of the damage that Leary, and
like-minded LSD apologists, could inflict if allowed to run amok. Closely
monitoring the activities of the high priest, the two psychiatrists anticipated what was to come. As Osmond pointed out to Hoffer,

> I don't think he has any intention of giving in.... At the moment he is
> a hero to and commands a loyalty among many thousands of young
> people.... If I have gauged Leary accurately he is much interested in
> revolutionary activity..., a protest and anti-establishment ideology

and this may blend subtly with the draft protest....Psychedelics are the most potent instrument in this sort of setting and the climate of intergenerational friction is certainly set right. [A lot of it] depends how heavy handed the FBI, FDA etc. become and how well organized Leary is in his "non-organization."[96]

Hoffer had his first face-to-face exchange with Leary at an LSD conference at Berkeley in June 1966. As Hoffer recalled, Leary was beyond reproach, driving the student population into a frenzy during the lecture that the two men gave at the university. As Hoffer sarcastically remarked, Leary could have read from a telephone directory, and it would not have mattered; the students would have reacted in the same manner.[97] Shortly after the Berkeley engagement, Hoffer confided to Osmond that

when I first met Leary I told my friends he would organize a new religion....It will be a curious religious group composed of types like Blewett, Leary, Alpert etc.—the problem will be for Leary to develop a code of religious ethics which will stabilize his religion and minimize the undesirable effects....Perhaps he will have to encode the use of nicotinic acid five days a week to prevent LSD toxicity.[98]

Hoffer used the Berkeley stage to advise attendees that, if they were insistent on doing LSD, it would be good to have some niacin on hand in the event of a bad trip, a nugget that eventually found its way into Abbie Hoffman's 1968 book *Revolution for the Hell of It*.[99]

As Hoffer and Osmond predicted, Leary continued down the revolutionary path, catapulting his LSD religion to international attention. Osmond made his fear of Leary's tactics well known both privately and publicly. In a September 1966 *Playboy* interview with Leary, Osmond was quoted as referring to Leary as "Irish and revolutionary, and to a good degree reckless," to which Leary responded "I plead guilty to the charges of being an Irishman and a revolutionary. But I don't think I'm careless about anything that's important."[100] In another attempt to get Leary to reconsider what he was doing, Osmond opted for a more direct approach. Writing to him in December 1966, Osmond took exception to some of the information that his group was circulating, which suggested that LSD might be helpful to a person suffering from schizophrenia. As Osmond emphasized to Leary, "I think you still underestimate the damage that can be done by this....Schizophrenia people don't need more openness but less....They need not be turned on but turned off, taught about their illness, helped to use it [LSD] only then with much caution and concern should [the] possibility of a psychedelic experience be considered with appropriate medical consultation."[101] But Leary was not listening, and the LSD movement continued unabated, as did legal restrictions on the drug.

So Hoffer, Osmond, and Blewett could only write and speak about LSD to all who would listen. Osmond emphasized the importance of understanding the younger generation of psychedelic adventurers: "These young people consider that it is neither possible nor desirable to prevent them from employing these substances in this way, and in fact they are challenging lawmakers, lawgivers and law enforcers to stop them."[102] He admitted that "some of my colleagues hope and indeed believe that this is just a fad which will soon die out. This is possible but I would not bet on it."[103] As Osmond saw it, authorities

> can either suppress psychedelics and punish those who make, distribute and use them or they can seek ways of incorporating these innovations in the mainstream of our society....Wholly different policies must be devised to ensure that safer substances and methods are developed....The worst possible solution would be to prohibit the substances with a ban that did not work.[104]

Blewett reiterated the value of LSD and other psychedelics, for both psychology and humankind, and he echoed many of Osmond's points. Reflecting on the LSD environment at the end of 1966, he wrote that

> agencies of government...have begun attempts to prohibit the use and eliminate the spreading influence of the psychedelics....[It has] long been evident to those working with the psychedelics that suppression will not be successful and that governments should...aim

at education, which would be relatively easy, instead of attempting prohibition, which would be costly and unavailing.[105]

Unfortunately, calls by all three psychedelic researchers, and others like them, for respectful, level-headed discussion, and a more honest and balanced appraisal of LSD by all, fell on deaf ears. There would be no truce over LSD, and the social movement based upon it and governments' responses to it were about to reach an explosive plateau.

SUMMER OF LOVE: PSYCHEDELIA PEAKS

In the so-called Summer of Love in 1967, the psychedelic social movement reached its apex. The drugs, and the concepts associated with them, had filtered into the heart of popular culture and could be seen in art, fashion, film and television, and music. The world's most popular band, the Beatles, whose members had all been turned on to LSD, released the psychedelic-infused *Sgt. Pepper's Lonely Hearts Club Band*, which instantly became the soundtrack to the movement. For many, psychedelics inspired a new way of life, a new social ethos, bound by feelings of unity, peace, love, and understanding. This attitude was exemplified by, and put on full public display at, the psychedelic epicentre in San Francisco's Haight-Ashbury district in 1967.

Most observers would agree that the official launch of the psychedelic revolution was on January 14, 1967, in San Francisco's Golden Gate Park, framed as a "Gathering of the Tribes for a Human Be-In." There had been other psychedelic get-togethers of sorts, some on a smaller scale and others on a grander scale, such as the acid tests, but the 1967 be-in signified an entirely new phase of the movement. Bringing together some of the era's most renowned countercultural icons and psychedelic movers and shakers, the event promised to be a day full of celebration, music, speeches, love, and, of course, psychedelic revelry (a slap in the face of authorities who had declared a ban on LSD in October 1966). As the posters read, "a new nation has grown inside the robot flesh of the old"; those interested in being a part of the revolution need only "hang [their] fear at the door and join the future."[106] The be-in was intended to be, as Lee and Shlain called it, "a spiritual occasion of otherworldly dimensions that would raise the vibrations of the entire planet."[107] The event also signified a certain politicization of the psychedelic movement. "One of the main purposes [of the organizers] was to bring together cultural and political rebels who did not always see eye to eye on strategies for liberation. In effect the goal was to psychedelicize the radical left."[108] Whether people were willing to admit it or not, and whether they liked it or not, LSD and other psychedelics *were* transforming America. For some, the be-in, and events leading up to it, were symbolic of a special time in American history: "Some three or four years of social experimentation, touched off by mass use of LSD, can be credited with having sparked a host of liberation movements. This period changed American attitudes toward work, toward the police and the military, and toward such groups as women and gays. It began our now established concern with consciousness raising and personal growth."[109]

The illustrious host of the be-in was the Haight-Ashbury community. The Haight, as it came to be known, was a quaint, Victorian-style neighbourhood that, in a few years, metamorphosed into the psychedelic capital of the nation. It was, as many would admit, the closest that anything would ever come to a psychedelic Eden, with its own newspaper (the *Oracle*), resident acid rock bands, "psychedelic" shop, "free" store, free medical clinic, and care network (i.e., the Diggers). In 1967, the call went out to those interested to make their way to the West Coast to join what was happening there. And the call was answered. According to various sources, the population of the Haight had swelled within a year (June 1966–June 1967) from 15,000 to between 75,000 and 100,000, many making the pilgrimage to the district for the Summer of Love. The mainstream press flocked to the Haight in droves as well, quickly turning the psychedelic community into a carnivalesque spectacle. With every major media outlet reporting on what was happening in the Haight, and with the strange new breed of cultural rebel, the "hippie" (a term popularized by *San Francisco Chronicle* reporter Herb Caen), much of America and the world became fully transfixed on the psychedelic revolution.

The stories, pictures, and live television feeds of the scene, and the appearance of Haight-like communities in other cities, were enough to shock the American populace. Many considered the gratuitous use of psychedelic drugs, and the lifestyle associated with it, morally repugnant. The spiritual and mystical language employed by many psychedelic movement members and LSD users was equally disturbing to mainstream society. On top of everything else were the physical and mental harms of using psychedelics, regularly cited by the media and opponents of the drugs. Inundated with reports of flashbacks, suicidal tendencies, and deaths brought on by the use of LSD, and stories of hospital emergency rooms overwhelmed by people having bad trips, many American citizens were understandably frightened and appalled. It did not matter that some of the stories were fabrications, as in the story of the six college students from Pennsylvania who reportedly went blind after staring at the sun while under the influence of LSD.[110] The most-quoted reports on the dangers of LSD in 1967 were on the irreparable genetic damage to chromosomes (though follow-up studies revealed the finding to be scientifically invalid and that a cup of coffee, in fact, did more damage).[111]

While the bombardment of medical alerts on LSD seemed to have the desired shock-and-awe effect on many in society, it had the opposite effect on those drawn to the drug, and "LSD acquired the emotional pull and magnetism of the taboo, and as a result…more and more people decided to try the drug."[112] Baumeister and Placidi later observed that "there are indeed genuine risks to the unsupervised use of LSD, and some concern (especially of parents regarding their children) was justified. However, it is clear that the objections to LSD use were greatly disproportionate to the actual dangers of using the drug."[113] The real problem was its use by the mainstream: "Middle America would tolerate minority and ghetto youngsters taking drugs, but LSD was a phenomenon of white middle- and upper-class youth, which made it salient."[114]

The LSD movement peaked during the highly touted Summer of Love, but it began a downward spiral from that moment on, and soon the psychedelic revolution was corrupted, co-opted, and commercialized. The sudden and irrevocable change in the atmosphere was particularly noticeable in the psychedelic heartland of the Haight.[115] Many blamed the demise of the LSD movement on the overexuberance and miscalculations of its members: "Inspired by a few good experiences, the movement's leaders adopted the attitude that the drug would be good for anyone, anytime.

By encouraging indiscriminate use of LSD, the movement evolved into a pathetic caricature of itself and made itself vulnerable to the wrath of the establishment. The mismanaged movement lost its credibility, its leaders were jailed, and it collapsed."[116] Others rightly suspected additional factors, such as negative media publicity or the sinister and hidden underbelly of the movement.[117] The death of the world that the Haight sought to realize was marked by a mock funeral led by the Diggers in the streets of the neighbourhood in October 1967 under the banner "Death of Hippie, Son of Media."

The movement had no place to go but down. After 1967, a series of events sealed its fate. In 1968, President Lyndon Johnson tackled the LSD menace in his State of the Union address, hoping to deal once and for all with what was deemed a threat more dangerous than the atom bomb. Additional changes to the Drug Abuse Act that year made possession of the drug a misdemeanour and sale of the drug a felony. By the end of the decade, LSD would be classified under the Schedule 1 category of the Controlled Substances Act, a label typically reserved for drugs deemed to have no medical value. Although LSD, and the movement surrounding it, would have a few last gasps, such as the Woodstock Festival in 1969, the scene was all but dead, a point hammered home by the senseless violence and mayhem that erupted at another music festival, headlined by the Rolling Stones at Altamont Speedway on the West Coast, the same year. Not long after the Altamont affair, Charles Manson and members of his cult family, all expats from the Haight, were arrested and sentenced for the brutal slayings of actress Sharon Tate and others.

As the LSD movement dissipated, so did mass use of the drug. The combination of scare tactics, anti-LSD propaganda, and stiff legal repercussions for LSD use, it seems, was enough to do the trick. LSD did not disappear, though, but went further underground since few users were willing to risk arrest and imprisonment. And the reasons for taking LSD changed:

> As the counterculture lost its momentum, the point of taking the drug shifted from spiritual exploration to "getting high"—a chemical form of instant gratification, reproducing larger culture's reliance on tranquilizers, alcohol and other sanctioned drugs....Despiritualized

and (after 1966) illegalized, LSD took its place among other potent drugs—cocaine, amphetamines, barbiturates—to which young people resorted principally for recreation, relaxation and escape and to which many became physiologically and psychologically addicted.[118]

According to Baumeister and Placidi, "LSD use subsided not because the arguments of its opponents prevailed but because its competitive appeal vis-à-vis other drugs faded. What distinguished LSD and the psychedelics from other drugs was the potential for self-encounter.... But drug-induced self-encounter was mismanaged and discredited collectively, undermined individually, and rendered obsolete and irrelevant socially."[119]

The jury is still out on whether the LSD movement had social value or not.[120] Whether LSD had scientific and medical value also remains debatable. Regardless, LSD and other psychedelics became symbolic of the turbulent 1960s.

THE PSYCHEDELIC IN-CROWD (PART 3)

Just as 1967 was a peak year for the psychedelic movement, so too it was a peak year for the careers of Hoffer, Osmond, and Blewett. Many both within and outside the psychedelic community continued to look to them as experts on psychedelics, as revealed by their frequent contributions to articles, books, and films on the subject and their participation in countless television and radio interviews, seminars, and conferences. Hoffer and Osmond's massive book *The Hallucinogens* was published in 1967. When the 600-page book was released, it immediately stood out as one of the most comprehensive books on the subject, describing the chemistry, biochemistry, pharmacology, and toxicology of all of the known classes of hallucinogens. LSD figured prominently in the book, with an in-depth discussion of its effects, its uses as a psychotomimetic and in psychotherapy, the psychedelic experience, and its sociological implications. As the two psychiatrists wrote in the preface,

the use of hallucinogens has been described as one of the major advances of this century. There is little doubt that they had a massive impact upon psychiatry and may produce marked changes in our society. The violent reaction for and against the hallucinogens suggests that, even if these compounds are not universally understood and approved of, they will neither be forgotten nor neglected.[121]

Many applauded the book as the definitive work in the field. Reviewing it, world-renowned Harvard University botanist Richard Evans Schultes wrote that "nothing like this book has ever appeared" and predicted that it would "take its place amongst the classics of hallucinogen literature."[122] The book sold many copies and went into second and third printings.

For Blewett as well, 1967 was dominated by the psychedelic issue. Much of his time was spent completing his own book, *The Frontiers of Being* (released in 1969). The book, which took him nearly ten years to complete, presented his thoughts on his years in the field of psychology. *The Frontiers of Being* brought together conventional psychological theories, the collected wisdom of some of the world's greatest philosophers, and the findings of psychedelic research. The result, Blewett hoped, was "a theory of personality based upon concepts and formulations currently in the field but moving beyond them by drawing upon (a) the insights discovered and recorded in the religious writing of the past four thousand years and (b) the remarkable wealth of psychological data acquired through the use of the psychedelic drugs."[123] In many respects, the book was his take on Huxley's *The Perennial Philosophy* with a psychedelic twist. Blewett believed that the psychedelic experience held important implications for personality theory.

Naturally, the psychedelic notoriety of Osmond, Hoffer, and Blewett caught them up in the LSD movement, Osmond perhaps more than the other two. In February 1967, Hoffer mailed Osmond a *Wall Street Journal* clipping on psychedelics with a message to him: "See what your word [*psychedelic*] is up to."[124] The article indicated how interconnected the word had become with the counterculture and mainstream culture in general. As the article pointed out, *psychedelic* was also "developing into a magnetic sales word." It was being used to label all sorts of things. Psychedelic shops were opening, and special night clubs were featuring psychedelic effects. Osmond was apparently taken aback and not sure what to make of the sudden popularity of the word. As he later confessed to his partner, "widespread use of the word probably surprises no one more than me....I had expected it at the most to be an afternote alongside such resonant terms as Lewin's fantastica."[125]

One of Osmond's main concerns at the beginning of 1967 remained the black market trade in LSD and illicit use of the drug in mainstream society. Osmond wanted to get to the bottom of the black market production and distribution of LSD. He warned Jonathon Cole,[126] the chief of psychopharmacology for the U.S. National Institute of Health, that the "alleged LSD on the black market may not be what it claims to be and yet is apparently psychologically active."[127] Osmond then requested Cole's assistance in collecting samples of the illicit LSD and having them analyzed by recognized specialists to determine which other substances they might contain.[128] Help from Cole and other U.S. government authorities never materialized, which exasperated Osmond.

His fear was that professional inaction, and the recent decision to bar researchers from continued studies of the drug, would only result in experts in the field falling behind black market amateurs. Osmond expressed his frustration in a memo of March 5, 1967, to his New Jersey colleague sociologist Frances Cheek:

> The absurd banning of LSD 25 in research at a time when its clandestine manufacture was the problem is an example of how not to do it. This alienated many well-known and able professionals while in no way coping with the problem of clandestine use....This is the worst kind of administrative blunder. The problem is too serious, too complex and too immediate for such clowning.[129]

Osmond opted to do some detective work of his own, at times inserting himself directly into the growing psychedelic movement. Writing to Hoffer on March 13, 1967, he laid out his concerns:

> New substances, combinations of substances, contaminated substances and new settings for use of old substances have become available. In addition they are being used mainly by very young

people....Most of us are quite ignorant as to what exactly is happening....This ignorance probably includes many members of the psychedelic movement....I am hoping to meet one of the big clandestine makers and to find out at least what he has in mind.[130]

Over April and May 1967, Osmond arranged to meet with some of the established and up-and-coming leaders of the LSD movement to keep his finger on its pulse and ascertain what might come next. He recounted some of these anonymous meetings to Hoffer. Detailing his meeting with "the alchemist,"[131] he wrote that he

> claim[s] to have supplied something like 90% of LSD in the US in recent times...through the use of temporary factories. [He] boasts [that] his LSD is purer than that of Sandoz...[and that he is] using/ spreading psychedelics for the betterment of humankind....He is organized, determined and ruthless....[The alchemist] is allegedly the greatest endogenous producer of psychedelics in the US....[His] opinion is that the drugs are not just mind changers but society changers....I think we can expect stirring and lively times ahead.[132]

Following his meeting with "the poet" (Ginsberg?), Osmond noted that he "does not think the alchemist is exaggerating....Clearly, further soundings of this sort are needed and I will attempt to undertake these in coming weeks."[133] Through his meeting with "the scholar," Osmond also came to learn of big plans for the summer in Haight-Ashbury.

Hoffer and Blewett found themselves pulled more into the debate on the psychedelic movement as the phenomenon spread into Canada. Readying himself for an appearance on a CBC film about LSD in March, Hoffer reaffirmed his and Osmond's approach to growing use of the drug on college and university campuses:

> [I] am going to educate people who have had bad trips to use nicotinic acid and vitamin c....Many people may construe this as an invitation to take LSD in the same way they consider the pill an invitation to promiscuity. These are the puritans who are out of touch

with the modern world....We have a sound position when we advise LSD intoxicated people to take nicotinic acid as [an] antidote.[134]

Hoffer discovered the extent of the LSD movement in Saskatchewan on March 20, 1967, when he was interviewed by a panel of three students from the University of Saskatchewan—Regina Campus for the LSD special on CBC TV. He was told by a female student that over five years at least 3,000 students on the Regina campus had taken LSD. As Hoffer wrote to Osmond,

this is 600 per year....If this is true this is 1/3 of the student body there....If she's exaggerating it 5 fold it is still enormous. Yet police issue statements that they know of no illegal use and I have heard of none....[It] must be remarkably discrete....I suspect we are losing the battle for controlled use.[135]

On many occasions, Hoffer was asked for his views on the often-hyped physical dangers of LSD. He typically bristled under the collar when responding to claims that LSD produced as much tragedy as thalidomide. On reading an article published by Abbott Labs on the effects and dangers of LSD, he penned a letter to the medical director, asking that it be published:

I am disturbed that you would publish such an inaccurate and biased account of LSD. Admittedly there are serious dangers when used in an uncontrolled setting but this applies to tranquilizers etc. But [William] Frosch's thesis is that the dangers are so powerful it overwhelms the therapeutic potential and this leads to the inference that LSD, like Thalidomide, should be banned.[136]

Another issue was the reputed genetic damage done to chromosomes by LSD, which Hoffer insisted was unfounded. He was aware of two studies with high-dose, long-term LSD users that found no damage; he also heard that the scientific community had refused at first to publish these reports and only did so after a controversy ensued with the authors.[137] Two additional reports appeared in early 1968, further damaging the credibility of the touted LSD-chromosome connection.

Blewett was drawn into many discussions with faculty members and students regarding the surge in psychedelic use and its effects on campuses and wider society. As he came to see it, the psychedelic movement was part of a larger transformation of society: "The young are no longer satisfied with the hypocrisies of their societies....Young people have led, and will continue to lead, the demand for social justice throughout the world. They consider this to be their world, and they wish and intend to take possession of it while it is still in livable condition."[138] Like other social commentators of the time, Blewett believed that rapid technological progress and a machine-dominated society had resulted in alienation and ultimately a "revolution in individuality" among many, particularly the young: "The degree of dissatisfaction and unrest...became manifest when young people attempted to follow Timothy Leary's advice to 'Turn on, tune in and drop out.' In large part the hippie movement gathered its initial momentum and maintained it as a result of social fears and frustrations from which Leary's dictum seemed to offer an escape."[139] Of course, for Blewett, psychedelics were an important part of societal transformation.[140]

Like Hoffer and Osmond, he was disheartened by the less than factual reporting on the harms caused by LSD, especially purported chromosomal damage: "It is indicative of the level of social anxiety surrounding the psychedelics that the authors of studies showing contradictory findings (that is, an absence of genetic damage) have had difficulty in getting their reports accepted for publication. Comparative data tends to be suppressed."[141] And, like his former colleagues, Blewett thought that "LSD, like atomic energy, is here to stay," so efforts should be directed toward finding the best way to manage it, whether through proper education or establishment of centres for controlled use of the drug.[142]

All three men blamed the campaign of fear over LSD waged by the scientific community, mass media, and government agencies as being largely responsible for provoking social hysteria in the first place; widespread dissemination of semi- or non-factual reports on the drug, as they saw it, only exacerbated social instability. In Osmond's words, "the most effective propagandizers for psychedelics were not those who advocated them...but those who condemned them....It was they who insisted that [the drugs] were forbidden fruits and so of necessity tastier than anything that was not forbidden.... [They] provided much of the impetus for [Leary's] ideas."[143] Blewett believed that authorities' focus only on negative reports on psychedelic drugs was bound to fail: "When the cry of 'Wolf' is too often used inappropriately, it loses its social utility.... It inhibits any process of educating young people on how to deal with these compounds that more and more of them are using."[144] For him, "the social establishment is becoming enmeshed in its own problem. Neither more severe legal sanctions nor harsher enforcement practice will reduce [the] hazards. Such measures will in fact add to them."[145]

Despite such protests, and the hope held by some that LSD in Canada might be saved from the same legal fate as that in the United States, Canadian government authorities escalated their clampdown on psychedelics, eventually following the lead of their American counterparts. In November 1968, Frank Ogden, a therapist at Hollywood Hospital in New Westminster, British Columbia, advised Osmond of a bill before the Senate to prohibit the use of LSD for all purposes, even in psychotherapeutic procedures, still used by recognized medical professionals (e.g., Hoffer) in Canada. Osmond immediately sent a letter to local BC Member of Parliament Douglas Hogarth, telling him that the move was ill-advised, particularly given recent research in Spring Grove (see Chapter 7) that confirmed earlier work in Saskatchewan and British Columbia:

> [It] would appear to me to be both absurd and unfair to prevent a
> useful and safe treatment being used, simply due to [a] state of public
> panic.... It should be perfectly possible to find simple ways for
> controlling use of this substance, and similar substances, and so not
> depriving science of a valuable tool and medicine of a useful...and
> now apparently proven treatment.[146]

Last-ditch attempts by Hoffer, Osmond, Blewett, and others to persuade the federal government to readjust its course failed. By 1969, possession of LSD without government authorization was a criminal offence. Soon enough the use of psychedelics as instruments in both scientific research and medical treatments in Canada would be shelved completely.

PSYCHEDELIC OBITUARY

From the end of 1967 on, authoritative statements disputing the value of LSD became the norm, and most professionals eventually agreed that the use of LSD and other psychedelics in scientific research and medicine was a dead issue, in spite of lingering evidence that supported its continuation. As Charles Grob noted, "a source of embarrassment and shame, hallucinogenic research became a non-issue, virtually disappearing from the professional literature and educational curriculums. By the early 1970s, psychiatric researchers and academicians had perceived that to continue to advocate for human research with hallucinogens, or even to be identified with past interest in their therapeutic potential, might seriously jeopardize their future careers."[147] In the United States, clinical studies testing the therapeutic potential of LSD had more or less come to a standstill by 1970.[148] In Canada, University of Toronto psychiatrist S. J. Holmes summed up the feeling of many in the Canadian medical community when he said in 1970 that "the curative powers of even a single dose of LSD...for alcoholism have proved to be disappointing. Follow-up experience and experiments with controls have shown no real benefits from this risky therapeutic procedure."[149] In its final report, the Le Dain Commission similarly noted that "the general medical effectiveness of LSD has not been adequately demonstrated."[150]

No doubt the LSD movement influenced scientific debate and killed any remaining possibility of the continuation of psychedelics in scientific research and medicine. The movement intensified, and in many ways obfuscated, debate on the drug, as played out in many scientific and medical journals of the time. On the subject of LSD, scientific, sociocultural, and political lines became blurred. A case in point was a 1967 review of Harold Abramson's *The Use of LSD in Psychotherapy and Alcoholism*; although the reviewer was "not that knowledgeable of LSD aside from reports...read in newspapers...regretfully I must report that except for a few highs it was mostly a bad trip."[151] His main concern was the "diminished professional judgment shown by so many of the contributors"; indeed, "a number of the reactions attributed to the drug find their expression in the therapist, to the extent I began to feel that in many cases the patient is given LSD in order to liberate the therapist, to enable him to become, in effect, a hippie in white."[152] As public fears about LSD increased and the movement kicked into high gear, even more moderate opinions on the scientific potential of the drug were called into question. Psychedelic investigators who remained in the field, and continued to support its use, soon came to be equated with the most extreme pro-LSD voices (e.g., Leary) and hence without morals or scruples. Recalling the sudden shift, American psychedelic pioneer and Californian psychiatrist Oscar Janiger commented that he had come to be perceived as "a villain who was, you know, trying to seduce people in taking it. It was absolutely bizarre! From the heroes, we were suddenly some creatures who were seducing people into changing their consciousness."[153]

What if the LSD movement had not materialized? Would its absence, and that of the illicit use of LSD and figures such as Leary, have made any difference to use of the drug in the scientific community? It is difficult to say with certainty what might have happened, especially given the additional variables at play (e.g., the CIA), but it is plausible that the use of LSD in research and even clinical practice might have continued. And what of research on its more psychedelic applications, specifically the psychedelic model of therapy? By the end of the 1960s, the debate about whether the psychedelic model, and psychedelic therapy for alcoholism, held scientific value remained at a crossroads, with evidence both for and against its use. This deadlock would likely have been the case had there been a psychedelic social movement or not.

The whole psychedelic approach, with its emphasis on spirituality, mysticism, and transcendence, had always been contentious within scientific ranks.[154] From the mid-1960s on, the odds were stacked against psychedelic approaches to the scientific use of LSD. In its 1966 position statement on LSD, the APA called for firmer regulations on and controls of the drug in the face of growing illicit use, but it did not recommend stopping scientific research with the drug, even though it was less comfortable with its psychedelic connotations: "Indiscriminate consumption of this hazardous drug can, and not infrequently does, lead to destructive physiological and personality changes....The association most particularly deplores the use by some persons in this way as a 'mind-expanding' or 'consciousness-expanding' experience."[155] Similarly, the head of the FDA declared any notion of psychedelic-inspired creativity and problem solving or other benefits to be "pure bunk."[156] Many within the scientific community undertaking different work with the drugs also blamed the more psychedelically inclined investigators for the social hysteria that occurred and its result, the complete ban on research with LSD.

Although Hoffer, Osmond, and Blewett were respected by many as pioneers in the field of psychedelic drugs, with the advent of the psychedelic social movement things changed considerably for them as well—and not for the better. Their outspokenness on issues related to LSD (its illicit use, its dangers, government regulations of it, or the psychedelic movement) struck a chord with some but became the target of scorn for others, no doubt because it challenged and often ran counter to the official lines of governments and their various agencies. As 1967 drew to a close, Hoffer admitted to Osmond that "there are no moderates left in the field....We now only hear extremist views....We will now find the major confrontation between the public who insist they have [a] right to use drugs and extremist physicians who jump from isolated events to broad conclusions."[157]

Like other professionals who remained openly committed to the idea that psychedelic drugs were beneficial to society, Hoffer, Osmond, and Blewett became marginalized in their respective fields. Hoffer and Osmond became personae non gratae in the psychiatric community, thorns in the sides of organizations such as the APA for their controversial theories on schizophrenia and their orthomolecular treatment of it. Their views on LSD only added to their less than scientific status in the eyes of many medical professionals and made it easier for them to write off their work. Some, like Sankar, made a direct parallel between the approaches of Leary and the two psychiatrists, writing that, just as Leary was "regarded with the same special love and respect as was reserved by the early Christians for Jesus, by the Moslems for Mohammed, etc...[so, too,] the partial magnetism of a leader is also perhaps responsible for the miraculous psychiatric cures effected by some of our orthomolecular psychiatrists."[158] One might surmise from statements such as these that Hoffer, Osmond, and Blewett's popularity had to do more with their guru-like qualities and anti-establishment stances than with their groundbreaking scientific research. But soon they and their work fell victim to the growing social hysteria over psychedelic drugs. In one of the more ridiculous instances, pictures of Hoffer and Blewett appeared in the *National Enquirer* under the caption "How American virgins are being sacrificed to the new sex drug smuggled from Red China."[159]

In many ways, then, the three men, in both scientific and social circles, came to be seen as acid quacks or hippies in white. One could argue that to some extent they contributed to their negative reputations simply because of unwavering commitment to their views and some of the people with whom they associated. Since all three were members of the psychedelic "in-crowd" and openly admired and fraternized with notorious mystics, outlaws, and philosophers such as Hubbard and Huxley, it became easier for critics to discredit them and their work. But even when they distanced themselves from people like Leary, they were still considered part of the lunatic fringe of the psychedelic movement.

So Hoffer, Osmond, and Blewett and their research in Saskatchewan with psychedelic drugs became tinged with countercultural references and all of the myths and misconceptions therein. Unfortunately, serious scientific questions about them and their work were obscured and, to a large extent, remain unresolved.

EPILOGUE

More than half a century has passed since the conclusion of the decade-long research program in Saskatchewan with psychedelic drugs, and Abram Hoffer, Humphry Osmond, and Duncan Blewett have all passed on now. Many people have never heard of them and know nothing of this episode in the history of psychiatry or that of Saskatchewan. Among those who do have some knowledge of it, opinions have been mixed over the years. In many ways, the research and the researchers remain controversial.

By the early 1970s, psychedelic research had ground to a halt, a casualty of both the transformations occurring in psychiatry and the countercultural revolution and explosive LSD social movement in the 1960s. Osmond and Hoffer returned to trying to establish their theories on and treatment methods for schizophrenia.

Following his tenure as director of New Jersey's Bureau of Research in Neurology and Psychiatry, Osmond spent his remaining professional years as a ward psychiatrist in Tuscaloosa, Alabama, and a University of Alabama Medical School faculty member, finally settling into retirement with his family in Appleton, Wisconsin, where he passed away on February 6, 2004. He was remembered as "a roman candle of new ideas," "a guru of the psychedelic movement," and "a grandfather of environmental psychiatry."[1]

Blewett remained in his position in the Department of Psychology at the University of Regina until his retirement in 1986, whereupon he moved to Gabriola Island on the West Coast with his wife, occasionally travelling to the Department of Psychology at the University of West Virginia as a visiting professor in humanistic psychology in the 1990s. Blewett passed away on February 24, 2007. In a glowing tribute to the psychedelic "trickster, magician, and alchemist," colleagues Larry Schor and Mike Arons wrote that, "for half a century," Blewett "led the expedition toward the further reaches of human consciousness. He truly believed that, one day soon, people of the world would live in peace and harmony. He dwelled in the space between what is and what is possible, and he always leaned toward the latter."[2]

Hoffer maintained his private practice in Saskatoon until 1976 and then moved to Victoria, where he continued to see and treat patients until his retirement in the early 2000s. On May 27, 2009, Hoffer passed away at the age of ninety-one. *Globe and Mail* columnist Sandra Martin referred to him as a "psychiatric contrarian, a maverick who waged a scientific war of attrition against the medical establishment and Big Pharma Goliaths; he will be remembered by many as an innovator and a founding father to naturopathic medicine and the health food movement."[3]

Nonetheless, Osmond, Blewett, and Hoffer have had outspoken critics. Although some have conceded that the Saskatchewan researchers produced some incredible findings, the verdict has been that their accomplishments ultimately failed to stand up to serious scientific scrutiny. At the more extreme end of opposition, their theories and experiments continue to be held up as examples of quackery, the men themselves as pseudo-scientific eccentrics and psychedelic cultists.[4]

So what conclusions can now be made about their research with psychedelic drugs half a century ago? More specifically, what conclusions can be made about the research as an episode in the history of science? Scholarship on the subject is not as sparse as it was in past years. Since I began work on this book, analyses of the province's psychedelic experiments, and the roles of the primary researchers, have grown considerably. Two of the biggest contributors to this evolving historiography, Erika Dyck and John Mills, have weighed in on opposite sides of the debate over the scientific value of the research.

For Dyck, who has provided the most thorough examination of the research, the psychedelic work spawned discoveries that deserve a place in psychiatric and psychopharmacological history. However, her analyses have been read by some as suggesting that the psychedelic innovations were ephemeral in their scientific importance.[5]

Mills has been much more explicit in his negative assessment of the Saskatchewan research: focusing on the theories of schizophrenia espoused by Hoffer and Osmond, he believes that their research largely failed the test of science. He even resorts to what might be construed as character assassination in his analyses to further discredit their research, revealing a particular disregard for Hoffer as a "remorseless advocate" of his own views:

> Hoffer pursued his programmes with the most ruthless disregard for the opinions of others....Above all he consistently attempted to discount the considerable body of evidence demonstrating that nicotinic acid therapy was ineffective....He retreated into a world of his own, with an idiosyncratic diagnostic system and therapeutic method. As a result he alienated himself not just from Saskatchewan's medical fraternity but even from his fellow PSB psychiatrists.[6]

Discussing the empirical evidence brought against the work of Hoffer and Osmond, Mills insists that "most of those familiar with [their] work would say that they were eager to promote the merits of their approach and were therefore ideologically rather than medically or scientifically driven."[7]

I have tried to show in this book that, though the two psychiatrists might have been subject to ideological whims and engaged in the occasional ideological battle, the same could be said of their scientific opponents. Moreover, I have argued that their work was indeed medically and scientifically driven. Mills takes the studies of their scientific adversaries at face value, downplaying, and in some cases missing altogether, some of their more blatant methodological shortcomings and personal biases. He thus fails to put these studies on the same level of scrutiny as the Saskatchewan research, while also neglecting to cite research in support of the latter. Mills is quick to castigate Hoffer and Osmond's unconventional diagnostic methods (e.g., the HOD Test) and their rumoured use of non-schizophrenic patients in many of their successful treatment outcomes, yet he does not even touch on some of the more questionable aspects of the CMHA studies, such as when they used chronic schizophrenic patients in purporting the inefficacy of niacin in treating acute schizophrenic patients.

This book has also attempted to demonstrate the revolutionary nature of the psychedelic experiments conducted by Hoffer, Osmond, and Blewett. Many within mainstream psychiatric, psychological, and larger scientific communities would not concur with such a view. An underlying question in this book has been what, exactly, constitutes legitimate science? As Paul Feyerabend maintained, "the history of science, after all, does not just consist of facts and conclusions drawn from facts. It also contains ideas, interpretations, mistakes, and so on."[8]

So, if we put aside the popular scientific and sociocultural myths and misconceptions about the Saskatchewan research and its three pioneers, it would be hard to ignore how much they accomplished from its beginning in 1951 to its rather abrupt end in the early 1960s. Chief among their achievements was the betterment of mental health in the province. During their tenure with the PSB, Hoffer, Osmond, and Blewett contributed greatly to improving an abysmal mental health situation. Of course, they did not do this alone and were among a host of others—such as Griff McKerracher, T. C. Douglas, Stan Rands, and F. S. Lawson—whose dedication and perseverance deserve just as much credit. Usually not mentioned in analyses of this period in Saskatchewan history, however, is the important role of psychedelic drug research in the transformation of mental health and how individuals such as Osmond, Hoffer, and Blewett made much of this change possible. Because of their willingness to take risks when and where no one else would, they made unique discoveries.

The work on schizophrenia resulted in the first specific biochemical theory of the disease, the adrenochrome hypothesis. Hoffer and Osmond proved that adrenochrome, and a similar adrenaline derivative, adrenolutin, could be classified as psychotomimetics, synthesized the first pure samples of these substances in the lab, invented new diagnostic tests that could be used to distinguish schizophrenics from other types of functional psychoses (the mauve factor and HOD), and originated a new form of treatment (orthomolecular) for mental illness. The model psychosis experiments also influenced mental hospital design and led to a host of multidisciplinary studies examining the world of the mental patient.

Additionally, Hoffer, Osmond, and Blewett pioneered the psychedelic concept, helping scientists to understand the human psyche. They can be credited with mapping out much of the psychedelic experience when little was known about it. And their successful use of LSD as a therapeutic adjunct, especially in treating alcoholism, was a watershed achievement. The LSD program in Saskatchewan became one of the first of its kind to graduate from experimental stage to preferred method for the treatment of severe alcohol addiction.

Although many observers of the psychedelic research in Saskatchewan recognized the above accomplishments, they were unwilling to concede that psychedelics had any utility or validity in actually treating mental illness or addiction, or any wider applicability for humankind. Detractors repudiated claims that adrenochrome is present in the human body and might play a role in the etiology of schizophrenia, that niacin is an effective treatment for schizophrenia and safer than tranquilizers/neuroleptics, and that psychedelic treatments produce lasting positive results.

Were the Saskatchewan findings simply too good to be true? Many detractors of the work point to the research of the NIMH, and of prestigious scientific researchers such as Seymour Kety and Julius Axelrod, as well as the niacin therapy studies carried out by the CMHA, as evidence of the non-scientific status of the work on schizophrenia carried out by Hoffer and Osmond. In the course of my research, I came across several former patients who maintained that they had been treated successfully with the megavitamin approach; hundreds of case histories and testimonials left behind by other former patients, as well as documentaries on the subject, also seem to confirm the effectiveness of the theories and treatment approaches espoused by Hoffer and Osmond.[9] Obviously, such anecdotal reports will never suffice in the high court of scientific opinion, nor should they, but neither should they be completely discounted. Moreover, the results obtained by Osmond and Hoffer with niacin and ascorbic acid received considerable support from two-time Nobel Prize winner and inventor of the term "orthomolecular," Linus Pauling; Hoffer's finding in 1955 that niacin was an effective anti-cholesterol agent was confirmed by the world-renowned Mayo Clinic; and several orthomolecular member countries have appeared since their early findings, with practitioners in India, Korea, Japan, Sweden, Finland, the United States, Canada, and elsewhere. Again, these are remarkable achievements even if they cannot be regarded as hard scientific proof of the theories or treatment approaches in question.

Psychiatry has come a long way since the days of the Saskatchewan research, with marked advances in understanding and treating mental illness. Magnetic resonance imaging (or MRI) techniques have enabled researchers to map the brain and measure its activity in ways unimaginable in the 1950s and 1960s.[10] Clinical techniques and methods have matured. New medicines have proliferated. And theories and approaches once dominant have fallen by the wayside.

Yet the debate over schizophrenia, its etiology, and its treatment remains fierce. Today's thinking on the disease remains firmly grounded in the concept of a chemically imbalanced brain and the use of pharmacological substances to treat that imbalance. The model guiding research on and treatment of schizophrenia is centred on the neurotransmitter dopamine, in spite of research in the past number of years that has raised serious questions about its assumed scientific validity and whether or not the treatment prescribed produces better outcomes.[11] Hewitt has shown how Woolley's serotonin hypothesis provided a rich source for subsequent pharmacological, psychiatric, neuroscientific, and other research. She also acknowledges how this research has been immensely profitable to pharmaceutical companies, how it "in turn produced many medications, but few answers and no cures."[12]

Perhaps, then, the adrenochrome hypothesis and the orthomolecular approach to treating schizophrenia and other mental disorders might still have something valuable to offer. Many critics have been steadfast in their disavowal of the orthomolecular school of thought, pointing to randomized controlled trials as the true test of the scientific validity of a treatment, a test that the Saskatchewan findings on schizophrenia apparently failed. Unfortunately, the chances of large-scale trials of niacin therapy occurring again are slim, as Hoffer's son John, a specialist in internal medicine, admits, and "the validation or refutation of...orthomolecular therapy will eventually come in the form of fundamental breakthroughs in our understanding of the biology of schizophrenia."[13]

Yet evidence in other areas of research lends credence to the findings of Hoffer and Osmond. Their former colleague John Smythies emphasized in 2002 that oxidized metabolites of catecholamines (e.g., noradrenochrome and nordopamine) have been shown to occur in the human brain, which leaves a large number of unanswered questions in the neurological field. He stated that empirical evidence supporting the adrenochrome hypothesis now exists, so both detection of and fuller understanding of the action of adrenochrome in the brain might not be that far off. Smythies believed that adrenochrome was a proven psychotomimetic and that the hypothesis had untapped potential:

> The recent evidence that adrenochrome may occur in strategic areas of the brain related to anxiety and to basic limbic functions suggests that further research in this area is indicated....Very little is known about the basic neuropharmacology of adrenochrome....An enormous amount of attention has been paid to the dopamine and noradrenergic systems of the brain—very little to the adrenergic system. This imbalance needs to be corrected.[14]

Recent research has also added to scientific knowledge of niacin in brain functioning. Much has been written about the blunted niacin flush response in individuals with schizophrenia, a phenomenon, some suggest, that provides a valuable diagnostic test of and biomarker for the disease.[15] Additionally, Johns Hopkins University researchers have revealed some of the operation of niacin in the neural pathways of postmortem schizophrenic brains and suggested that "early clinical studies by Hoffer...should now be reevaluated in the context of the limitations imposed by a deficient receptor."[16] It is too early to say whether these developments suggest that the adrenochrome hypothesis and orthomolecular treatment for schizophrenia will garner further acceptance and use within mainstream medical and psychiatric communities, but there are promising signs: "Although...adrenochrome has not been measured and compared between schizophrenic, non-schizophrenic and normal control groups of population, we can draw conclusions in favour of niacin and hence, indirectly for the adreno-chrome hypothesis, from Hoffer's studies and reports that spanned many decades."[17] Other researchers are revisiting the possibility of niacin augmentation to assist in the treatment of certain subsets of patients with schizophrenia.[18] If Hoffer and Osmond's theories on schizophrenia are indeed revealed to have scientific merit, as some of the recent research seems to suggest, then they could have profound implications not only for how we view psychiatric history but also for the present and future of psychiatry. Given the somewhat shaky ground on which psychiatry once again finds itself, it will be interesting to see which developments occur with challenges to the reigning biological paradigm.

What about psychedelic therapy and the claims made about LSD as an effective adjunct in the treatment of alcoholism? The idea of spiritual transcendence evoked by psychedelic therapy left a sour taste in the mouth of many a scientist, and it still rankles many scientific and medical authorities today. Blewett was frank about the matter in 2002, writing to me that "it is still the case that the areas of spiritual awareness and spiritual development have remained foreign to the thinking of most psychologists and social scientists." He continued to believe, however, that in time there would be "a scientific rationale for the study of self and social awareness and responsibility."[19] It is too bad that he did not live to see that day.

Over the past two decades, there has been a return to animal and human testing of the effects and applications of not just LSD but also older and newer psychedelic substances (e.g., ayahuasca, psilocybin, ibogaine, MDMA,[20] and ketamine). Non-profit organizations such as the Multidisciplinary Association for Psychedelic Studies (MAPS) and the Heffter Institute have done much to promote education on the drugs and secure regulatory approval for their use "in thoughtfully designed and carefully conducted scientific experiments."[21] Model psychosis experiments have resurfaced, the drugs now being used to explore neurotransmitter systems. Psychedelic-assisted therapy has also had a revival of sorts in the scientific community, and there are promising new studies of the drugs and their uses in the treatment of alcoholism and other addictions, post-traumatic stress disorder, cluster headaches, treatment-resistant depression, and anxiety in cases of terminal cancer.[22] Psychedelics, as UK psychiatrist Ben Sessa has noted, "have reentered the mainstream with impressive force, pushing themselves to the forefront of contemporary brain research."[23]

Of course, resistance from orthodox medical and psychiatric circles to the gradual resurgence of psychedelic therapy remains considerable, but there is more openness "towards a science of spiritual experience" not seen since the early 1960s.[24] Another indication of the psychedelic comeback within the scientific community has been the plethora of psychedelic-themed international academic forums—among them the 2013 Psychedelic Science and 2016 Horizons conferences in the United States, and the annual Breaking Convention and Beyond Psychedelics conferences in Europe—which have brought together thousands of professionals from around the world to discuss the role of the drugs in science, healing, culture, and spirituality.

In Canada, too, medical professionals are beginning to reassess the potential of psychedelic-assisted therapy.[25] As reported in a 2015 *Canadian Medical Association Journal* article, "continued medical research and scientific inquiry into psychedelic drugs may offer new ways to treat mental illness and addiction in patients who do not benefit from currently available treatments. The re-emerging paradigm of psychedelic medicine may open clinical and therapeutic doors long closed."[26] The most notable Canadian endeavour in the field in recent years is that of Vancouver physician and addictions specialist Gabor Maté, whose work with Peruvian indigenous shamans to treat local addicts with the hallucinogenic Amazonian ayahuasca is attracting international acclaim.[27] Maté believes that "complex unconscious psychological stresses underlie and contribute to all chronic medical conditions, from cancer to addiction, from depression to multiple sclerosis. Therapy that is assisted by psychedelics, in the right context and with the right support, can bring these dynamics to the surface and thus help a person liberate themselves from their influence."[28]

Undoubtedly, the subject of psychedelics will remain tinged with references to the "swinging '60s," hippies, acidheads, and Leary-like LSD gurus. These references will obfuscate discussions on the drugs' scientific value. Or are the scientific community and Western society at a turning point? Will they move beyond the view of psychedelics as either substances of danger and abuse or means to a new world order? Although psychedelics virtually disappeared from scientific labs and medical clinics by the 1970s, their recreational use has continued. A recent American survey estimated the use of drugs such as LSD to be about 22 million people (9.1 percent of the population).[29] With the recent spurt of research, there might be more scientific credibility attached to claims about the value of such drugs. As San Francisco physician David Smith and his colleagues report, "LSD is not the poison it has been pictured as by the U.S. judicial and medical establishment. What has been lost for the last 45 years is the opportunity for a tremendous amount of research on LSD and other psychedelics' potential for ameliorating brain disease."[30]

Psychedelics, then, have been given a second chance in the scientific world. Sensationalized media reports on the drugs will continue to surface, such as the recent one about the hospitalization of a Florida family poisoned by LSD-tainted meat from Walmart,[31] and they will spark the moral outrage and fearmongering that appeared in the 1960s. Psychedelics will continue to have claims both for and against their medical and scientific value.[32] Yet there is now an opportunity for both the scientific community and society in general to deal with the matter differently, learning from previous excesses, avoiding the same mistakes going forward, understanding the more spiritual elements associated with the drugs, and coming to grips with their wider social use.

As Sessa sees it, "we need a new way forward, a renaissance as some are calling it, a reformation of the relationship between the people using these drugs, their society, science and the greater spiritual consciousness. This is the future. We are going back to it."[33] Researchers can now pick up where the psychedelic investigators of the 1950s and 1960s left off, testing hypotheses, refining scientific methods, and exploring new avenues. For the true potentials of these consciousness-altering substances to be realized, we need to proceed in an intelligent, calm, collaborative, and safe manner. As Sessa notes, "doctors need randomized, double-blind, placebo-controlled clinical studies, not reports of anecdotal drug experiences, no matter how convincing they may seem to the users."[34]

A number of significant hurdles will have to be overcome, such as "convincing a profession funded by the pharmaceutical industry to spend millions on researching and developing a therapy that could end up challenging the need for anti-depressants."[35] Until now, psychedelic projects have moved forward only because they have "fit into the conservative paradigm," and tamed references to the "mystico-spiritual elements" of the drugs, but the medical model alone is insufficient.[36] Langlitz has pointed out that much of the contemporary hallucinogenic drug research has been more in the vein of Kuhn's "normal science" than a paradigm shift because most research projects have fit themselves into established, mainstream, neuroscientific/psychopharmacological research models instead of "overtly disputing the status quo."[37]

As Sessa argues, if psychedelic therapy is to be effective, "the full healing element of the experience" has to be embraced; "a watered down, capsulated version will not suffice....All healing substances used by all healers ought to have equal footing in medicine. Doctors may still have something to learn from the hippies."[38] He even suggests that progression in the field might require a change in semantics:

> We need a new language with which to talk about the psychedelic experience and perhaps...a new name for these compounds. Psycholytic, entheogen, or entactogen are all viable alternatives to the now too-negatively-biased psychedelic. These substances are not recreational drugs; they are medical agents, pharmacological compounds designed in the main part in laboratories by and for the medical profession. That is where they started and that is where they deserve to return. We owe that much to the population of patients with intractable mental health disorders who may benefit from their effects.[39]

Echoing the advice proffered by Osmond, Hoffer, Huxley, and others before him, Sessa emphasized that "everything possible must be done to avoid the past promises of chemical utopia.... Those of us who see benefits of psychedelic drugs have a much better chance of infiltrating into mainstream consciousness if we adopt a cautious approach."[40]

In light of these recent developments in psychedelic therapy, what can be said of the Saskatchewan research many decades ago with LSD in the treatment of alcoholics? At the end of the 1960s and into the 1970s, there was a sizable body of research both for and against the efficacy of LSD-assisted therapy for alcoholism. Further testing of the claims from Saskatchewan and Spring Grove Hospital in Maryland did not materialize for the reasons cited above. Thus, LSD-assisted therapy for alcoholism faded into history. Today most psychedelic proponents have some familiarity with the Saskatchewan research, and Hoffer, Osmond, and Blewett have been credited as pioneers in the field, though some practitioners still lack in-depth knowledge of their contributions. In the 1990s, psychiatrist John Beresford referred to the *Handbook for the Therapeutic Use of Lysergic Acid Diethylamide-25* as an "unjustly neglected and now lost masterpiece."[41] With the psychedelic approach now being dusted off and used once again, further testing and validation of the findings of the Saskatchewan researchers might be forthcoming.

Saskatchewan had one of the highest rates of success in the world for treating alcoholism. Yet, as with any other treatment, some cases failed, and some succeeded. Among those treated with LSD, many experienced positive changes to their lifestyles that lasted for many years. Yet Kenneth Bell attempted to dispel any notion that the treatment was successful when he interviewed some former patients and their family members and brought to attention one of the more unfortunate episodes in Saskatchewan, the suicide of one of the patients involved in the LSD program. Bell gathered that "few found [the experience] pleasant or beneficial."[42] Other patients reported that psychedelic drug experiences were transformative. Some still credit their LSD treatment and the dedication of people like Hoffer, Osmond, and Blewett for helping them to maintain sobriety. As one patient recalled, "I had an experience that has been with me till today. I've never had a drink since. And it turned my life around completely."[43] One weakness of the Saskatchewan program was its lack of long-term follow-up of patients, which makes it almost impossible to reach conclusions about the treatment. The question of whether the successes were lasting might never be fully answered.

Norwegian researchers recently conducted a meta-analysis of all known controlled trials of LSD treatment for alcoholism.[44] In comparing the various studies (which included those by the Saskatchewan researchers), they indicated that, methodological inconsistencies and weaknesses aside, there is enough evidence to support the use of LSD in treating alcoholism. Indeed, the "effectiveness of a single dose of LSD compares well with [the] effectiveness of daily naltrexone, acamprosate, or disulfiram."[45] Moreover, "as an alternative to LSD, it may be worthwhile to evaluate shorter-acting psychedelics, such as mescaline, psilocybin, or dimethyltryptamine."[46] Perhaps, then, the scientific psychedelic trip is not over yet.

The psychedelic research in Saskatchewan half a century ago was indeed revolutionary. Many professionals and laypersons alike have believed that the research, though full of promise and good intentions, produced little of value. I am inclined to side with a more optimistic appraisal. That the research has not been picked up by the mainstream scientific community, that it has not reached paradigmatic status as "psychedelic psychiatry," or whatever else one might wish to call it, should not be equated with failure.

At the height of the research, Osmond was acutely aware of the resistance and hostility that he and his Saskatchewan colleagues faced as innovators in such a controversial field. Fondly quoting sage advice from Carl Jung, Osmond wrote to Hoffer that "'a sheep that is two yards ahead of the flock is a leader, but if he is 100 yards ahead then he becomes a wolf.'...We are nearer 100 yards ahead than 2 yards and this leads to much agitation among the sheep."[47] Arrogant as this claim might sound, it is an accurate reflection of what the Saskatchewan researchers were doing and what they had to contend with as a result. It also reflects Vine Deloria Jr.'s point about "the rank and file of academia [or those in medicine] usually [being] a generation behind the original thinkers within its peer group."[48] Hoffer, Osmond, and Blewett were just such "original thinkers." Perhaps the new generation, following finally in their footsteps, will understand that they were indeed psychedelic revolutionaries.

NOTES

INTRODUCTION

1   Archer, *Saskatchewan*, 301.
2   http://www.cbc.ca/news/canada/saskatchewan/did-you-know-saskatchewan-is-psychedelic 1.3120921.
3   See Crockford, "Dr. Yes"; Littlefield, *Hofmann's Potion* and *Feed Your Head*; MacDonald, "Peaking on the Prairies"; and McLennan, *Psychedelic Pioneers*.
4   Bell, *Acid Tests*.
5   For a description of the psychedelic drug scene on Regina's campus in the 1960s and the reaction of Hoffer and Blewett to student experimentation with the drugs, see Pitsula, *New World Dawning*, 211–18.
6   Although *Feed Your Head* includes an overview of the Saskatchewan psychedelic research, the focus is on the adrenalin-based theories of Hoffer and Osmond concerning schizophrenia and their role in establishing megavitamin (or orthomolecular) chemotherapy to treat it and other mental illnesses.
7   Bell, *Acid Tests*.
8   See Dyck, "Flashback," "Hitting Highs at Rock Bottom," "Prairies, Psychedelics, and Place," and *Psychedelic Psychiatry*.
9   See Mills, "Hallucinogens as Hard Science" and "Lessons from the Periphery."
10   See De Bont, "Schizophrenia, Evolution, and the Borders of Biology"; Donaldson, "Psychomimesis"; and Edginton, "Architecture as Therapy." Historiography in this area has also been enriched by the recent contributions of a new group of Saskatchewan historians writing on psychiatry and psychedelic drugs in mainstream society and culture more generally. See Elcock, "The Fifth Freedom" and "From Acid Revolution to Entheogenic Evolution"; Montgomery, "Book Reviews, *Neuropsychedelia*"; and Richert, "Therapy Means Political Change, Not Peanut Butter."

11   The evidence-based medicine (or EBM) movement has been defined as one that "purports to eschew unsystematic and 'intuitive' methods of clinical practice in favour of a more scientifically rigorous approach." Goldenberg, "On Evidence and Evidence-Based Medicine," 2621.

12   See http://www.theglobeandmail.com/life/health-and-fitness/health/ psychedelic-drugs-may-be helpful-in-treating-addiction-anxiety-and-ptsd/ article26249575/; http://www.newyorker.com/magazine/2015/02/09/trip-treatment; and https://soundcloud.com/cmajpodcasts/141124-ana.

13   Langlitz, *Neuropsychedelia*, 240. Also see De Gregario et al., "D-Lysergic Acid Diethylamide (LSD) as a Model of Psychosis"; and Lieberman and Shalev, "Back to the Future."

14   Parsons, "The Institutionalization of Scientific Investigations," 7.

15   Feyerabend, "Democracy, Elitism, and the Scientific Method," 218.

16   Latour and Woolgar, *Laboratory Life*, 11.

17   Koyré, "Commentary on H. Gueclac's 'Some Historical Assumptions of the History of Science,'" 848.

18   Ibid.

19   Mackenzie and Barnes, "Scientific Judgement," 191.

20   Tullock, *The Organization of Inquiry*, 52.

21   Shapin, "Why the Public Ought to Understand Science-in-the-Making," 28.

22   Ibid., 28–29.

23   Shapin, "How to Be Antiscientific," 105–06.

24   Polanyi, *Personal Knowledge*, viii, 3.

25   Kuhn, *The Structure of Scientific Revolutions*, 1.

26   Ibid., viii.

27   Ibid., 24.

28   Ibid., 92.

29   Ibid.

30   See Barnes, *T.S. Kuhn and Social Science*, 120.

31   Ghaemi, "Paradigms of Psychiatry," 619.

32   Baumeister and Hawkins, "Continuity and Discontinuity," 200.

33   Tart, "Science, States of Consciousness, and Spiritual Experiences," 18.

34   Bloor, *Knowledge and Social Imagery*, 2.

35   Scull, "Somatic Treatments and the Historiography of Psychiatry," 12.

36   Whyte, *The Organization Man*, 7.

37   Ibid.

38   Ibid. 4.

39   Ibid. 12.

40   Ibid., 205.

41   Ibid., 206–07.

42   Whyte, ibid., 205, describes one study conducted in the United States in the 1950s in which it was estimated "that of the 600,000 people engaged in scientific work...no more than 5,000 [were] free to pick their own problems."

43 For a discussion of the potential implications of the increasing collectivization and bureaucratization of scientific research, see Hagstrom, "Traditional and Modern Forms of Scientific Teamwork," 241–42.

44 Polanyi, "The Autonomy of Science," 16.

45 Hagstrom, "Traditional and Modern Forms of Scientific Teamwork," 262.

46 For analyses of this era in psychiatry, see Grob, *From Asylum to Community*; Scull, "Somatic Treatments and the Historiography of Psychiatry"; and Shorter, *A History of Psychiatry*.

47 As Scull revealed, "there were more than a million admissions to US hospitals in the war years for neuropsychiatric problems. Among combat units in the European theatre in 1944, admissions were as high as 250 per thousand men, an extraordinary percentage. . . . The surge in the ranks of the psychiatrically impaired showed no signs of diminishing in the immediate aftermath of the conflict. In 1945 there were 50,662 neuropsychiatric casualties crowding the wards of military hospitals, and to those institutionalized we must add the 475,397 discharged servicemen who were receiving Veterans' Administration pensions for psychiatric disabilities by 1947." Scull, "The Mental Health Sector...Part 1," 7.

48 As Gerald Grob has contended, the move from "a concern with mental disease in institutional populations to the incidence in the general population and the role of socioenvironmental variables represented an extraordinary intellectual leap." Grob, "Psychiatry's Holy Grail," 212.

49 For a thorough discussion of the psychological profession during this period, see Capshew, *Psychologists on the March*.

50 Scull, "The Mental Health Sector...Part 1," 9.

51 Mental illness was viewed along a continuum, with neurotic disorders at one end and more serious psychoses at the other.

52 Hale, "American Psychoanalysis since World War II," 81.

53 Baumeister and Hawkins, "Continuity and Discontinuity," 200.

54 Popper, "Science," 38.

55 See Farreras, "The Historical Context for NIMH Support."

56 Polanyi, "From Copernicus to Einstein," 113.

57 Kety, "Recent Biochemical Theories of Schizophrenia," 123.

58 Giorgi, *Psychology as a Human Science*, 14.

59 Maslow, *Religions, Values, and Peak-Experiences*, 5.

60 Maslow, *The Psychology of Science*, 33.

61 Tart, "Science, States of Consciousness, and Spiritual Experiences," 20.

62 Rogers, "Towards a Science of the Person," 112.

63 Scull, "The Mental Health Sector...Part 2," 269.

64 Whether the psychopharmacological revolution was the primary motivator of the deinstitutionalization era has been questioned by some historians of psychiatry. Scull, for example, has maintained that "antipsychotics played at best a secondary role in the demise of the asylum. Deinstitutionalization was driven

far more by fiscal concerns, and by conscious shifts in state policy." Scull, "The Art of Medicine," 1247.

65   See Grob, "Psychiatry's Holy Grail"; and Valenstein, *Blaming the Brain*.

66   Emil Kraepelin, a nineteenth-century German psychiatrist noted for his somatic approach to mental disorders, is widely regarded as the founder of modern scientific psychiatry and psychiatric nosology, hence the term "neo-Kraepelinian" applied to biologically oriented psychiatric reform from the mid-1960s on.

67   *DSM-I* (1952) and *DSM-II* (1968) were embedded within the psychoanalytical paradigm and steered away from formal, empirically based systems of classification.

68   See Grob, "Origins of *DSM-I*"; and Wilson, "The Transformation of American Psychiatry."

69   See Szasz, *The Manufacture of Madness*. The view of mental illness as a myth was strengthened in light of psychiatry's continuing classification of homosexuality as a mental disorder and the famous Rosenhan experiment in the early 1970s in which a group of "pseudopatients" presented themselves to hospitals with imaginary symptoms and were diagnosed with schizophrenia.

70   For an analysis of the *DSM-III* and its impact on psychiatry, see Faust and Miner, "The Empiricist and His New Clothes"; and Galatzer-Levy and Galatzer-Levy, "The Revolution in Psychiatric Diagnosis."

71   In Shorter's assessment, "the biological approach to psychiatry—treating mental illness as a genetically influenced disorder of brain chemistry—has been a smashing success. Freud's ideas, which dominated the history of psychiatry for the past half century, are now vanishing like the last snows of winter." Shorter, *A History of Psychiatry*, vii.

72   See Richert and Reilly, "American Psychiatry Scholarship."

73   Rasmussen, "Making the First Anti-Depressant," 291.

74   Pickersgill, "From Psyche to Soma?," 305.

75   Metzl, *Prozac on the Couch* and "Psychoanalysis and the Miltown Revolution."

76   Referring to Metzl's history of the minor tranquilizing drugs and his own history of amphetamines, Rasmussen writes that "the psychoanalytic approach was not inconsistent with psychiatric drug prescription—although in such situations the drugs were thought of as acting on anxiety symptoms without affecting the neurotic illness itself." Rasmussen, "Making the First Anti-Depressant," 290.

77   As Valenstein writes, the "evidence and arguments supporting all these claims about the relationship of brain chemistry to psychological problems and personality and behavioral traits are far from compelling and are most likely wrong." Valenstein, *Blaming the Brain*, 3. Also see Healy, *The Creation of Psychopharmacology*.

78   Grob, "Psychiatry's Holy Grail," 215.

79   See Pilecki, Clegg, and McKay, "The Influence of Corporate and Political Interests," 198.

80 Rasmussen, "Making the First Anti-Depressant," 323.

81 Whitaker, *Anatomy of an Epidemic*, 65.

82 Ibid.

83 Fortunately, this situation is now changing for the better with the contributions of historians such as Hewitt, "Rehabilitating LSD History in Postwar America"; and Oram, "Efficacy and Enlightenment" and "Prohibited or Regulated?"

84 Grof, *Realms of the Human Unconscious*, 32.

85 Healy, *The Creation of Psychopharmacology*, 194.

86 The "memory hole" is a term employed by Noam Chomsky and other authors to demonstrate how certain facts and events have been discarded in official histories so as to manufacture citizens' consent. The term first appeared in George Orwell's dystopian novel *1984* and referred to a file transfer system used by the ruling party to destroy documents that contradicted or challenged its version of history. Healy has indicated a similar assessment in his work, noting that theories attached to LSD and other psychedelics, including those of the Saskatchewan researchers, have been largely "written out" of the history of psychopharmacology. See ibid., 185.

87 Some historians have questioned the overreliance on external, namely countercultural, factors as the principal reason for the downfall of psychedelic science in the 1960s. In "Prohibited or Regulated?" Oram traces the decline of LSD psychotherapy to U.S. food and drug regulatory amendments in the early 1960s.

88 "Subjectivity," as Goldenberg writes in "Iconoclast or Creed?," 169, is "popularly understood to be a pejorative intrusion on scientific investigation," whereas "objectivity is an epistemic virtue in science that stands for an aperspectival 'view from nowhere,' certainty, and freedom from bias, values, interpretation, and prejudice."

89 Osmond and Smythies, "Schizophrenia," 6.

90 Jung, "The Psychology of Dementia Praecox," 37. Kraepelin's term "dementia praecox," meaning "prematurely out of one's mind," was later changed by psychiatrist Eugen Bleuler to "schizophrenia," implying a "splitting of one's personality."

91 Fromm-Reichmann, "Notes on the Development of Treatment of Schizophrenics," 163.

92 Freud, as Whyte notes, never thought himself to be a "Freudian" in the sense that so many others have applied the term; he "never maintained that man was forever hostage to childhood traumata," and "with resolution and intelligence...the individual could by understanding these factors...perhaps surmount them." Whyte, *Organization Man*, 22.

93 Hoffer and Osmond, *The Chemical Basis of Clinical Psychiatry*, 24–25.

94 Ibid., 26.

95 Braden, *The Private Sea*, n. pag.

96 Ibid.

97   Blewett, *The Frontiers of Being*, 20.

98   Healy, "Psychedelic Psychiatry," 699–700.

99   Healy, *The Creation of Psychopharmacology*, 185. Hewitt challenges this interpretation of the adrenochrome hypothesis and Hoffer and Osmond, claiming the serotonin hypothesis of American LSD researcher Dilworth Wayne Woolley to be "more sophisticated and more fruitful." Hewitt, "Rehabilitating LSD History in Postwar America," 317.

100   Ibid., 186.

101   For a recent discussion, see Langlitz, "The Persistence of the Subjective in Neuropsychopharmacology."

102   Kety, "The Biological Roots of Schizophrenia," 36.

103   Maslow, *Religions, Values, and Peak-Experiences*, 6.

104   "Molecules and Mental Health," 296.

105   Baekeland, "Evaluation of Treatment Methods in Chronic Alcoholism," 424.

106   Dyck, "Flashback," 381.

107   Dyck, "Prairie Psychedelics," 233.

108   Dyck, *Psychedelic Psychiatry*, 8.

109   Dyck, "Prairie Psychedelics," 228. Also see Dyck, "Prairies, Psychedelics, and Place" and "Land of the Living Sky with Diamonds"; and Mills and Dyck, "Trust Amply Recompensed."

110   For a contemporary rationale in favour of freedom in scientific research, see Wilholt, "Scientific Feedom."

111   Hoffer, interview with Barber and Dyck, June 27, 2003.

112   Although it is true that Hoffer and Osmond's approach to diagnosing and treating schizophrenia (i.e., orthomolecular) did not involve the use of psychedelics per se, historians would be remiss to exclude or downplay it as a major component of Saskatchewan's psychedelic research story. Hoffer stressed this repeatedly: "All the work we did, the many hypotheses of which the adrenochrome hypothesis is the best, all of our work with hallucinogens were a means to an end and that was the prevention and cure of schizophrenia." Hoffer, correspondence with the author, February 26, 2008.

113   Mills, "Psychiatry in Saskatchewan," 187.

114   Kuhn, *The Structure of Scientific Revolutions*, 92.

115   See Langlitz, *Neuropsychedelia*.

116   Rouse, "Kuhn's Philosophy of Scientific Practice," 115.

PART ONE

1   SAB, Hoffer Papers, A207, XVIII 10(c), Hoffer-Osmond Correspondence, December 30, 1961.

2   Quoted in Marsh, "Meaning and the Mind-Drugs," 426.

3   Hoffer and Osmond, *The Clinical Basis of Psychiatry*, 5.

CHAPTER 1

1   Again, the main reason for the cold reception of the findings of model psychosis research was the predominance of the psychodynamic paradigm. The introduction of neuroleptic drugs such as chlorpromazine in the early 1950s to treat mental illness helped to soften the resistance to biochemical theories, but few understood how these medications worked, and none was originally based upon etiological theories of mental illness backed by empirical evidence. See Hewitt, "Rehabilitating LSD History in Postwar America," 310.

2   Hewitt has made a similar argument in her evaluation of the model psychosis work of Rockefeller Institute chemist Dilworth Wayne Woolley with LSD and serotonin. As she put it, Woolley's serotonin hypothesis was instrumental in leading the paradigm shift in psychiatry from psychogenic to biochemical models of mental illness and "one of the first broad theories of biochemical origins of mental disturbances based on empirical evidence." Ibid., 327.

3   The CCF was "a party that combined high idealism, in the tradition of British-inspired Fabian socialism and North American social gospel, with a pragmatism born of grassroots populism and limited financial resources." Marchildon, "A House Divided," 307.

4   Whyte, *The Organization Man*, 209.

5   The term "orthomolecular" was coined in the late 1960s by two-time Nobel Prize–winner Linus Pauling to emphasize the importance of maintaining a proper balance of naturally occurring minerals and vitamins in the human body.

6   Dickinson, *The Two Psychiatries*, 77.

7   Kahan, *Brains and Bricks*, 21.

8   Ibid.

9   CMHA, *Ten Giant Steps*, 12.

10 Provincial health officer Low was quoted as saying as far back as 1907 that insane people ought to be treated not as criminals but as sick people.

11 Responsibility for the asylum had been shared between the Department of Public Works and the Department of Public Health.

12 Rands, "Community Psychiatric Services in a Rural Area," 406. Also see Dooley, "The End of the Asylum (Town)"; Marchildon, "A House Divided"; and Mills, "Lessons from the Periphery."

13 Details of Osmond's early years in England come from Dorothy Gale, his sister, in a letter in February 2008 forwarded to me by his daughter.

14 Osmond and Smythies, "Schizophrenia," 2.

15 Ibid., 3.

16 Ibid., 6.

17 Ibid., 7.

18 Ibid.

19 Hoffer, letter to the author, July 26, 2003.

20 Hoffer, *Adventures in Psychiatry*, 3.

21 Ibid., 4.

22 Saskatchewan Archives Board (hereafter SAB), Hoffer Papers, A207, X12, Ford Foundation submission.

23 SAB, Hoffer Papers, A207, III 194(a), D. G. McKerracher Correspondence, April 20, 1950.

24 SAB, Hoffer Papers, A207, XVIII 1(a), Hoffer-Osmond Correspondence, copy of the research plan, November 1951.

25 Ibid.

26 Ibid.

27 Hoffer later pointed out that what they "actually got [was] about $27,000 [and that] the grant was approved only after it was submitted to professor Nolan D.C. Lewis, the most prominent American psychiatrist, a friend of Freud's, a pathologist and biochemist and head at Columbia University New York." He added that "all three chairmen from the departments of psychiatry from McGill, Toronto, and Western University rejected [their] proposal." Hoffer, letter to the author, 2004.

28 Janiger, "The Use of Hallucinogenic Agents in Psychiatry," n. pag.

29 Ibid.

30 Ibid.

31 Hoffer and Osmond, *The Chemical Basis of Clinical Psychiatry*, 212.

32 See Hoffer, Osmond, and Smythies, "Schizophrenia...II," 30.

33 Osmond, "Inspiration and Method in Schizophrenia Research," 7.

34 "The indole nucleus, known to chemists as benzpyrrole, is the parent member of a broad spectrum of nitrogen heterocyclic biochemicals commonly found in nature. Indole derivatives occur in flower oils such as jasmine and orange blossom, and in less pleasant substances such as coal and fecal matter. Indoles also exist in melanin-related organics and indigoid pigments." www.microscopy. fsu.edu/phytochemicals/pages/ indole.html.

35 Hoffer, interview with Barber and Dyck, June 2003.

36 The "asthmatic historian," a subject in Osmond and Smythies' early mescaline experiments, is referenced in many of the early accounts of the adrenaline-based theories but never identified. See Hoffer, Osmond, and Smythies, "Schizophrenia...II," 23.

37 Osmond, "Inspiration and Method in Schizophrenia Research," 8.

38 Hoffer, Osmond, and Smythies, "Schizophrenia...II," 33.

39 Hofmann, *LSD—My Problem Child*, 15.

40 Ibid., 45.

41 SAB, Hoffer Papers, A207, XVIII 1(b), Hoffer-Osmond Correspondence, April 2, 1952.

42 Hoffer and Osmond, *The Chemical Basis of Clinical Psychiatry*, 71.

43 University of Regina Archives (hereafter URA), Duncan Blewett Papers, 93-19, "Psychological Study LSD-25—Humphry Osmond and Ben Stefaniuk," August 1955, 455.

44 Osmond, "On Being Mad," 28.

45 Ibid.

46 McKellar, "Scientific Theory and Psychosis," 172.

47 For a list of the earliest research reports of the psychological and physiological effects of LSD on schizophrenic and other patient populations, see Cholden, Kurland, and Savage, "Clinical Reactions."

48 Hoffer and Osmond, *The Hallucinogens*, 105.

49 Agnew and Hoffer, "Nicotinic Acid Modified Lysergic Acid Diethylamide Psychosis," 18.

50 Agnew described the reasoning for the experiments as such: "One of my jobs was rounding up volunteers to take LSD so we could test various antidotes—if they worked with LSD—a psychotomimetic—the same antidote might work with schizophrenics. In order to tell potential volunteers what the LSD experience was like I, of course, had to take it. As far as I know, I was the first of the group to do so. At the peak of the LSD 'trip' Abe injected me with nicotinic acid—the supposed antidote. Nicotinic acid makes you feel very warm and...flush red. I felt like my skin was made of cellophane about to burst into flame at any moment." Agnew, letter to the author, November 27, 2003.

51 SAB, Hoffer Papers, A207, XVIII 1(b), Hoffer-Osmond Correspondence, April 7, 1953.

52 Agnew and Hoffer, "Nicotinic Acid Modified Lysergic Acid Diethylamide Psychosis," 19.

53 Ibid., 13.

54 Ibid., 19.

55 Much of the information on Blewett's early years comes from his archives or discussions and correspondence with his daughter, sister-in-law, and second wife.

56 Blewett, letter to Irene Blewett (n.d.), provided by Blewett's daughter, Mary Lowe.

57 Hoffer and Osmond, *The Hallucinogens*, 83.

58 SAB, Hoffer Papers, A207, XVIII 1(b), Hoffer-Osmond Correspondence, October 9, 1952.

59 Hoffer, Osmond, and Smythies, "Schizophrenia...II," 42.

60 Ibid., 39.

61 Ibid., 35.

62 The parallels between Hewitt's history of Woolley's serotonin-LSD work and Saskatchewan's model psychosis experiments, and between Hoffer and Woolley in particular, are remarkable. For example, both Hoffer and Woolley had backgrounds in agricultural chemistry and a preoccupation with B vitamins such as niacin and their role in nutritional health. Unfortunately, Hewitt offers no information on what Woolley, who died in 1966, thought about the Saskatchewan research findings or the use of niacin in treating schizophrenia.

63 Barber, "Resistance by Scientists to Scientific Discovery," 596–97.

64 Here, too, Barber referred to how methodological concepts serve as a "second cultural source of resistance to scientific discovery." Ibid., 598.

65 Kety, "The Biological Roots of Schizophrenia," 35.

66 Ibid., 36.

67 Langlitz, "The Persistence of the Subjective in Neuropsychopharmacology," 45.

68 Ibid.

69 Ibid., 43.

70 Ibid.

71 Katz, "My 12 Hours as a Madman," 9.

72 Ibid., 55.

73 SAB, Hoffer Papers, A207, XVIII 1(b), Hoffer-Osmond Correspondence, January 2, 1953.

74 With Osmond as superintendent, mental health services in the Weyburn hospital underwent a period of modernization and "liberalization," for example in the placement of female nurses on male wards and the parole and discharge of patients into the community. See Dooley, "The End of the Asylum (Town)," 341–43.

75 SAB, Hoffer Papers, A207, XVIII 1(b), Hoffer-Osmond Correspondence, December 23, 1953.

CHAPTER 2

1 See Kuhn, *The Structure of Scientific Revolutions*, 7.

2 Polanyi, "The Nature of Scientific Convictions," 52.

3 Hoffer and Osmond, *The Chemical Basis of Clinical Psychiatry*, 10–11.

4 Smith, "The Unscientific Bases of Psychiatry," 68.

5 SAB, Hoffer Papers, A207, XVIII 1(b), Hoffer-Osmond Correspondence, letter from Kluver, August 30, 1953.

6 "Catechol indoles," or "catecholamines," is the term often used to describe the group of hormones comprising dopamine, noradrenaline, and adrenaline, all of which share a close resemblance in chemical structure and derive from the amino acid tyrosine.

7 Hoffer and Osmond, "Schizophrenia...III," 654.

8 Hoffer and Osmond, *The Hallucinogens*, 268.

9 Ibid.

10 Hoffer and Osmond, "Schizophrenia...III," 656.

11 The term "double-blind" describes experiments in which neither subjects nor researchers know whether the agent administered, say adrenochrome, is real or a placebo. In such experiments, the bias of the experimenters and the suggestibility of the subjects are supposedly kept to a minimum.

12 Hoffer and Osmond, *The Hallucinogens*, 348.

13 See Osmond, "Oliloqui."

14 Osmond, "Models of Madness," 779.

15 Hoffer, "Epinephrine Metabolites," 129–30.

16 For a brief description of the taraxein work, see Segal, Yager, and Sullivan, *Foundations of Biochemical Psychiatry*, 60–62.

17 Hoffer et al., "Psychological Response to d-Lysergic Acid Diethylamide," 125.

18 Ibid.

19 A letter from an asthmatic in the United States lent further credibility to their theories. The individual had unknowingly inhaled some discoloured adrenaline and subsequently suffered schizophrenic symptoms for a few weeks. This led the two scientists to carry out tests on thousands of commercial samples of deteriorated adrenaline. See Osmond and Hoffer, "The Case of Mr. Kovish."

20 Schwarz et al., "Effects of Mescaline, LSD-25, and Adrenochrome"; Taubmann and Jantz, "Untersuchung über die dem Adrenochrom Zugeschniebenen Psychotoxinschen Winkungen."

21 Grof et al., "Clinical and Experimental Study of Effects of Adrenochrome," 210.

22 Hoffer and Osmond, *The Chemical Basis of Clinical Psychiatry*, 8.

23 Ibid., 8–9.

24 Polanyi, "The Autonomy of Science," 21–22.

25 Hoffer et al., "Psychological Response to d-Lysergic Acid Diethylamide," 126.

26 Hoffer and Osmond, "Schizophrenia...III," 660.

27 Hoffer et al., "Psychological Response to d-Lysergic Acid Diethylamide," 132.

28 Ibid., 130.

29 Hoffer and Osmond, *The Hallucinogens*, 174.

30 Hoffer et al., "Psychological Response to d-Lysergic Acid Diethylamide," 132.

31 See Hoffer and Osmond, *How to Live with Schizophrenia*, 65.

32 The present-day term is "kryptopyrole."

33 SAB, Hoffer Papers, A207, X, HOD File (1961–63).

34 Experts in the area (e.g., biochemists) said that stable/crystalline adrenochrome could not be synthesized, that the oxidation of adrenaline to adrenochrome could not occur in the body, and that adrenochrome was not hallucinogenic. The Saskatchewan findings proved these objections wrong.

35 SAB, Hoffer Papers, A207, XVIII 5(a), Hoffer-Osmond Correspondence, August 8, 1958.

36 Feyerabend, "Science," 180.

37 Bleuler, "Comparison of Drug-Induced and Endogenous Psychoses in Man," 162.

38 Ibid., 163.

39 The adrenochrome phenomenon provides an interesting parallel with another important scientific discovery, that of X-rays. Their discovery was "greeted not only with surprise but with shock. Lord Kelvin at first pronounced them as an elaborate hoax. Others, though they could not doubt the evidence, were clearly staggered by it. Though X-rays were not prohibited by established theory, they violated deeply entrenched expectations." Kuhn, *The Structure of Scientific Revolutions*, 59.

40 Ibid., 103.

41 Ibid., 49.

42 Ibid., 10.

CHAPTER 3

1   Osmond, "A Review of the Clinical Effects of Psychotomimetic Agents," 419.
2   SAB, Hoffer Papers, A207, XVIII 1(b), Hoffer-Osmond Correspondence, April 7, 1953.
3   Huxley, *The Doors of Perception*, 69.
4   See Busch and Johnson, "LSD-25 as an Aid in Psychotherapy"; Frederking, "The Use of LSD and Mescaline in Psychotherapy"; and Sandison, "Psychological Aspects of the LSD Treatment of the Neuroses."
5   Hoffer and Osmond, *The Hallucinogens*, 148.
6   Hofmann, *LSD—My Problem Child*, 50.
7   Grof, *Realms of the Human Unconscious*, 2.
8   Smith, "Some Reflections on the Possible Therapeutic Effects of the Hallucinogens," 292–93.
9   Osmond, "A Review of the Clinical Effects of Psychotomimetic Agents," 421.
10 Ibid.
11 MacLean et al., "The Use of LSD-25 in the Treatment of Alcoholism," 43.
12 Grof, *Realms of the Human Unconscious*, 20.
13 Campbell, *The Grand Vision*, xi.
14 Hoffer and Osmond, *New Hope for Alcoholics*, 15.
15 "In the city of Regina, which has a population of over 130,000, court and law enforcement facilities are strained to the limit by offences which involve a high percentage of active alcoholics. Dr. [Saul] Cohen estimates that, if you discount traffic offences such as speeding and parking violations, 92.5 percent of prosecutions under Regina's bylaws in 1963 involved abusive or pathological use of beverage alcohol." Quoted in ibid., 21.
16 Campbell, *The Grand Vision*, 18–19.
17 This venture resulted in the first double-blind controlled experiments (using niacin) in the country.
18 Hoffer, "A Program for the Treatment of Alcoholism," 343–44.
19 James, *The Varieties of Religious Experience*; Tiebout, "The Ego Factors in Surrender in Alcoholism."
20 Smith, "A New Adjunct to the Treatment of Alcoholism," 407.
21 Attaching concrete definitions to the various diagnostic categories is complicated. Typically, definitions change every twenty to thirty years and might differ slightly from one medical school/practitioner to the next. For full descriptions of diagnostic categories, see *DSM-V*.
22 Smith, "A New Adjunct to the Treatment of Alcoholism," 408.
23 Ibid.
24 Ibid., 415.
25 Ibid., 416.
26 Ibid.
27 Huxley, *The Doors of Perception*, 12–13.

28   SAB, Hoffer Papers, A207, XVIII 1(b), Hoffer-Osmond Correspondence, July 21, 1953.

29   Ibid.

30   Huxley, *The Perennial Philosophy, vii.*

31   SAB, Hoffer Papers, A207, XVIII 3, Hoffer-Osmond Correspondence, September 15, 1956.

32   URA, Blewett Papers, Miscellaneous Writings, 2005-05, Box 2.

33   Blewett, *The Frontiers of Being,* 74.

34   Blewett, "The Psychedelic Experience in the Native American Church," 240.

35   Osmond, "A Review of the Clinical Effects of Psychotomimetic Agents," 427.

36   Jung, *Modern Man in Search of a Soul,* 173.

37   Blewett, *The Frontiers of Being,* 11.

38   Watts, "Psychedelics and Religious Experience," 74.

39   Osmond, "A Review of the Clinical Effects of Psychotomimetic Agents," 428.

40   For an in-depth account of the Native American Church and its peyote ceremony, see Slotkin, *The Peyote Religion.* For a review of peyote's therapeutic use among Indigenous peoples, see Albaugh and Anderson, "Peyote in the Treatment of Alcoholism among American Indians"; and Calabrese, "Clinical Paradigm Clashes." For reports of the church in western Canada and Saskatchewan, see Dyck and Bradford, "Peyote on the Prairies"; and Sagi, "White Men Witness Indian Peyote Rites." As well, Kahan's rejected "The Road to Manitou" manuscript (which contains essays by Blewett, Osmond, and Weckowicz) has recently been published as *A Culture's Catalyst,* edited and introduced by Dyck.

41   Crow Dog, quoted in Lame Deer and Erdoes, *Lame Deer,* 218–19.

42   Campbell, in *Debates of the House of Commons,* Third Session, Twenty-Second Parliament, January 24, 1956, 2474.

43   Ibid., April 23, 1956, 3163.

44   Hoffer, letter to the author, July 5, 2004.

45   Dyck and Bradford, "Peyote on the Prairies," 47.

46   Ibid., 47–48.

47   Sagi, "White Men Witness Indian Peyote Rites."

48   Osmond, "Peyote Night," 84.

49   Weckowicz, "Peyotl Ceremony and Jungian Archetypes," 219.

50   Calabrese, "Clinical Paradigm Clashes," 334.

51   Blewett, "The Psychedelic Experience in the Native American Church," 233.

52   Ibid.

53   Dyck, Introduction to Kahan, *A Culture's Catalyst.*

54   Feyerabend, "'Science,'" 173.

55   Ibid., 174.

56   Stolaroff, "LSD."

CHAPTER 4

1   See http://www.hofmann.org/Reviews/BLEWETT%20HANDBOOK.htm.

2   SAB, Hoffer Papers, A207, XVIII 10(b), Hoffer-Osmond Correspondence, "A New Approach 1950–56."

3   Giorgi, *Psychology as a Human Science*, 52–53.

4   Like their humanistic counterparts, however, psychoanalytical psychologists saw merit in subjective experience, which behaviourists did not.

5   Riesman, *The Lonely Crowd*.

6   Ibid., 25.

7   Ibid., 31.

8   Whyte, *The Organization Man*, 211–12.

9   Smith, "Some Reflections on the Possible Therapeutic Effects of the Hallucinogens," 296.

10 Hoffer and Osmond, *The Chemical Basis of Clinical Psychiatry, 105*.

11 Osmond, "A Review of the Clinical Effects of Psychotomimetic Agents," 428.

12 SAB, Hoffer Papers, A207, XVIII 3, Hoffer-Osmond Correspondence, March 23, 1956. Apparently, there is some confusion over whether the words are spelled with an *o* or an *e* (i.e., psychodelic or psychedelic). Even though the latter spelling is now dominant, the two spellings have often been used interchangeably in various articles and personal correspondence.

13 Ibid., March 28, 1956.

14 Huxley, *Moksha*, 107.

15 Ibid.

16 Lee and Shlain, *Acid Dreams*, 53.

17 Ibid., 44.

18 SAB, Hoffer Papers, S-A1101, II. A. 109, Dr. A. M. Hubbard (1955–59), October 22, 1955.

19 As Hoffer dryly noted, Hubbard did not have a bogus MD: he had paid twenty-five dollars for an official PhD. That Hoffer often referred to him by the term "doctor" annoyed many people. Hoffer, letter to the author, June 13, 2005.

20 Blewett, *Frontiers of Being*, 253–54.

21 This was conducted in the presence of three psychiatrists, Hoffer, Smith, and Chwelos.

22 Hoffer and Osmond, *The Hallucinogens*, 158.

23 Blewett, "Interim Report on the Therapeutic Use of LSD-25," 5.

24 URA, Blewett Papers, 88-29, File 46, Box 3, Miscellaneous Notes, n.d.

25 Blewett, "Interim Report on the Therapeutic Use of LSD-25," 4–5.

26 Blewett and Chwelos, *Handbook for the Therapeutic Use of Lysergic Acid Diethylamide-25*, 10.

27 Ibid.

28 Ibid., 11–12.

29 Ibid.

30 Ibid., 12.
31 Ibid., 13.
32 Ibid., 14.
33 Grof, *Realms of the Human Unconscious*, 219.
34 Chwelos et al., "Use of D-Lysergide Acid Diethylamide in the Treatment of Alcoholism," 3.
35 Blewett, *The Frontiers of Being*, 90.
36 Blewett and Chwelos, *Handbook*, 6.
37 See Masters and Houston, *The Varieties of Psychedelic Experience*, 133.
38 Blewett, "Interim Report on the Therapeutic Use of LSD-25," 31.
39 Carl Rogers also emphasized the value of empathic understanding or "interpersonal knowledge," as he liked to call it. See Rogers, "Towards a Science of the Person," 115–16.
40 Blewett and Chwelos, *Handbook*, 55.
41 Lee and Shlain, *Acid Dreams*, 75. Rick Doblin later revealed that the published claims of this experiment were false but should not be taken as "proof of the lack of value of psychedelics as adjuncts to psychotherapy in criminals." Doblin, "Dr. Leary's Concord Prison Experiment," 425.
42 Chwelos et al., "Use of D-Lysergide Acid Diethylamide in the Treatment of Alcoholism," 4.
43 Ibid., 9.
44 Jensen, "A Treatment Program for Alcoholics in a Mental Hospital," 4.
45 Ibid.
46 Ibid., 2.
47 Ibid.
48 Ibid.
49 Ibid.
50 O'Reilly and Reich, "Lysergic Acid and the Alcoholic," 1.
51 Ibid., 2.
52 Ibid., 4.
53 Hoffer and Osmond, *The Hallucinogens*, 66.
54 Ibid., 167.
55 MacLean et al., "The Use of LSD-25 in the Treatment of Alcoholism," 36.
56 Masters and Houston, *The Varieties of Psychedelic Experience*, 134.
57 URA, Blewett Papers, 88-29, File 46, Box 3, Miscellaneous Notes, n.d.
58 As documented in an unpublished Bureau of Alcoholism report, roughly 50 percent of alcoholics (69 of 145) treated between 1957 and 1962 and followed up over intervals ranging from two months to five years improved after LSD treatment. Most cases, as the report noted, "had the last treatment from two to four years before the check was made," and fifty of the sixty-nine cases were totally dry. The other nineteen "have occasional relapses but contrive to try and find sobriety. Further, in some of these cases their relapses are becoming fewer and of shorter duration, e.g. one day of intoxication compared to previous

bout pattern of one week; gainfully employed as compared to former chronic unemployment." As quoted in Costello, "An Evaluation of Aversion and LSD Therapy," 37.

59   Smith, "Some Reflections on the Possible Therapeutic Effects of the Hallucinogens," 298–99.

60   Ibid., 299.

61   Ibid.

62   Quoted in Hoffer and Osmond, *The Hallucinogens*, 168.

63   As quoted in Unger, "Mescaline, LSD, Psilocybin, and Personality Change," 122.

64   Hoffer and Osmond, *New Hope for Alcoholics*, 36.

65   SAB, Hoffer Papers, A207, XVIII 13(a), Hoffer-Osmond Correspondence, June 4, 1963.

66   SAB, Hoffer Papers, S-A1101, II. A. 106/107, Mr. M. Stolaroff International Foundation for Advanced Study #1 (1960–63) and #2 (1963–66), June 6, 1961.

67   SAB, Hoffer Papers, A207, XVIII 10(c), Hoffer-Osmond Correspondence, December 27, 1961.

68   For an example, see "The Drug that Brings the Alcoholic Face to Face with Himself."

69   As quoted in Anderson, "Safe from Utopia?," 15. Likely, as Anderson asserts, the support of the CCF had more to do with political advantage at the time than anything else. Popular support for the party was waning by the 1960s; attaching itself to the psychedelic research and its innovative findings when they were relatively well received inside and outside the province could have been solely a matter of political convenience. Perhaps aside from T. C. Douglas himself, few in the party closely followed the Saskatchewan research, knew what it was about, or regularly supported it.

70   Buckman, "Theoretical Aspects of LSD Therapy," 137.

71   As quoted in Anderson, "Safe from Utopia?," 19.

72   Ibid., 21.

73   Tyhurst's campaign resulted in Maclean, Hubbard, and the rest of the group operating out of Hollywood Hospital in New Westminster having to appear before an investigative committee of doctors. Although the group received compliments on their success, they were reprimanded for allowing leaks to the press. Doubts were also raised about Hubbard's professional credentials.

74   SAB, Hoffer Papers, A207, III 271, Correspondence Sandoz Pharmaceuticals Ltd.—Montreal, September 22, 1959.

75   The Josiah Macy Foundation was a secret CIA conduit that funded two conferences on LSD (1959, 1965).

76   SAB, Hoffer Papers, A207, III 271, Correspondence Sandoz Pharmaceuticals Ltd.—Montreal, September 28, 1959.

77   Ibid.

78   As Oram notes, premarket clinical research in the 1950s was largely controlled by the drug manufacturer: "The manufacturer was free to distribute new drugs

to qualified researchers so long as they were labelled for investigational use. The manufacturer simply had to obtain a written statement from the researcher that they had adequate facilities to perform research with the drug, and that all research would be under their direction." This type of regulation became a concern for organizations such as the U.S. Food and Drug Agency because it "left them unable to prevent widespread distribution of investigational drugs to physicians for the ulterior purpose of establishing their place in the market prior to official release." Oram, "Prohibited or Regulated?," 4.

79 URA, Blewett Papers, 88-29, File 46, Box 3, Miscellaneous Notes, n.d.

80 SAB, Hoffer Papers, A207, XVIII 9, Hoffer-Osmond Correspondence, June 19, 1961.

CHAPTER 5

1 SAB, Hoffer Papers, A207, XVIII 10(c), Hoffer-Osmond Correspondence, December 30, 1961.

2 SAB, Hoffer Papers, A207, XVIII 7(a), Hoffer-Osmond Correspondence, February 23, 1960.

3 Hoffer and Osmond, *The Chemical Basis of Clinical Psychiatry,* 5.

4 Tart, "Science, States of Consciousness, and Spiritual Experiences," 21.

5 Rogers, "Towards a Science of the Person," 114.

6 Goffman, "The Moral Career of the Mental Patient—Inpatient Phase," 301.

7 Kahan, *Brains and Bricks*, 28.

8 An American psychiatrist and founding member of the APA, Kirkbride is regarded by many as one of the pre-eminent specialists in mental health in the nineteenth century. He is best remembered for his humane approach to the mentally ill and the construction of asylums that would best suit their needs and treatment. Kirkbride advocated buildings with a maximum population of 250 patients.

9 Low was a provincial health officer who, in 1907, was tasked by Premier Walter Scott to tour and assess mental hospitals in the eastern United States and Canada and report on his findings. His recommendations, one of which was the construction of smaller, cottage-type institutions, were not followed, and the province subsequently built two large mental hospitals in North Battleford and Weyburn.

10 Rands, "Community Psychiatric Services in a Rural Area," 405.

11 Osmond noted that "a hospital, whether a general or mental hospital, is one of the more complicated pieces of social machinery. For although its goal may appear to be fairly clear, in order to achieve it all sorts of people with different skills have to cooperate and finally their cooperative effort must be joined with an effort by the sick person, his family, friends, and not infrequently his community." Quoted in Canadian Mental Health Association, *Ten Giant Steps: Strides in Mental Health*, 10.

12 Hoffer and Osmond were committed to the Saskatchewan Plan as originally envisioned: that is, as a plan for the expansion and improvement of community mental health services and not a program solely for emptying institutions. As historian Chris Dooley notes, by the mid-1960s the plan remained the same in name only, and "much of what had set [it] apart from the more general pattern of deinstitutionalization had vanished." Dooley, "The Older Staff," 105.

13 SAB, Hoffer Papers, A207, XVIII 2(b), Hoffer-Osmond Correspondence, April 1953.

14 Ibid.

15 Osmond, "Function as the Basis of Psychiatric Ward Design," 23.

16 Ibid.

17 Ibid.

18 As contained in ibid.

19 As contained in ibid.

20 Ibid.

21 Quoted in Kahan, *Brains and Bricks*, 48.

22 Izumi, "LSD and Architectural Design," 387.

23 URA, Blewett Papers, 91-87, Osmond, "Hallucinogenic Drugs in Psychiatric Research," 4.

24 For a similar assessment of Saskatchewan's psychedelic explorations of mental hospital design, see Dyck, "Spaced-Out in Saskatchewan." Her thesis is that, "although the Izumi-Osmond-Sommer design was never fully realized, their in-depth discussions about the role of the mental institution in modern society tapped into broader international trends: namely, modernism, anti-psychiatry, and deinstitutionalization" (664).

25 Quoted in "Sociopetal Building Arouses Controversy," 30.

26 Quoted in ibid., 27.

27 SAB, Hoffer Papers, A207, XVIII 5(a), Hoffer-Osmond Correspondence, February 12, 1958.

28 Sommer and Osmond, "Autobiographies of Former Mental Patients," 660.

29 In one study comparing normal people and schizophrenic patients, it was found that "schizophrenics drew statistically smaller hands than normals and non-schizophrenics when asked to draw the upper part of the body in natural size." Weckowicz and Sommer, "Body Image and Self-Concept in Schizophrenia," 26.

30 Ibid., 17.

31 Ibid.

32 Weckowicz, "Depersonalization," 158.

33 Izumi, "LSD and Architectural Design," 395.

34 Sommer, "Floor Designs Can Be Therapeutic," 54.

35 Sommer, "Personal Space," 81.

36 Quoted in CMHA, *Ten Giant Steps*, 44.

37 Izumi, "LSD and Architectural Design," 389.

38 According to Arthur Allen, Izumi's friend and associate, the "success of the [sociopetal] concept was limited, but it...did influence psychiatric facility

planning in Toronto where architects Somerville, McMurrich, and Oxley credited the work of Izumi and Osmond in their 1970 renovations to that city's hospital on Queen Street." Allen also insisted that the YPC buildings "do not in any way incorporate the social essence of the circular sociopetal plan." Allen, letter to the author, September 2005.

39 Quoted in Osmond, "A Comment on Some of the Uses of Psychotomimetics in Psychiatry," 430.

40 Ibid., 431.

41 This practice is highlighted in Huxley's *Island* when the matriarchal character Lakshmi receives the Moksha medicine on her deathbed. Huxley did the same prior to his own death. This mode of psychedelic therapy was later picked up by researchers such as Stanislav Grof, following his arrival in the United States, and Eric Kast, whose extensive work with LSD and cancer patients revealed a "lessening of the patients' physical distress and a lifting of their mood and outlook that lasted about ten days." Kast, "A Concept of Death," 381.

42 Blewett, "Psychedelic Drugs in Parapsychological Research," 54.

43 SAB, Hoffer Papers, A207, XVIII 5(a), Hoffer-Osmond Correspondence, February 10, 1958.

44 Huxley, *Moksha*, 132–33.

45 URA, Blewett Papers, 88-29, File 92, No. 5, Memorandum by Dr. Abram Hoffer—Parapsychology Foundation, Inc.—Conference on Parapsychology and Psychedelics—1958.

46 Ibid.

47 Fogel and Hoffer, "The Use of Hypnosis to Interrupt and to Reproduce an LSD-25 Experience," 12.

48 URA, Blewett Papers, 88-29, File 92, No. 5, Memorandum by Dr. Abram Hoffer— Parapsychology Foundation, Inc.—Conference on Parapsychology and Psychedelics—1958.

49 Polanyi, *Personal Knowledge*, 167.

50 URA, Blewett Papers, 88-29, Blewett, "Interim Report on the Therapeutic Use of LSD-25," 60.

51 Ibid., 67.

52 Sommer, "Psychology in the Wilderness," 26.

53 SAB, Hoffer Papers, A207, XVIII 7(a), Hoffer-Osmond Correspondence, March 2, 1960.

54 Whyte, *Organization Man*, 222.

55 Ibid., 222–23. Wilholt reinforces this view in arguing that "individualized freedom of research will lead to a diversity of scientific approaches that can be expected to surpass the epistemic yield of centralized forms of research organization." Wilholt, "Scientific Freedom," 177.

56 SAB, A207, XVIII 9, Hoffer-Osmond Correspondence, May 25, 1961.

57 URA, Blewett Papers, 88-29, File 46, Box 3, Miscellaneous Notes, n.d.

PART TWO

1   For example, see Kershner, "Leeches, Quicksilver, Megavitamins, and Learning Disabilities."

2   Popper, "Science," 37.

3   Feyerabend, *Against Method*, 27.

4   Ibid., 23–24.

5   Polanyi, *Personal Knowledge*, 163–64.

6   Ibid.

7   Barnes, "Catching Up with Robert Merton," 180.

8   Barnes, Bloor, and Henry, "Scientific Knowledge," 140.

9   Ibid., 142.

10   Ibid., 168.

11   Polanyi, *Personal Knowledge*, 216.

12   Ibid.

13   Shapin, *The Scientific Revolution*, 10.

14   Merton, *The Sociology of Science*, 326.

15   Tullock, *The Organization of Inquiry*, 152.

16   Ibid.

17   Ibid., 54.

18   Scheffler, *Science and Subjectivity*, 2.

19   For a thought-provoking critique of evidence-based medicine's reliance on randomized controlled trials as the "gold standard," see Goldenberg, "Iconoclast or Creed?"

CHAPTER 6

1   Arieti, *Interpretations of Schizophrenia*, 3.

2   "The UCLA Interdepartmental Conference," 108.

3   Peters, "Concepts of Schizophrenia after Kraepelin and Bleuler," 96.

4   Shorter, *A History of Psychiatry*, 296.

5   See Kirk and Kutchins, *The Selling of DSM*, 219.

6   Lehmann, "Chlorpromazine in Psychiatric Conditions," 97.

7   Ibid.

8   Ban, "Fifty Years Chlorpromazine," 497.

9   Rudnick, "The Molecular Turn in Psychiatry," 292.

10   Kety, "The Academic Lecture," 393.

11   Ibid.

12   Roueché, "Placebo," 86.

13   Shepherd, "The Evaluation of Treatment in Psychiatry," 89.

14   See Barsa, "The Fallacy of the Double Blind," 1175; and Tuteur, "The 'Double Blind' Method," 922.

15 Guy, Gross, and Dennis, "An Alternative to the Double Blind Procedure," 1505.

16 SAB, Hoffer Papers, A207, XVIII 9(b), Hoffer-Osmond Correspondence, September 7, 1961.

17 Benjamin, "Biological Research in Schizophrenia," 428.

18 Ibid., 436.

19 Ibid.

20 Sourkes, "Review Article," 487.

21 Quoted in ibid., 488.

22 Hoffer and Osmond, "Letter to the Journal," 1309.

23 Ibid., 1311.

24 Ibid.

25 Snyder, "Julius Axelrod (1912–2004)," 4.

26 Axelrod, "Reflections," 7.

27 Axelrod, interview with National Institutes of Health, November 25, 2003, http://history.nih.gov/archives/downloads/Axelrod%20Julius.pdf.

28 Kety, "The Biochemical Theories of Schizophrenia," 1594.

29 See Axelrod, "O-Methylation of Epinephrine and Other Catechols in Vitro and in Vivo"; and Labrosse, Mann, and Kety, "The Physiological and Psychological Effects of Intravenously Administered Epinephrine."

30 Kety, "The Biochemical Theories of Schizophrenia," 1594.

31 Ibid., 1595.

32 Ibid., 1596.

33 Hoffer, "Abnormalities of Behavior," 363–64.

34 Ibid.

35 Labrosse, Mann, and Kety, "The Physiological and Psychological Effects of Intravenously Administered Epinephrine," 68.

36 Smythies, "The Biochemistry of Schizophrenia," 27.

37 Baldessarini and Snyder, "Schizophernia," 1111–12.

38 Weil-Malherbe and Szara, *The Biochemistry of Functional and Experimental Psychoses*, 150.

39 Ibid., 155.

40 Hoffer, "Remembering Humphry Osmond," 76.

41 SAB, Hoffer Papers, A207, XVIII 1(b), Hoffer-Osmond Correspondence, June 18, 1952.

42 See Hoffer et al., "Treatment of Schizophrenia with Nicotinic Acid and Nicotinamide."

43 SAB, Hoffer Papers, A207, X-12, Ford Foundation Submission, 1955, 19–20.

44 Ibid.

45 Ibid., 21.

46 See Gach, "Biological Psychiatry in the Nineteenth and Twentieth Centuries," 401.

47 Goldberger and Wheeler, "The Experimental Production of Pellagra," 81.

48 Sydenstricker, "The History of Pellagra," 412.

49 For examples of this research, see Elvehjem, "Tryptophan and Niacin Relations"; Frazier and Friedemann, "Pellagra, a Study in Human Nutrition"; and Goldsmith et al., "Studies of Niacin Requirement in Man."

50 See Cleckley, Sydenstricker, and Geeslin, "Nicotinic Acid in the Treatment of Atypical Psychotic States"; Gould, "The Use of Vitamins in Psychiatric Practice"; and Sydenstricker and Cleckley, "Effect of Nicotinic Acid in Stupor, Lethargy, and Various Other Psychotic Disorders."

51 Lehmann, "Case Report."

52 Osmond and Hoffer, "Massive Niacin Treatment in Schizophrenia."

53 Ibid., 8.

54 Ibid., 10.

55 Osmond and Hoffer, "Massive Niacin Treatment in Schizophrenia," 320.

56 Hoffer and Osmond, "Treatment of Schizophrenia with Nicotinic Acid," 171.

57 Osmond, interview with McEnaney, 1960, SAB Tape S-135 a and b.

58 Hoffer, "Letters to the Journal Re: Adverse Effects of Niacin in Emergent Psychosis," 1355.

59 Although some preliminary clinical experiments had been carried out on schizophrenic patients with various preparations of NAD, made from niacin, it never became widely available for extensive testing.

60 See CMHA, obituary of Griffin, July 12, 2001.

61 SAB, Hoffer Papers, A207, XVIII 21, Hoffer-Osmond Correspondence, January 10, 1966.

62 SAB, Hoffer Papers, A207, File 71, Press Relations 1961, April 11 and 19, 1966.

63 Ibid.

64 Kline et al., "Controlled Evaluation of Nicotinamide Adenine Dinucleotide."

65 Baihly, "The Professional Woos the Layman," 2.

66 SAB, Hoffer Papers, A207, XVIII 27, Hoffer-Osmond Correspondence, letter to Robert Robinson, APA, May 23, 1967.

67 Ibid., letter to Walter Barton, APA, May 9, 1967.

68 Ibid., Osmond to Hoffer, May 21, 1967.

69 Tullock, *The Organization of Scientific Inquiry*, 140–41.

70 SAB, Hoffer Papers, A207, XVIII 27, Hoffer-Osmond Correspondence, May 2, 1967.

71 Ibid., May 7, 1967.

72 In the early 1970s, the Saskatchewan Schizophrenia Foundation expanded to become a national organization, the Canadian Schizophrenia Foundation (CSF). The ASA and CSF subsequently merged to become the International Schizophrenia Foundation (ISF).

73 Pauling, "Orthomolecular Psychiatry," 265.

74 Oken and Pauling, "Vitamin Therapy for the Mentally Ill," 1181.

75 Goertzel, Goertzel, and Goertzel, "Linus Pauling," 374.

76 Some of the other investigative units involved were headed by other well-known American psychopharmacological experts, such as Leo Hollister,

Jonathon Cole, and Jerome Levine, all involved in the American LSD research in the 1950s and '60s.

77 Quoted in Ban, *Recent Advances in the Biology of Schizophrenia*, 80.

78 Ibid. See also Ban, "Pharmacotherapy of Schizophrenia," 178.

79 Recounted in Ban, *Recent Advances in the Biology of Schizophrenia*, 81–82; and Hoffer, "The Controversy over Orthomolecular Therapy," 168.

80 SAB, Grant Papers, R-45, File 80, memo from Smith, PSB, to Grant, December 17, 1968.

81 Ibid., Ban to Smith, January 2, 1969.

82 SAB, CMHA Papers, Saskatchewan Division, R 1265-V-K-169, Griffin to Peacock, April 9, 1969. Griffin made a similar plea to Robert Weil, president of the Canadian Psychiatric Association. See ibid., Griffin to Weil, April 10, 1969.

83 SAB, Hoffer Papers, A207, XVIII 34, Hoffer-Osmond Correspondence, letter from Garber, April 14, 1969.

84 Ibid., letter to Garber, April 25, 1969.

85 Ibid., Osmond to Hoffer, April 21, 1969.

86 Ibid., Hoffer to Osmond, May 1, 1969.

87 In 1968, Hawkins, then director of the North Nassau Mental Health Center in New York state, claimed to have successfully treated 2,000 schizophrenic patients using orthomolecular therapy.

88 Hawkins, "Letter to the Editor Re: Adverse Effects of Niacin in Emergent Psychosis," 1010.

89 SAB, Hoffer Papers, A207, XVIII 31(a), Hoffer-Osmond Correspondence, May 21, 1968.

90 Pauling, "Preface," vi.

91 Whitaker, *Anatomy of an Epidemic*, 57, notes that the "new marketplace for drugs proved profitable for all involved. Drug industry revenues topped $1 billion in 1957, the pharmaceutical companies enjoying earnings that made them 'the darlings of Wall Street,' one writer observed. Now that physicians controlled access to antibiotics and all other prescription drugs, their incomes began to climb rapidly, doubling from 1950 to 1970 (after adjusting for inflation). The [American Medical Association's] revenues from drug advertisements in its journals rose from $2.5 million in 1950 to $10 million in 1960, and not surprisingly...these advertisements painted a rosy picture. A 1959 review of drugs in six major medical journals found that 89 percent of the ads provided no information about the drugs' side effects."

92 SAB, Hoffer Papers, A207, XVIII 35(a), Hoffer-Osmond Correspondence, June 12, 1969.

93 Ibid., June 16, 1969.

94 Ibid., December 11, 1969.

95 Ramsay et al., "Nicotinic Acid as Adjuvant Therapy," 942.

96 Ananth et al., "Nicotinic Acid in the Treatment of Newly Admitted Schizophrenic Patients."

97 Ban, *Recent Advances in the Biology of Schizophrenia*, 89–90.

98 SAB, CMHA Papers, Saskatchewan Division, R 1265-VI-A-114, reprint of the introduction to "Nicotinic Acid in the Treatment of Schizophrenia Progress Report I"; emphasis added.

99 SAB, CMHA Papers, Saskatchewan Division, R 1265-VI-H-100, "The Megavitamin Therapy," CMHA committee report prepared for the National Scientific and Planning Council, 1973.

100 SAB, CMHA Papers, Saskatchewan Division, R 1265-VI-H-171, quoted in "The Schizophrenias," July 1973, 2.

101 Mills, "Hallucinogens as Hard Science," 188.

102 Ibid.

103 SAB, CMHA Papers, Saskatchewan Division, R 1265-VI-H-100, "The Megavitamin Therapy," CMHA committee report prepared for the National Scientific and Planning Council, 1973.

104 See McGrath et al., "Short Report"; and Sehdev and Olson, "Nicotinic Acid Therapy in Chronic Schizophrenia."

105 Pauling, "Preface," ix.

106 Ibid., vii.

107 Ibid., vii–viii. As another example of "misrepresentation," Pauling referred to a double-blind study by University of Toronto psychiatrist Gerald Greenbaum that failed to show any benefit of massive doses of niacinamide with a group of fifty-seven children with schizophrenia. As Pauling pointed out, though Greenbaum did not find any statistically significant difference between niacin and placebo, his findings still indicated a 54 percent improvement in the niacinamide group compared with the placebo group.

108 Hoffer Jr., "Letters to the Editor," 493.

109 Lehmann and Ban, "Author's Reply," 494.

110 Hoffer Jr., "Megavitamin Treatment in Schizophrenia," 134.

111 Brown, "Megavitamin and Orthomolecular Therapy of Schizophrenia," 99.

112 Wittenborn, "Niacin in the Long-Term Treatment of Schizophrenia."

113 Wittenborn, "A Search for Responders to Niacin Supplementation," 552.

114 Ban, "Nicotinic Acid and Psychiatry," 427.

115 Ban, *Recent Advances in the Biology of Schizophrenia*, 90.

116 Cathcart, "Nutrition and Schizophrenia," 50.

117 McFarlane, "And Another Reader Comments," 51.

118 Autry, "Workshop on Orthomolecular Treatment of Schizophrenia," 94.

119 Ibid., 97.

120 Ibid., 101.

121 SAB, Hoffer Papers, A207, XVIII 36, Hoffer-Osmond Correspondence, Osmond to Hardin Branch, October 1969.

122 See Whitaker, *Anatomy of an Epidemic*.

123 Mosher, "Nicotinic Acid Side Effects."

124 APA Task Force on Vitamin Therapy in Psychiatry, "Megavitamin and Orthomolecular Therapy in Psychiatry," 10.

125 Ibid.

126 Ibid., 48.

127 Ibid.

128 Their full response can be viewed online at http://www.iahf.com/orthomolecular/reply_to_apa_tfr_7.pdf.

129 Pauling, "On the Orthomolecular Environment of the Mind," 1256.

130 SAB, Hoffer Papers, A207, III 82, Douglas to Hoffer, November 1, 1961.

131 American Academy of Pediatrics—Committee on Nutrition, "Megavitamin Therapy for Childhood Psychoses and Learning Disabilities."

132 "Statement on Current Status of Megavitamin and Orthomolecular Therapies (1976-2)," http://www.cpa.apc.org/publications/position_papers/status.asp.

133 Hoffer, *Adventures in Psychiatry*, 253.

134 Klein, "Comment," 1265.

CHAPTER 7

1 Levine, "The Discovery of Addiction," 151.

2 Quoted in ibid., 152.

3 Ibid., 153.

4 White, "The Rebirth of the Disease Concept of Alcoholism in the 20th Century," 62.

5 Levine, "The Alcohol Problem in America," 116.

6 Kurtz, "Alcohol Problems in the United States," 6.

7 Ibid.

8 Page, "E. M. Jellinek and the Evolution of Alcohol Studies," 1627.

9 Page, "The Origins of Alcohol Studies," 1096.

10 Glatt, "Alcoholism Disease Concept and Loss of Control Revisited," 142.

11 Kurtz, "Alcohol Problems in America," 18.

12 Ibid.

13 See Davies, "Normal Drinking in Recovered Alcoholic Addicts."

14 Quoted in Kurtz, "Alcoholics Anonymous and the Disease Concept of Alcoholism," 7.

15 Page, "E. M. Jellinek and the Evolution of Alcohol Studies," 1626.

16 In Canada, Jellinek was instrumental in the formation of Toronto's Addictions Research Foundation in 1958 and had a rotating professorship in the Departments of Psychiatry of the University of Toronto and the University of Alberta in the early 1960s.

17 Quoted in White, "The Rebirth of the Disease Concept of Alcoholism in the 20th Century," 66.

18 White, "A Disease Concept for the 21st Century," 1.

19 In 1998, the ARF merged with two psychiatric hospitals and the Donwood Institute, an addictions treatment centre, to become what is now known as the Centre for Addiction and Mental Health.

20 Archibald, *ARF*, 18.

21 See Alcoholics Anonymous, *"Pass It On,"* 370–71.

22 Campbell, *The Grand Vision*, 59–60.

23 Hoffer and Osmond, "Outside AA," n. pag.

24 Blewett defined psychotronics as the "science which studies fields of interaction between people and their environment—both internal and external—and the energies involved in those interactions; it recognizes that matter, energy and consciousness are interconnected and makes a study of these interactions." URA, Blewett Papers, 91-87, Box 3, File 20.

25 URA, Blewett Papers, Miscellaneous Writings 2005-5, Box 2.

26 URA, Blewett Papers, 88-29, Box 1, File 14, Blewett, "LSD."

27 SAB, Hoffer Papers, A207, XVIII 13(a), Hoffer-Osmond Correspondence, January 10, 1963.

28 Ibid., June 6, 1963.

29 Grinker, "Lysergic Acid Diethylamide," 425.

30 Ibid.

31 Grinker, "Bootlegged Ecstasy," 768.

32 Ibid.

33 Oram, who has written extensively on the impact of the drug amendments of 1962 in the United States, notes how the new regulations, with their emphasis on the randomized controlled methodology, resulted in LSD researchers changing "their focus to the scientific rigor of their clinical trial design, often to the detriment of their therapeutic method." Oram, "Efficacy and Enlightenment," 234.

34 See URA, Blewett Papers, 91-87, Box 3, File 14.

35 SAB, Hoffer Papers, A207, XVIII 13(a), Hoffer-Osmond Correspondence, January 7, 1963.

36 Hoffer, "A Program for the Treatment of Alcoholism," 396.

37 Smith, "Exploratory and Controlled Studies of Lysergide in the Treatment of Alcoholism," 742–43.

38 Cohen, "Lysergic Acid Diethylamide," 30.

39 Ibid.

40 Smith, "Exploratory and Controlled Studies of Lysergide in the Treatment of Alcoholism," 743.

41 Ibid.

42 Cole and Katz, "The Psychotomimetic Drugs," 758.

43 Ibid.

44 Ibid., 761.

45 Ibid., 759.

46 Ibid.

47 Along with other researchers, Unger would go on to validate psychedelic therapy in the double-blind LSD treatment of alcoholic patients in the NIMH-sponsored Spring Grove experiment in the mid-1960s.

48 Unger, "Mescaline, LSD, Psilocybin, and Personality Change," 122–23. Other researchers experimenting with LSD therapy acknowledged some of the methodological weaknesses of Jensen's study but considered it a "major progression from previous therapy research efforts." See Salzman, "Controlled Therapy Research with Psychedelic Drugs," 23.

49 Unger, "Mescaline, LSD, Psilocybin, and Personality Change," 123.

50 See Smart, interview in Edwards, *Addiction*.

51 Smart and Storm, "The Efficacy of LSD in the Treatment of Alcoholism," 333.

52 Ibid.

53 See ibid., 334, for their list of requirements.

54 Ibid. At the time of writing, Smart and Storm addressed only the earlier Saskatchewan reports (e.g., Smith, Chwelos, et al.) and not those of Jensen, which purportedly had some controls and in which the LSD therapy performed better compared with standard therapies. In their subsequent writings, Smart and Storm insisted that the control group in Jensen's study (individual, out-patient therapy) was too different from the LSD-treated group (long-term, in-patient group therapy).

55 Ibid.

56 Ibid.

57 Ibid., 336.

58 American Psychiatric Association, "Position Statement on LSD," 353.

59 Smart et al., "A Controlled Study of Lysergide in the Treatment of Alcoholism," 476.

60 Ibid., 481.

61 Ibid., 479.

62 Ibid.

63 Ibid.

64 Ibid., 471.

65 Kurland et al., "LSD in the Treatment of Alcoholics," 84.

66 Surprisingly, Smart and his researchers managed to achieve marked improvements using this approach with some of the LSD patients. Referring to tactics used by Baker in earlier LSD treatments, Savage noted that "Baker has been chided...for the seeming barbarity of strapping the patients down and swacking them out....Using this particular method of therapy he had achieved therapeutic results of, let's say, 60%." Hicks and Fink, *Psychedelic Drugs*, 47.

67 Johnson, "LSD in the Treatment of Alcoholism," 483.

68 See ibid., 486, for some of the "transcendental" and "affective" changes noted by Johnson.

69 Ibid., 485.

70 Ibid.

71 Ibid.

72 Savage et al., "Research with Psychedelic Drugs," 17.

73 SAB, Hoffer Papers, A207, XVIII 32(c), Hoffer-Osmond Correspondence, November 30, 1968.

74 Hoffer, "Treatment of Alcoholism with Psychedelic Therapy," 359.

75 Blewett, "Psychedelic Drugs in Parapsychological Research," 51.

76 URA, Blewett Papers, 88-29, Box 3, File 46, Miscellaneous Notes, n.d.

77 "LSD in the Treatment of Alcoholism," 41.

78 Levine, as mentioned earlier, was also an expert on the APA Task Force that issued a formal report in the 1970s condemning Hoffer and Osmond's controversial megavitamin therapy for schizophrenia.

79 Levine, Ludwig, and Lyle, "The Controlled Psychedelic State," 163–64.

80 Ibid., 164.

81 For an example of these works, see Ludwig, "Studies in Alcoholism and LSD (1)"; and Ludwig and Levine, "A Controlled Comparison of Five Brief Treatment Techniques."

82 Ludwig et al., "A Clinical Study of LSD Treatment in Alcoholism," 59.

83 In the last group, patients received neither drugs nor psychotherapy; instead, they were put into a room and told to contemplate their alcoholism and write down their thoughts.

84 As an additional component of the study, disulfiram was given to half of the patients post-treatment to test whether it enhanced any of the four treatment approaches. Results indicated that the drug had no added value.

85 Ludwig et al., "A Clinical Study of LSD Treatment in Alcoholism," 66.

86 Ibid., 68.

87 Kantor, "Letters to the Editor," 140.

88 Ludwig, "Letters to the Editor," 140–41.

89 Hicks and Fink, *Psychedelic Drugs*, 48.

90 Grob, "Psychiatric Research with Hallucinogens," 14.

91 Ibid., 18.

92 Hicks and Fink, *Psychedelic Drugs*, 51.

93 For another such study, see Hollister et al., "A Controlled Comparison of Lysergic Acid Diethylamide (LSD) and Dextroamphetamine in Alcoholics."

94 Savage also had ties to Myron Stolaroff, Al Hubbard, and others at the IFAS.

95 Spring Grove researchers also replicated the work of Dr. Eric Kast in their use of LSD as an analgesic to treat chronic pain in terminally ill cancer patients. For more information on the analgesic properties of LSD, see Kast, "Attenuation of Anticipation."

96 Kurland et al., "Psychedelic Therapy in the Treatment of the Alcoholic Patient," 1202.

97 Ibid., 1208.

98 Earlier attempts to use a placebo were of little use: "Within one hour the cat was out of the bag and the study was neither double- nor single-blind." Savage et al., "Research with Psychedelic Drugs," 17.

99 Kurland et al., "LSD in the Treatment of Alcoholics," 84.

100 Ibid., 93.

101 Ibid., 92.

102 Salzman, "Controlled Therapy Research with Psychedelic Drugs," 27.

103 Ibid., 28.

104 Savage et al., "Research with Psychedelic Drugs," 21.

105 Salzman, "Controlled Therapy Research with Psychedelic Drugs," 28. For Spring Grove's response to the issue of therapist bias and the question of independent rating of psychedelic reaction, see Hicks and Fink, *Psychedelic Drugs*, 43.

106 Salzman, "Controlled Therapy Research with Psychedelic Drugs," 29.

107 SAB, Hoffer Papers, A207, XVIII 32(c), Hoffer-Osmond Correspondence, November 30, 1968.

108 Savage et al., "Research with Psychedelic Drugs," 16.

109 Maslow, *Religions, Values, and Peak Experiences*, 27.

110 Ibid.

111 For examples, see Bowen et al., "Lysergide Acid Diethylamide as a Variable in the Hospital Treatment of Alcoholism"; Cheek et al., "Observations Regarding the Use of LSD-25 in the Treatment of Alcoholism"; and Kurland et al., "LSD in the Treatment of Alcoholics."

112 Kurland et al., "LSD in the Treatment of Alcoholics," 93.

113 Cheek et al., "Observations Regarding the Use of LSD-25 in the Treatment of Alcoholism," 56–57.

114 Ibid., 73.

115 Bowen, Soskin, and Chotlos, "Lysergic Acid Diethylamide as a Variable in the Treatment of Alcoholism," 118.

116 Ibid.

117 See Grinspoon and Bakalar, "Can Drugs Be Used to Enhance the Psychotherapeutic Process?," 397.

118 Halpern, "The Use of Hallucinogens in the Treatment of Addiction," n. pag.

119 Ibid.

120 Savage et al., "Research with Psychedelic Drugs," 21.

CHAPTER 8

1 Retrospectives of "the '60s" are fraught with oversimplifications and nostalgic references (e.g., hippies). The period was notoriously complex and has been subject to various interpretations by both participants and historians, with disagreements over when it started and ended, what occurred and why, and what it implied. Rather than a monolithic movement of dissent, the counterculture has been viewed by many historians as representing "an inherently unstable collection of attitudes, tendencies, postures, gestures, 'lifestyles,' ideals, visions, hedonistic pleasures, moralisms, negations and affirmations." Braunstein and Doyle, *Imagine Nation*, 10. Arthur Marwick, too, insists that "there was no unified, integrated counterculture, totally and consistently in opposition to mainstream culture....What is called

the counterculture was in reality made up of a large number of very varied subcultures." Marwick, *The Sixties*, 12–13.

2   Affluence was "the most distinctive feature" of post–Second World War America (1945–70), with the average citizen "command[ing] 50 percent more real income at the end of the period than at the beginning." Matusow, *The Unraveling of America*, xx. Also see Farber, *The Sixties*.

3   Anderson, *The Sixties*, 7.

4   The theme of America in the process of splitting apart is common in many analyses of the 1960s. See Matusow, *The Unraveling of America*; and O'Neill, *Coming Apart*.

5   Most historians of the 1960s agree that the decade radically transformed America. Whether the changes were positive or negative remains open to debate. According to Todd Gitlin, "there is a specific reason...why 'the Sixties' are still so heated a subject. To put it briefly, the genies that the Sixties loosed are still abroad in the land, inspiring and unsettling, making trouble." Gitlin, *The Sixties*, xiv.

6   Peck, *Uncovering the Sixties*, xiii.

7   See Goodman, "Crisis and New Spirit."

8   Roszak, *The Making of a Counter Culture*, 7.

9   Ibid., 9.

10   Ibid., 27.

11   Ibid., 34.

12   In 1953, William S. Burroughs and Allen Ginsberg were documenting their experiences with hallucinogens such as ayahuasca. See Burroughs and Ginsberg, *The Yage Letters*.

13   The role of LSD and other psychedelics in the counterculture of the era remains understudied. In-depth sources most cited continue to be Lee and Shlain, *Acid Dreams*; and Stevens, *Storming Heaven*. More recent examinations include Cottrell, *Sex, Drugs, and Rock 'n' Roll*. Also see Farber, "The Intoxicate State/Illegal Nation."

14   Quoted in Peck, *Uncovering the Sixties*, 34.

15   Ibid.

16   For a more in-depth discussion of the debate about religious experience and psychedelics, see Clark, "Do Drugs Have Religious Import?"

17   Mayhew, *Time to Explain*, 50–51.

18   Leary, as the main advocate of the use of LSD and other psychedelics, was "an influence on sixties attitudes we cannot ignore." Marwick, *The Sixties*, 310.

19   Lee and Shlain, *Acid Dreams*, 73.

20   Leary, "The Religious Experience," 324.

21   Lee and Shlain, *Acid Dreams*, 75.

22   One of Leary's colleagues in the Harvard project who went on to become president of the American Psychiatric Association.

23   SAB, Hoffer Papers, A207, XVIII 9, Hoffer-Osmond Correspondence, Leary to Osmond, January 17, 1961.

24 Lee and Shlain, *Acid Dreams*, 76.

25 Ironically, it was Osmond who referred Ginsberg to Leary following Ginsberg's reading of the poem "Lysergic Acid" at a Boston psychiatric conference. Whereas most conference attendees scoffed at the performance, thinking the poet to be morally repugnant, perhaps even schizophrenic, Osmond was intrigued and suggested to Ginsberg that he look into the Harvard experiments and contact Leary. See Stevens, *Storming Heaven*, 144. Ginsberg also crossed paths with Hoffer through various LSD speaking engagements and was intrigued by the psychiatrist's use of niacin in the treatment of schizophrenia. See Hoffer, *Adventures in Psychiatry*, 114–15.

26 Stevens, *Storming Heaven*, 145.

27 "Statement of Purpose," 6.

28 Braden, "LSD and the Press," 403–04.

29 Ibid.

30 SAB, Hoffer Papers, A207, XVIII 13, Hoffer-Osmond Correspondence, March 17, 1963.

31 Ibid., Osmond to Metzner, June 25, 1963.

32 Ironically, one of the most respected biological psychiatrists of the twentieth century, Nathan Kline, was serious about putting psychopharmacological substances in drinking water, declaring that, "since we are already putting chlorine and fluorine in the water supply, maybe we should also put in a little lithium. It might make the world a little better place to live in for all of us." Quoted in Szasz, *Cruel Compassion*, 167.

33 Blewett was known to occasionally take his LSD patients out of the formal treatment setting and into the countryside so that they could be in nature and watch the sun come up.

34 SAB, Hoffer Papers, A207, XVIII 7, Hoffer-Osmond Correspondence, May 17, 1960.

35 SAB, Hoffer Papers, S-A1101, II. A. 106/107, Mr. M. Stolaroff International Foundation for Advanced Study #1 (1960–63) and #2 (1963–66).

36 Calder, a close friend and supporter of Blewett, was director of the Saskatchewan Bureau of Alcoholism.

37 SAB, Hoffer Papers, A207, XVIII 13, Hoffer-Osmond Correspondence, June 3, 1963.

38 Blewett Papers, 88-29, Box 3, File 46, Miscellaneous Notes, n.d.

39 Ibid.

40 Ibid., August 31, 1963.

41 SAB, Hoffer Papers, A207, XVIII 13, Hoffer-Osmond Correspondence, May 20, 1963.

42 Ibid.

43 Hofmann had noted how Sandoz LSD deteriorated when exposed to air, which is why samples were kept in air-tight vials under nitrogen. See ibid.

44 Ibid., Osmond to Hubbard, May 24, 1963.

45 Ibid., Osmond to Hoffer, May 20, 1963.

46 Ibid., Osmond to Huxley, May 25, 1963.

47 Ibid., Osmond to Hoffer, August 29, 1963.

48 Ibid., August 31, 1963.

49 Ibid.

50 Ibid., Hoffer to Osmond, September 5, 1963.

51 Ibid., XVIII 16, June 18, 1964.

52 Ibid., XVIII 17, Osmond to Hubbard, Stolaroff, and Wilson, November 3, 1964.

53 Lee and Shlain, *Acid Dreams*, 91.

54 For in-depth discussions of the U.S. Army's experiments with LSD and other drugs as possible agents of chemical warfare, see Ketchum, *Chemical Warfare*; and Khatchadourian, "Operation Delirium."

55 Marks, *The Search for the "Manchurian Candidate,"* 58.

56 Buckman, "Brainwashing, LSD, and CIA," 10–11.

57 Stevens, *Storming Heaven*, 82.

58 Lee and Shlain, *Acid Dreams*, 26.

59 Ibid.

60 Marks, *The Search for the "Manchurian Candidate,"* 72–73.

61 Lee and Shlain, *Acid Dreams*, 19.

62 Those on the CIA and U.S. Army Chemical Corps payrolls included Boston psychiatrist Max Rinkel (the first to conduct scientific studies of LSD in the United States), Harris Isbell (director of the NIMH), and Paul Hoch (one of the first to theorize on the ability of hallucinogens to model psychoses).

63 Cameron, a former president of the Canadian, American, and World Psychiatric Associations, was contracted by the CIA under its MK-ULTRA program to test several drugs on patients. For a detailed account of his CIA-backed experiments, see Collin, *In the Sleep Chamber*.

64 "Acid Tests," *Canada Now*, CBC Regina, 1992.

65 Ibid.

66 Lee and Shlain, *Acid Dreams*, 45.

67 See SAB, Hoffer Papers, A207, XVIII 1(b), Hoffer-Osmond Correspondence, October 28, 1952.

68 SAB, Hoffer Papers, A207, Dr. H. Abramson III 3, Hoffer to Abramson, December 23, 1954.

69 Aaronson, who went on to co-author the psychedelics anthology with Osmond, unknowingly had his research on hypnosis and altered states of consciousness paid for by a CIA conduit. Pfeiffer, a Princeton University pharmacologist, was known to have close links to the agency and tested LSD on inmates in federal penitentiaries.

70 Lee and Shlain, *Acid Dreams*, 52.

71 Ibid., 53.

72 Dallas, *Pink Floyd*, 32.

73 At the time, the FBI was conducting its top-secret counterintelligence program, or COINTELPRO, to keep tabs on, and disrupt through whatever means

necessary (agents provocateurs, blackmail, and assassination), other domestic political and radical threats to the American status quo, such as the civil rights/Black Power, American Indian, and New Left movements.

74 Gitlin, "On Drugs and Mass Media in American Consumer Society," 50.

75 Ibid.

76 Baumeister and Placidi, "A Social History and Analysis of the LSD Controversy," 25.

77 Historian Chris Elcock has analyzed the patriotic elements inherent in the psychedelic movement used by Leary and others as a means to legitimize the psychedelic experience. As he asserts, the movement's "revolutionary agenda needs to be tempered and described more fittingly as a reformist movement that may have departed from mainstream American culture in many ways, but not from several of its cornerstones." Elcock, "The Fifth Freedom," 32.

78 Dallos, "Dr. Leary Starts 'New' Religion with 'Sacramental' Use of LSD."

79 Stevens, *Storming Heaven*, 223.

80 Cassady provided the amphetamine-fuelled inspiration for Kerouac's central character, Dean Moriarty, in *On the Road*.

81 For footage of the adventures, see Gibney and Ellwood, *Magic Trip*.

82 Stevens, *Storming Heaven*, 301.

83 See Greenfield, "The King of LSD."

84 Marsh, "Meaning and the Mind Drugs," 408–09.

85 Ibid., 410.

86 Alpert, Cohen, and Schiller, LSD.

87 Ibid., 10.

88 Ibid.

89 Kennedy's wife, Ethel, was one of a number of prominent Americans to have gone through LSD-assisted therapy under Dr. Ross McLean and company at Hollywood Hospital in New Westminster, British Columbia. Kennedy himself was also reputed to have had experiences with psychedelics in New York.

90 Baumeister and Placidi, "A Social History and Analysis of the LSD Controversy," 37.

91 Lee and Shlain, *Acid Dreams*, 94.

92 URA, Blewett Papers, 88-29, Correspondence—Professional, Blewett to Katz, February 15, 1965.

93 SAB, Hoffer Papers, A207, XVIII 20, Hoffer-Osmond Correspondence, November 5, 1965.

94 Blewett, *The Frontiers of Being*, 262.

95 SAB, Hoffer Papers, A207, XVIII 20, Hoffer-Osmond Correspondence, December 14, 1965.

96 Ibid.

97 Discussion with the author, July 2004.

98 SAB, Hoffer Papers, A207, XVIII 25, Hoffer-Osmond Correspondence, October 6, 1966.

99 Hoffman, *Revolution for the Hell of It*, 28.

100 Leary, "Interview with Timothy Leary," 93.
101 SAB, Hoffer Papers, A207, XVIII 25, Hoffer-Osmond Correspondence, Osmond to Leary, December 10, 1966.
102 Stafford and Golightly, *LSD*, 11–12.
103 Ibid., 12.
104 Ibid., 13.
105 Ibid., 16–17.
106 Weller, "Suddenly That Summer."
107 Lee and Shlain, *Acid Dreams*, 159.
108 Ibid.
109 Stafford, *Psychedelics Encyclopedia*, 54.
110 "Governor Shafer Calls LSD Blindings a Hoax."
111 For a brief discussion on how the scientific claims regarding LSD and genetic damage were politicized, see Shapin, "Cordelia's Love," 264–65.
112 Lee and Shlain, *Acid Dreams*, 155.
113 Baumeister and Placidi, "A Social History and Analysis of the LSD Controversy," 37.
114 Ibid., 38.
115 See Peck, *Uncovering the Sixties*, 51.
116 Baumeister and Placidi, "A Social History and Analysis of the LSD Controversy," 51.
117 See Lee and Shlain, *Acid Dreams*, 188.
118 Gitlin, "On Drugs and Mass Media in America's Consumer Society," 44.
119 Baumeister and Placidi, "A Social History and Analysis of the LSD Controversy," 54.
120 See ibid., 47.
121 Hoffer and Osmond, *The Hallucinogens*, v–vi.
122 SAB, Hoffer Papers, A207, XVIII 28, Hoffer-Osmond Correspondence, December 11, 1967.
123 Blewett, *The Frontiers of Being*, 7.
124 SAB, Hoffer Papers, A207, XVIII 26, Hoffer-Osmond Correspondence, February 13, 1967.
125 Ibid., XVIII 32, October 25, 1968.
126 Cole had come out strongly against the psychedelic concept and its application in psychotherapy in the early 1960s. He had questioned its scientific value, emphasized the psychotomimetic qualities and dangers of the drugs, and chastised early investigators (e.g., Osmond) for their overenthusiasm and contribution to the start of the black market problem and illicit use of drugs such as LSD.
127 SAB, Hoffer Papers, A207, XVIII 26, Hoffer-Osmond Correspondence, Osmond to Cole, February 9, 1967.
128 Ibid.
129 Ibid., Osmond to Cheek, March 5, 1967.
130 Ibid., Osmond to Hoffer, March 13, 1967.
131 From the account provided by Osmond, the alchemist seems to have been none other than Owsley, the LSD king himself.

132 Ibid., XVIII 27, April 30, 1967.

133 Ibid.

134 Ibid., March 13, 1967.

135 Ibid., March 20, 1967.

136 SAB, Hoffer Papers, IX, Papers for Publication (1964–78), Hoffer to Nash, May 16, 1967.

137 See Houston, "Review of the Evidence and Qualifications Regarding the Effect of Hallucinogenic Drugs."

138 Blewett, *Frontiers of Being*, 245.

139 Ibid.

140 Ibid., 251–52.

141 Ibid., 258.

142 Ibid., 253.

143 SAB, Hoffer Papers, A207, XVIII 28, Hoffer-Osmond Correspondence, September 21, 1967.

144 Blewett, *Frontiers of Being*, 259.

145 Ibid., 260.

146 SAB, Hoffer Papers, A207, XVIII 32, Hoffer-Osmond Correspondence, Osmond to Hogarth, November 26, 1968.

147 Grob, "Psychiatric Research with Hallucinogens."

148 The team based in the Maryland Psychiatric Research Center in the 1970s had perhaps the only research project left in the United States, examining the uses of LSD-assisted psychotherapy in the treatment of alcoholism and cancer. See Rhead, "Psychedelic Medicine."

149 Holmes, "Treatment of Alcoholism," 48.

150 *Final Report of the Commission of Inquiry into the Non-Medical Use of Drugs*, 359.

151 Schneider, "Book Review," 199.

152 Ibid.

153 Quoted in Stafford, *Psychedelics Encyclopedia*, 20.

154 See Watts, "Psychedelics and Religious Experience," 141.

155 American Psychiatric Association, "Position Statement on LSD," 353.

156 Quoted in Stafford, *Psychedelics Encyclopedia*, 20.

157 SAB, Hoffer Papers, A207, XVIII 28, Hoffer-Osmond Correspondence, December 11, 1967.

158 Sankar, *LSD—A Total Study*, 787.

159 Quoted in Blewett, "LSD—A Learning Process."

EPILOGUE

1 Sommer, "In Memoriam," 257–58.

2 Schor and Arons, "Tribute to Duncan Blewett," 541.

3 Martin, "A Psychiatric Contrarian."

4   See http://www.quackwatch.org/01QuackeryRelatedTopics/ortho.html.

5   See Kroker, "Book Review of *Psychedelic Psychiatry*," 225.

6   Mills, "Lessons from the Periphery," 192.

7   Mills, "Hallucinogens as Hard Science," 186.

8   Feyerabend, *Against Method*, 19.

9   See, for example, Ford, *Schizophrenia Cured*; and Littlefield, *Feed Your Head*.

10 For an analysis of the first modern neuroimaging study of LSD, see Carhart-Harris et al., "Neural Correlates of the LSD Experience Revealed by Multimodal Neuroimaging."

11 See Montcrieff, "A Critique of the Dopamine Hypothesis of Schizophrenia and Psychosis" and "Why Is It So Difficult to Stop Psychiatric Drug Treatment?" Recent admissions by NIMH Director Tom Insel and other influential psychiatric leaders before and after the release of DSM-V in 2013 have also highlighted the immaturity of "scientific psychiatry" and the biological paradigm for mental illness. See http://www.newyorker.com/tech/elements/the-rats-of-n-i-m-h and https://www.psychologytoday.com/blog/side-effects/201305/the-nimh-withdraws-support-dsm-5.

12 Hewitt, "Rehabilitating LSD History," 328.

13 Hoffer, "Orthomolecular Psychiatry," 16.

14 Smythies, "The Adrenochrome Hypothesis of Schizophrenia Revisited," 149. Also refer to Smythies, "Oxidative Reactions and Schizophrenia."

15 See Liu et al., "Absent Response to Niacin Skin Patch"; Messamore, "Niacin Subsensitivity"; Ward et al., "Niacin Skin Flush in Schizophrenia"; and Yao et al., "Prevalence and Specificity of the Abnormal Niacin Response."

16 Miller and Dulay, "The High-Affinity Niacin Receptor HM74A," 41.

17 Singh, "Correspondence," 198.

18 Xu and Jiang, "Niacin-Respondent Subset of Schizophrenia."

19 Correspondence with the author, August 7, 2002.

20 MDMA (3,4-methylenedioxymethamphetamine), popularly known as ecstasy, is an amphetamine known for its empathic effects (hence the term "empathogen"). Although it has some psychedelic qualities, MDMA is generally assumed to possess unique attributes that separate it from classic psychedelics such as LSD. Nevertheless, some include it under the psychedelic umbrella.

21 See www.heffter.org.

22 See Bogenschutz et al., "Psilocybin-Assisted Treatment for Alcohol Dependence"; CBC News, "Psilocybin and Cancer Anxiety"; Griffiths et al., "Psilocybin Can Occasion Mystical-Type Experiences"; Grob et al., "Pilot Study of Psilocybin Treatment for Anxiety"; Krupitsky and Grinenko, "Ketamine Psychedelic Therapy (KPT)"; *Lancet*, "Reviving Research into Psychedelic Drugs"; Rhead, "Psychedelic Medicine"; Sessa, "Is There a Case for MDMA-Assisted Psychotherapy in the UK?"; and Sewell, Halpern, and Pope, "Response of Cluster Headache to Psilocybin and LSD."

23 Sessa, "Why Psychiatry Needs Psychedelics," 57.

24   See de Wit, "Towards a Science of Spiritual Experience."

25   "Psychedelics Could Help Addicts, Say B.C. Drug Officials."

26   Tupper et al., "Psychedelic Medicine," 5–6.

27   For a review of Maté's use of ayahuasca-assisted therapy, see Johnston et al., *The Jungle Prescription*, which debuted on CBC Television's *The Nature of Things* on June 28, 2012.

28   Maté, "Postscript—Psychedelics in Unlocking the Unconscious," 217.

29   See Smith, Raswyck, and Dickerson Davidson, "From Hofmann to Haight Ashbury," 6.

30   Ibid., 7.

31   See http://www.tampabay.com/news/publicsafety/tampa-police-close-lsd-investigation-with-no-answers-about-tainted-meat/2206752.

32   See, for example, http://reset.me/story/howpsychedelicssavedmylife/; and http://www.theguardian.com/commentisfree/2006/oct/12/acidandalcoholdontmix.

33   Sessa, "Why Psychiatry Needs Psychedelics," 57.

34   Ibid., 60.

35   Ibid. While some groups such as MAPS have managed to get around some of the funding and other regulatory hurdles to conduct psychedelic research, there remain major barriers that prevent the therapeutic potential of "controlled" psychedelics such as LSD from being fully realized. See Nutt, King, and Nichols, "Effects of Schedule I Drug Laws on Neuroscience Research and Treatment Innovation."

36   Ibid.

37   Langlitz, *Neuropsychedelia*, 240. Randomized-controlled trials (or RCTs) remain the gold standard of the EBM paradigm, but, as Goldenberg writes, "in the interest of better science…open-ended critical inquiry should be encouraged, as should comparative clinical research and problem-specific methodology (which may include uncontrolled methods and even reliance on clinical judgment." Goldenberg, "Iconoclast or Creed?," 184.

38   Sessa, "Why Psychiatry Needs Psychedelics," 61.

39   Ibid., 62.

40   Ibid.

41   Beresford, "Introduction to First Edition," 15.

42   See Bell, *Acid Tests*.

43   Anonymous patient, interview with Barber and Dyck, June 22, 2003.

44   Krebs and Johansen, "Review—Lysergic Acid Diethylamide (LSD) for Alcoholism."

45   Ibid., 6.

46   Ibid., 7.

47   SAB, Hoffer Papers, A207, XVIII 7, Hoffer-Osmond Correspondence, February 25, 1960.

48   Deloria, "Perceptions and Maturity," 7.

REFERENCES

PERSONAL INTERVIEWS AND CORRESPONDENCE

Agnew, Neil. Correspondence. November 27, 2003.

Allen, Arthur. Correspondence. September 2005.

———. Personal interview. June 26, 2003, and July 2004.

Anonymous former patient. Personal interview. June 22, 2003.

Blewett, Duncan. Correspondence. August 7, 2002.

———. Personal interview. June 28, 2003.

Hoffer, Abram. Correspondence. May 2002–March 2009.

———. Personal interview. June 27, 2003.

GOVERNMENT REPORTS

Canada. Commission of Inquiry into the Non-Medical Use of Drugs. *Final Report of Commission of Inquiry into the Non-Medical Use of Drugs*. Ottawa: Queen's Printer, 1973.

———. *Debates of the House of Commons*. Third Session, Twenty-Second Parliament. Vol. 3: 2203–3286.

UNPUBLISHED THESIS

Anderson, Erik. "Safe from Utopia? The LSD Controversy in Saskatchewan, 1950–67." MA thesis, University of British Columbia, 1996.

ARCHIVAL SOURCES

Saskatchewan Archives Board (SAB)
Canadian Mental Health Association, Saskatchewan Division Papers

> R 1265-V-K-169.
> R 1265-VI-A-114.
> R 1265-VI-H-100.
> R 1265-VI-H-171.
> R 1265-VI-H-100.

T. C. Douglas Papers

> R-33.1. 573a.

Gordon Grant Papers

> R-45, File 80 (8-4), Psychiatric Services: Miscellaneous, January 1967–December 1969.
> R-45, File 203b, Correspondence, July 2, 1968.

Abram Hoffer Papers

> A207, III (3), Correspondence Dr. H. Abramson.
> A207, III (82), Correspondence T. C. Douglas.
> A207, III (109), Correspondence Al Hubbard.
> A207, III (194a), Correspondence D. G. McKerracher.
> A207, III (271), Correspondence Sandoz Pharmaceuticals Ltd.—Montreal.
> S-A1101, II. A. 106/107, Mr. M. Stolaroff, International Foundation for Advanced Study #1 (1960–62) and #2 (1963–66).
> A207, IX, Papers for Publication (1964–78), Dr. P. Nash, Medical Director, Abbott Labs Ltd.—Montreal.
> A207, X, 3, HOD File, 1961–63.
> A207, X, 12, Ford Foundation Submission.
> A207, XVIII, Hoffer-Osmond Correspondence, 1951–92, 1–36.
> A207, File 71, Press Relations, 1961, 1966.

Tape S-135a and b. Interview with Humphry Osmond by Marjorie McEnaney, 1960.
University of Regina Archives (URA)
Duncan Blewett Papers

> 88-29.
> 91-87.
> 93-19.
> 2005-05.

VIDEO ARCHIVES

Bell, Kenneth. *Acid Tests*. Canada Now, CBC, 1993.
Gibney, Alex, and Alison Ellwood. *Magic Trip: Ken Kesey's Search for a Kool Place*. Magnolia Pictures, 2011.

Johnston, Mark, et al. *The Jungle Prescription*. Documentary for CBC's The Nature of
    Things. Nomad Films, 2012.
Littlefield, Connie. *Hofmann's Potion*. Montreal: NFB, 2002.
———. *Feed Your Head*. Conceptafilm, 2010.
McLennan, Gordon. *Psychedelic Pioneers*. Saskatoon: Kahani Entertainment, 2005.
"Psilocybin and Cancer Anxiety." *The National*, CBC, November 5, 2007.

BOOKS AND ARTICLES

Aaronson, Bernard, and Humphry Osmond, eds. *Psychedelics: The Uses and
    Implications of Hallucinogenic Drugs*. New York: Doubleday and Company, 1970.
Agnew, Neil, and Abram Hoffer. "Nicotinic Acid Modified Lysergic Acid Diethylamide
    Psychosis." *Journal of Mental Science* 101 (1955): 12–27.
Albaugh, Bernard, and Philip Anderson. "Peyote in the Treatment of Alcoholism
    among American Indians." *American Journal of Psychiatry* 131, 11 (1974):
    1247–49.
Alcoholics Anonymous. *"Pass It On": The Story of Bill Wilson and How the AA Message
    Reached the World*. New York: Alcoholics Anonymous World Services, 1984.
Alpert, Richard, Sidney Cohen, and Lawrence Schiller. *LSD*. New York: New American
    Library, 1966.
American Academy of Pediatrics, Committee on Nutrition. "Megavitamin Therapy for
    Childhood Psychoses and Learning Disabilities." *Pediatrics* 58, 6 (1976): 910–12.
American Psychiatric Association. "Position Statement on LSD." *American Journal of
    Psychiatry* 123, 3 (1966): 353.
———. *Task Force Report 7: Megavitamin and Orthomolecular Therapy in Psychiatry—A
    Report of the APA Task Force on Vitamin Therapy in Psychiatry*. Washington, DC:
    American Psychiatric Association, 1973.
Ananth, J. V., L. Vacaflor, G. Kekhwa, C. Sterlin, and T. A. Ban. "Nicotinic Acid in the
    Treatment of Newly Admitted Schizophrenic Patients: A Placebo Controlled
    Study." *International Journal of Clinical Pharmacology* 5 (1972): 406–10.
Anderson, Terry. *The Sixties*. 3rd ed. New York: Pearson and Longman, 2007.
Archer, John. *Saskatchewan, a History*. Saskatoon: Western Producer Prairie Books,
    1980.
Archibald, H. David. *The Addiction Research Foundation: Voyage of Discovery*. Toronto:
    ARF, 1990.
Arieti, Silvano. *Interpretations of Schizophrenia*. New York: Basic Books, 1974.
Autry, Joseph III. "Workshop on Orthomolecular Treatment of Schizophrenia: A
    Report." *Schizophrenia Bulletin* 1, 12 (1975): 91–103.
Axelrod, Julius. "O-Methylation of Epinephrine and Other Catechols in Vitro and in
    Vivo." *Science* 126 (1957): 400–01.
———. "Reflections: Journey of a Late Blooming Biochemical Neuroscientist."
    Journal of Biological Chemistry 278, 1 (2003): 1–13.

Baekeland, Frederik. "Evaluation of Treatment Methods in Chronic Alcoholism." In *The Biology of Alcoholism 5*, edited by Benjamin Kissin and Henri Begleiter, 385–440. New York: Plenum Press, 1977.

Baihly, Lee. "The Professional Woos the Layman: Psychiatrists Seek the Public's Support for Drug They Say Cures Schizophrenia." *Psychiatric News*, April 1967, 1–2.

Baldessarini, Ross, and Solomon Snyder. "Schizophrenia: A Critique of Recent Genetic-Biochemical Formulations." *Nature 204* (1965): 1111–12.

Ban, Thomas. "Fifty Years Chlorpromazine: A Historical Perspective." Neuropsychiatric Disease and Treatment 3, 4 (2007): 495–500.

———. "Nicotinic Acid and Psychiatry." *Canadian Psychiatric Association Journal 16*, 5 (1971): 413–31.

———. "Pharmacotherapy of Schizophrenia: Facts, Speculations, Hypotheses, and Theories." *Psychosomatics 15*, 4 (1974): 178–87.

———. *Recent Advances in the Biology of Schizophrenia*. Springfield, IL: Charles C. Thomas, 1973.

Barber, Bernard. "Resistance by Scientists to Scientific Discovery." *Science 134* (1961): 596–602.

———. "The Sociology of Science: A Trend Report and Bibliography." *Current Sociology* 5, 2 (1956): 91–111.

Barnes, Barry. "Catching Up with Robert Merton: Scientific Collectives as Status Groups." *Journal of Classical Sociology 7*, 2 (2007): 179–92.

———. *T.S. Kuhn and Social Science*. New York: Columbia University Press, 1982.

Barnes, Barry, David Bloor, and John Henry. *Scientific Knowledge: A Sociological Analysis*. Chicago: University of Chicago Press, 1996.

Barsa, Joseph. "The Fallacy of the Double Blind." *American Journal of Psychiatry 119* (1963): 1174–75.

Baumeister, Alan, and Mike Hawkins. "Continuity and Discontinuity in the Historical Development of Modern Psychopharmacology." *Journal of the History of the Neurosciences 14*, 3 (2005): 199–209.

Baumeister, Roy, and Kathleen Placidi. "A Social History and Analysis of the LSD Controversy." *Journal of Humanistic Psychology 23*, 4 (1983): 25–58.

Benjamin, John. "Biological Research in Schizophrenia." *Psychosomatic Medicine 20*, 6 (1958): 427–45.

Beresford, John. "Introduction." In *Psychedelics Encyclopedia*, 3rd ed., edited by Peter Stafford, 13–19. Berkeley: Ronin Publishing, 1992.

Bleuler, Manfred. "Comparison of Drug-Induced and Endogenous Psychoses in Man," in *Neuro-Psychopharmacology: Proceedings of the First International Congress of Neuro-Pharmacology* (Rome, September, 1958), edited by P. B. Bradley, P. Deniker, and C. Radoucho-Thomas, 161–65. Amsterdam: Elsevier Publishing Company, 1959.

Blewett, Duncan. *The Frontiers of Being*. New York: Award Books, 1969.

———. "Interim Report on the Therapeutic Use of LSD." Unpublished manuscript, 1958.

———. "LSD—A Learning Process: An Interview with Duncan Blewett." *Carillon*, March 6, 1967.

———. "Psychedelic Drugs in Parapsychological Research." *International Journal of Parapsychology* 5, 1 (1963): 43–73.

———. "The Psychedelic Experience in the Native American Church." In "The Road to Manitou," by Fannie Kahan, 233–49. Unpublished manuscript, 1958.

Blewett, Duncan, and Nicholas Chwelos. *Handbook for the Therapeutic Use of Lysergic Acid Diethylamide-25: Individual and Group Procedures*. http://www.maps.org/research-archive/ritesofpassage/lsdhandbook.pdf.

Bloor, David. *Knowledge and Social Imagery*. London: Routledge Direct Editions and Kegan Paul, 1976.

Bogenschutz, Michael, et al. "Psilocybin-Assisted Treatment for Alcohol Dependence: A Proof-of-Concept Study." *Journal of Psychopharmacology* 29, 3 (2015): 289–99.

"Book Reviews: Lysergic Acid Diethylamide (LSD) in the Treatment of Alcoholism by Smart, Storm, Baker, and Solurish." *Journal of the Royal Society for the Promotion of Health* 89, 1 (1969): 41.

Bowen, W. E., R. A. Soskin, and J. W. Chotlos. "Lysergic Acid Diethylamide as a Variable in the Treatment of Alcoholism. A Follow-Up Study." *Journal of Nervous Mental Disease* 150 (1970): 111–18.

Braden, William. "LSD and the Press." In *Psychedelics*, edited by Bernard Aaronson and Humphry Osmond, 400–18. Garden City, NY: Anchor Books, 1970.

———. *The Private Sea: LSD and the Search for God*. http://www.druglibrary.org/schaffer/lsd/braden3.htm.

Braunstein, Peter, and Michael William Doyle, eds. *Imagine Nation: The American Counterculture of the 1960s and '70s*. New York: Routledge, 2002.

Brown, W. T. "Megavitamin and Orthomolecular Therapy of Schizophrenia." *Canadian Psychiatric Association Journal* 20, 2 (1975): 97–100.

Buckman, John. "Brainwashing, LSD, and CIA: Historical and Ethical Perspective." *International Journal of Social Psychiatry* 23, 8 (1977): 8–19.

———. "Theoretical Aspects of LSD Therapy." In *The Use of LSD in Psychotherapy and Alcoholism*, edited by Harold Abramson, 83–101. New York: Bobbs-Merrill, 1967.

Burroughs, William S., and Allen Ginsberg. *The Yage Letters*. San Francisco: City Lights Books, 1963.

Busch, A. K., and W. C. Johnson. "LSD-25 as an Aid in Psychotherapy (Preliminary Report of a New Drug)." *Diseases of the Nervous System* 10 (1950): 241–45.

Calabrese, Joseph. "Clinical Paradigm Clashes: Ethnocentric and Political Barriers to Native American Efforts at Self-Healing." *Ethos* 36, 3 (2008): 334–53.

Campbell, Angus. *The Grand Vision: A History of the Bureau on Alcoholism and the Saskatchewan Alcoholism Commission*. Winnipeg: Kromar Printing, 1993.

Canadian Mental Health Association. *Ten Giant Steps: Strides in Mental Health*. Regina: CMHA—Saskatchewan Division, 1960.

Capshew, James. *Psychologists on the March: Science, Practice, and Professional Identity in America, 1929–1969*. New York: Cambridge University Press, 1999.

Carhart-Harris, Robin, et al. "Neural Correlates of the LSD Experience Revealed by Multimodal Neuroimaging." *PNAS* 13, 17 (2016): 4853–58.

Cathcart, L. M. "Nutrition and Schizophrenia: More on the Controversy." *Canadian Family Physician 21*, 6 (1975): 49–50.

Cheek, Frances, H. Osmond, M. Sarett, and R. Albahary. "Observations Regarding the Use of LSD-25 in the Treatment of Alcoholism." *Journal of Psychopharmacology 1*, 2 (1966): 56–74.

Chwelos, N., D. Blewett, C. Smith, and A. Hoffer. "Use of D-Lysergide Acid Diethylamide in the Treatment of Alcoholism." *Quarterly Journal of Studies on Alcohol 20*, 3 (1959): 577–90.

Clark, Huston. "Do Drugs Have Religious Import?" *Journal of Philosophy 61*, 18 (1964): 517–30.

Clarke, Arthur C. *Profiles of the Future (An Inquiry into the Limits of the Possible).* 5th ed. New York: Warner Books, 1985.

Cleckley, H. M., V. P. Sydenstricker, and L. E. Geeslin. "Nicotinic Acid in the Treatment of Atypical Psychotic States." *Journal of American Medical Association 112*, 21 (1939): 2107–10.

Cohen, Sidney. "LSD: Side Effects and Complications." *Journal of Nervous Disease 130* (1960): 30–40.

Cole, Jonathan, and Martin Katz. "The Psychotomimetic Drugs: An Overview." *Journal of the American Medical Association 187* (1964): 758–61.

Collins, Anne. *In the Sleep Chamber: The Story of the CIA Brainwashing Experiments in Canada.* Toronto: Lester and Orpen Dennys, 1988.

Costello, C. G. "An Evaluation of Aversion and LSD Therapy in the Treatment of Alcoholism." *Canadian Psychiatric Association Journal 14*, 1 (1969): 31–42.

Cottrell, Robert. *Sex, Drugs, and Rock 'N' Roll: The Rise of America's 1960s Counterculture.* Lanham, MD: Rowman and Littlefield, 2015.

Crockford, Ross. "Dr. Yes." *Western Living*, December 2001, 42–53.

Dallas, Karl. *Pink Floyd: Bricks in the Wall.* New York: SPI Books, 1994.

Dallos, Robert. "Dr. Leary Starts 'New' Religion with 'Sacramental' Use of LSD." *New York Times*, September 20, 1966.

Davies, D. L. "Normal Drinking in Recovered Alcoholic Addicts." *Quarterly Journal of Studies on Alcohol 23* (1962): 94–104.

De Bont, Raf. "Schizophrenia, Evolution, and the Borders of Biology: On Huxley et al.'s 1964 Paper in Nature." *History of Psychiatry 21*, 2 (2010): 144–59.

de Wit, Harriet. "Editorial: Towards a Science of Spiritual Experience." *Psychopharmacology 187*, 3 (2006): 267.

Deloria, Vine Jr. "Perceptions and Maturity: Reflections on Feyerabend's Point of View." In *Spirit and Reason: The Vine Deloria Jr. Reader*, edited by Barbara Deloria, Kristen Foehner, and Sam Scinta, 3–16. Golden, CO: Fulcrum Publishing, 1999.

———. *Red Earth, White Lies: Native Americans and the Myth of Scientific Fact.* Golden, CO: Fulcrum Publishing, 1997.

Dickinson, Harley D. *The Two Psychiatries: The Transformation of Psychiatric Work in Saskatchewan, 1905-1984*. Regina: Canadian Plains Research Center, 1984.

Doblin, Rick. "Dr. Leary's Concord Prison Experiment: A 34-Year Follow-Up Study." *Journal of Psychoactive Drugs 30*, 4 (1998): 419-26.

Donaldson, Elizabeth. "Psychomimesis: LSD and Disability Immersion Experiences of Schizophrenia." *Disability Studies Quarterly 33*, 1 (2013): n. pag. http://dsq-sds.org/article/view/3431/3203.

Dooley, Chris. "The End of the Asylum (Town): Community Responses to the Depopulation and Closure of the Saskatchewan Hospital, Weyburn." *Social History 44*, 88 (2011): 331-54.

Dyck, Erika. "Flashback: Psychiatric Experimentation with LSD in Historical Perspective." *Canadian Journal of Psychiatry 50*, 7 (2005): 381-88.

———. "Hitting Highs at Rock Bottom: LSD Treatment for Alcoholism, 1950-1970." *Social History of Medicine 19*, 2 (2006): 313-29.

———. "Land of the Living Sky with Diamonds: A Place for Radical Psychiatry?" *Journal of Canadian Studies 43*, 3 (2007): 42-66.

———. "Prairie Psychedelics: Mental Health Research in Saskatchewan, 1951-67." In *Mental Health and Canadian Society: Historical Perspectives*, edited by James E. Moran and David Wright, 221-44. Montreal: McGill-Queen's University Press, 2006.

———. "Prairies, Psychedelics, and Place: The Dynamics of Region in Psychiatric Research." *Health and Place 15* (2009): 657-63.

———. *Psychedelic Psychiatry: LSD from Clinic to Campus*. Baltimore: Johns Hopkins University Press, 2008.

———. "Spaced-Out in Saskatchewan: Modernism, Anti-Psychiatry, and Deinstitutionalization, 1950-1968." *Bulletin of the History of Medicine 84* (2010): 640-66.

Dyck, Erika, and Tolly Bradford. "Peyote on the Prairies: Religion, Scientists, and Native Newcomer Relations in Western Canada." *Journal of Canadian Studies 46*, 1 (2012): 28-52.

Edginton, Barry. "Architecture as Therapy: A Case Study in the Phenomenology of Design." *Journal of Design History 23*, 1 (2010): 83-97.

Edwards, Griffith, ed. *Addiction: Evolution of a Specialist Field*. Oxford: Blackwell Science, 2002.

Elcock, Chris. "The Fifth Freedom: The Politics of Psychedelic Patriotism." *Journal of the Study of Radicalism 9*, 2 (2015): 17-40.

———. "From Acid Revolution to Entheogenic Evolution: Psychedelic Philosophy in the Sixties and Beyond." *Journal of American Culture 36*, 4 (2013): 296-311.

Elvehjem, C. A. "Tryptophan and Niacin Relations and Their Implications to Human Nutrition." *Journal of the American Dietetic Association 24*, 8 (1948): 653-57.

Farber, David. "The Intoxicate State/Illegal Nation: Drugs in the Sixties Counterculture." In *Imagine Nation: The American Counterculture of the 1960s and '70s*, edited by Peter Braunstein and Michael William Doyle, 17-40. New York: Routledge, 2002.

————, ed. *The Sixties: From Memory to History*. Chapel Hill: University of North Carolina Press, 1994.

Farreras, Ingrid. "The Historical Context for NIMH Support of APA Training and Accreditation Efforts." In *Psychology and the National Institute of Mental Health: A Historical Analysis of Science, Practice, and Policy*, edited by Wade Pickren and S. F. Schneider, 153–79. Washington, DC: American Psychological Association, 2005.

Faust, David, and Richard Miner. "The Empiricist and His New Clothes: DSM-III in Perspective." *American Journal of Psychiatry 143*, 8 (1986): 962–67.

Feyerabend, Paul. *Against Method: Outline of an Anarchistic Theory of Knowledge*. London: Humanities Press, 1975.

————. "Democracy, Elitism, and the Scientific Method." In *Paul Feyerabend: Knowledge, Science, and Relativism*, edited by John Preston, 212–26. Cambridge, UK: Cambridge University Press, 1999.

————. "'Science': The Myth and Its Role in Society." *Inquiry 18*, 2 (1975): 167–81.

Fogel, S., and Abram Hoffer. "The Use of Hypnosis to Interrupt and to Reproduce an LSD-25 Experience." *Journal of Clinical and Experimental Psychopathology and Quarterly Review of Psychiatry and Neurology 23*, 1 (1962): 11–16.

Ford, Terra. *Schizophrenia Cured: A Case History and a Look at Orthomolecular Therapy*. Regina: Canadian Schizophrenia Foundation, 1978.

Frazier, E. I., and T. E. Friedemann. "Pellagra, a Study in Human Nutrition. The Multiple-Factor Principle of the Determination of Minimum Vitamin Requirements." *Quarterly Bulletin Northwestern University Medical School 20*, 24 (1946): 24–48.

Frederking, W. "The Use of LSD and Mescaline in Psychotherapy." *Psyche 7* (1953–54): 342.

Fromm-Reichmann, Frieda. "Notes on the Development of Treatment of Schizophrenics by Psychoanalytic Psychotherapy." In *Psychoanalysis and Psychotherapy: Selected Papers of Frieda Fromm-Reichmann*, edited by Dexter Bullard, 160–75. Chicago: University of Chicago Press, 1959.

Gach, John. "Biological Psychiatry in the Nineteenth and Twentieth Centuries." In *History of Psychiatry and Medical Psychology*, edited by Edwin Wallace and John Gach, 381–418. New York: Springer Science, 2008.

Galatzer-Levy, Isaac, and Robert Galatzer-Levy. "The Revolution in Psychiatric Diagnosis: Problems at the Foundations." *Perspectives in Biology and Medicine 50*, 2 (2007): 161–80.

Ghaemi, Seyyed Nassir. "Paradigms of Psychiatry: Eclecticism and Its Discontents." *Current Opinion in Psychiatry 19* (2006): 619–24.

Giorgi, Amedeo. *Psychology as a Human Science: A Phenomenologically Based Approach*. New York: Harper-Row, 1970.

Gitlin, Todd. "On Drugs and Mass Media in American Consumer Society." In *Youth and Drugs: Society's Mixed Messages*, edited by Hank Resnik, 31–52. Rockville, MD: U.S. Department of Health and Human Services, 1990.

————. *The Sixties: Years of Hope, Days of Rage*. New York: Bantam Books, 1993.

Glatt, M. M. "Alcoholism Disease Concept and Loss of Control Revisited." *British Journal of Addiction* 71, 2 (1976): 135–44.

Goertzel, Ted, Mildred Goertzel, and Victor Goertzel. "Linus Pauling: The Scientist as Crusader." *Antioch Review* 38, 3 (1980): 371–82.

Goffman, Erving. "The Moral Career of the Mental Patient—Inpatient Phase." In *The Sociology of Mental Illness*, edited by Oscar Grusky and Melvin Pollner, 301–08. New York: Holt, Rinehart and Winston, 1981.

Goldberger, Joseph, and G. A. Wheeler. "The Experimental Production of Pellagra in Human Subjects by Means of Diet." In *Goldberger on Pellagra*, edited by Milton Terris, 54–94. Baton Rouge: Louisiana State University Press, 1964.

Goldenberg, Maya. "Iconoclast or Creed? Objectivism, Pragmatism, and the Hierarchy of Evidence." *Perspectives in Biology and Medicine* 52, 2 (2009): 168–87.

———. "On Evidence and Evidence-Based Medicine: Lessons from the Philosophy of Science." *Social Science and Medicine* 62 (2006): 2621–32.

Goldsmith, Grace, et al. "Studies of Niacin Requirement in Man. I. Experimental Pellagra in Subjects on Corn Diets Low in Niacin and Tryptophan." *Journal of Clinical Investigation* 31, 6 (1952): 533–42.

Goodman, Paul. "Crisis and New Spirit." In *Utopian Essays and Practical Proposals*, by Paul Goodman, 274–89. New York: Random House, 1962.

Gould, Jonathon. "The Use of Vitamins in Psychiatric Practice." *Proceedings of the Royal Society of Medicine* 47 (1953): 215–20.

"Governor Shafer Calls LSD Blindings a Hoax." *New York Times*, 19 January 1968.

Greenfield, Robert. "The King of LSD." In *Grateful Dead: The Ultimate Guide—Rolling Stone Special Collector's Edition (2013):* 78–83.

Griffiths, R. R., et al. "Psilocybin Can Occasion Mystical-Type Experiences Having Substantial and Sustained Personal Meaning and Spiritual Significance." *Psychopharmacology* 187 (2006): 268–83.

Grinker, R. R. "Bootlegged Ecstasy." *Journal of the American Medical Association* 187, 10 (1964): 768.

---. "Lysergic Acid Diethylamide." *Archives of General Psychiatry* 8 (1963): 425.

Grinspoon, Lester, and James Bakalar. "Can Drugs Be Used to Enhance the Psychotherapeutic Process?" *American Journal of Psychotherapy* 40, 3 (1986): 393–404.

Grob, Charles. "Psychiatric Research with Hallucinogens: What Have We learned?" *Yearbook for Ethnomedicine and the Study of Consciousness* 3 (1994): n. pag. http://www.druglibrary.org/Schaffer/lsd/grob.htm.

Grob, Charles S., et al. "Pilot Study of Psilocybin Treatment for Anxiety in Patients with Advanced-Stage Cancer." *Archives of General Psychiatry* 68, 1 (2010): n. pag. http://archpsyc.jamanetwork.com/article.aspx?articleid=210962&maxtoshow= &hits=10&RESULTFORMAT=&fulltext=Charle%20S.%20Grob&searchid= 1&FIRSTINDEX=0&resourcetype=HWCIT.

Grob, Gerald. *From Asylum to Community: Mental Health Policy in Modern America.* Princeton, NJ: Princeton University Press, 1991.

———. "Origins of DSM-I: A Study in Appearance and Reality." *American Journal of Psychiatry* 148, 4 (1991): 421–31.

———. "Psychiatry's Holy Grail: The Search for the Mechanisms of Mental Disease." *Bulletin of Medicine* 77, 2 (1998): 189–219.

Grof, Stanislav. *Realms of the Human Unconscious.* New York: Viking Press, 1975.

Grof, Stanislav, M. Vojtechovsky, V. Vítek, and S. Franko-vá. "Clinical and Experimental Study of Central Effects of Adrenochrome." *Journal of Neuropsychiatry* 5 (1963): 33–50.

Guy, William, Martin Gross, and Helen Dennis. "An Alternative to the Double Blind Procedure." *American Journal of Psychiatry* 123, 12 (1967): 1505–11.

Hagstrom, Warren. "Traditional and Modern Forms of Scientific Teamwork." *Administrative Science Quarterly* 9, 3 (1964): 241–63.

Hale, N. G. Jr. "American Psychoanalysis since World War II." In *American Psychiatry after World War II 1944-1994,* edited by Roy Menniger and John Nemiah, 77–102. Washington, DC: American Psychiatric Press, 2000.

Halpern, J. H. "The Use of Hallucinogens in the Treatment of Addiction." *Addiction Research* 4, 2 (1996): 177–89. https://catbull.com/alamut/Bibliothek/1996_halpern_6646_1.pdf.

Hawkins, David. "Letter to the Editor re: Adverse Effects of Niacin in Emergent Psychosis." *Journal of American Medical Association* 204 (1968): 1010–11.

Healy, David. *The Creation of Psychopharmacology.* Cambridge, MA: Harvard University Press, 2002.

———. *Let Them Eat Prozac.* Toronto: James Lorimer and Company, 2003.

———. "Psychedelic Psychiatry: LSD from Clinic to Campus (Review)," *Bulletin of the History of Medicine* 84, 4 (2010): 699–700.

Hewitt, Kim. "Rehabilitating LSD History in Postwar America: Dilworth Wayne Woolley and the Serotonin Hypothesis of Mental Illness." *History of Science* 54, 3 (2016): 307–30.

Hicks, Richard, and Paul Fink, eds. *Psychedelic Drugs: Proceedings of a Hahnemann Medical College and Hospital Symposium Sponsored by the Department of Psychiatry.* New York: Grune and Stratton, 1969.

Hoffer, Abram. "Abnormalities of Behavior." *Annual Review of Psychiatry* 11 (1960): 351–80.

———. *Adventures in Psychiatry: The Scientific Memoirs of Dr. Abram Hoffer.* Caledon, ON: KOS Publishing, 2005.

———. "Epinephrine Metabolites: Relationship of Epinephrine Metabolites to Schizophrenia." In *Chemical Concepts of Psychosis: Proceedings of the Symposium on Chemical Concepts of Psychosis Held at the Second International Congress of Psychiatry in Zurich, Switzerland, September 1 to 7, 1957,* edited by Max Rinkel and Herman Denber, 127–40. New York: McDowell and Obolensky, 1958.

———. "Letters to the Journal re: Adverse Effects of Niacin in Emergent Psychosis." *Journal of the American Medical Association* 207, 7 (1969): 1355.

———. "A Program for the Treatment of Alcoholism: LSD, Malvaria, and Nicotinic Acid." In *The Use of LSD in Psychotherapy and Alcoholism*, edited by Harold Abramson, 357–66. Indianapolis: Bobbs-Merrill, 1967.

———. "Remembering Humphry Osmond: A Pioneer in Orthomolecular Medicine." *Journal of Orthomolecular Medicine* 19, 2 (2004): 71–84.

———. "A Theoretical Examination of Double Blind Design." *Canadian Medical Association Journal* 97, 1 (1967): 123–27.

———. "Treatment of Alcoholism with Psychedelic Therapy." In *Psychedelics*, edited by B. Aaronson and H. Osmond, 357–66. New York: Doubleday and Company, 1970.

Hoffer, Abram, and M. J. Callbeck. "Drug-Induced Schizophrenia." *Journal of Mental Science* 106 (1960): 138–59.

Hoffer, Abram, and Humphry Osmond. *The Chemical Basis of Clinical Psychiatry.* Springfield, IL: Thomas, 1960.

———. "Concerning an Etiological Factor in Alcoholism: The Possible Role of Adrenochrome Metabolism." *Quarterly Journal of Studies on Alcohol* 20, 4 (1959): 750–56.

———. *The Hallucinogens*. New York: Academic Press, 1967.

———. *How to Live with Schizophrenia*. London: Johnson, 1966.

———. "Letter to the Journal: The Biochemistry of Mental Disease." *Canadian Medical Association Journal* 85 (1961): 1309–11.

———. *New Hope for Alcoholics*. New Hyde Park, NY: University Books, 1968.

———. "On Critics and Research." *Psychosomatic Medicine* 21, 4 (1959): 311–20.

———. "Outside AA: Alcoholism and the Researcher." *AA Grapevine*, January 1960, n. pag. From a reprint.

———. "Treatment of Schizophrenia with Nicotinic Acid: A Ten Year Follow-Up." *Acta Psychiatrica Scandinavica* 40, 2 (1964): 171–89.

Hoffer, Abram, Humphry Osmond, and John Smythies, "Schizophrenia: A New Approach II. Result of a Year's Research." *Journal of Mental Science* 100, 18 (1954): 29–45.

Hoffer, A., C. Smith, N. Chwelos, M. Callbeck, and M. Mahon. "Psychological Response to D-Lysergic Acid Diethylamide and Its Relationship to Adrenochrome Levels." *Journal of Clinical and Experimental Psychopathology and Quarterly Review of Psychiatry and Neurology* 20, 2 (1959): 125–34.

Hoffer, L. John. "The Controversy over Orthomolecular Therapy." *Orthomolecular Psychiatry* 3, 3 (1974): 167–85.

———. "Letters to the Editor: Megavitamin Treatment of Schizophrenia," *Canadian Psychiatric Association Journal* 20 (1975): 492–94.

———. "Megavitamin Treatment in Schizophrenia." *Canadian Psychiatric Association Journal* 21 (1976): 133–34.

———. "Orthomolecular Psychiatry: What Would Abram Hoffer Do?" Unpublished manuscript.

Hoffman, Abbie. *Revolution for the Hell of It*. New York: Dial Press, 1968.

Hofmann, Albert. LSD—*My Problem Child: Reflections on Sacred Drugs, Mysticism, and Science*. Translated by Jonathon Ott. Los Angeles: J. P. Tarcher, 1983.

Holmes, S. J. "Treatment of Alcoholism." *Canadian Family Physician*, January 1970, 46–49.

Hollister, Leo E., Jack Shelton, and George Krieger. "A Controlled Comparison of Lysergic Acid Diethylamide (LSD) and Dextroamphetamine in Alcoholics." *American Journal of Psychiatry* 125, 10 (1969): 1352–57.

Houston, Kent. "Review of the Evidence and Qualifications Regarding the Effect of Hallucinogenic Drugs on Chromosomes and Embryos." *American Journal of Psychiatry* 126, 2 (1969): 137–40.

Huxley, Aldous. *The Doors of Perception and Heaven and Hell*. London: Grafton, 1977.

———. *Moksa: Aldous Huxley's Classic Writings on Psychedelics and the Visionary Experience (1931–1963)*. Edited by Michael Horowitz and Cynthia Palmer. Rochester, VT: Park Street Press, 1999.

———. *The Perennial Philosophy*. New York: Harper and Row, 1945.

"Interview with Timothy Leary." *Playboy*, September 1966, n. pag. https://archive.org/details/playboylearyinteooplayrich.

Izumi, Kiyo. "LSD and Architectural Design." In *Psychedelics*, edited by B. Aaronson and H. Osmond, 381–97. New York: Doubleday and Company, 1970.

James, William. *The Varieties of Religious Experience: A Study in Human Nature*. New York: Modern Library, 1908.

Janiger, Oscar. "The Use of Hallucinogenic Agents in Psychiatry." *Californian Clinician*, July–August 1959, n. pag. From URA, Blewett Papers, 91-87, Box 3, File 23, Publications LSD.

Jensen, Sven. "A Treatment Program for Alcoholics in a Mental Hospital." *Quarterly Journal of Studies on Alcohol* 23, 2 (1962): 1–6. From URA, Blewett Papers, 91-87, Box 5, File 35, Publications LSD.

Johnson, F. Gordon. "LSD in the Treatment of Alcoholism." *American Journal of Psychiatry* 126, 4 (1969): 481–87.

Jung, C. G. *Modern Man in Search of a Soul*. New York: Harcourt, Brace and World, 1933.

———. "The Psychology of Dementia Praecox." In *The Psychogenesis of Mental Disease*, by Carl Jung, 1–37. Translated by R. C. F. Hull. London: Routledge and Kegan Paul, 1960.

Kahan, Fanny H. *Brains and Bricks: The History of the Yorkton Psychiatric Centre*. Regina: White Cross Publications, 1965.

———. *A Culture's Catalyst: Historical Encounters with Peyote and the Native American Church in Canada*. Edited with an Introduction by Erika Dyck. Winnipeg: University of Manitoba Press, 2016.

———. "The Road to Manitou." Unpublished manuscript, 1959.

Kantor, Seymour. "Letters to the Editor: Drugs and Drink." *American Journal of Psychiatry* 126, 12 (1970): 1798.

Kast, Eric. "Attenuation of Anticipation: A Therapeutic Use of Lysergic Acid Diethylamide." In *The Psychopharmacology of the Normal Human*, edited by Wayne Evans and Nathan Kline, 206–18. Springfield, IL: Charles C. Thomas, 1969.

———. "A Concept of Death." In *Psychedelics*, edited by B. Aaronson and H. Osmond, 366–81. New York: Doubleday and Company, 1970.

Katz, Sidney. "The Heaven or Hell Drug." *Maclean's*, June 20, 1964, 9–13, 26–28.

———. "My 12 Hours as a Madman." *Maclean's*, October 1, 1953, 9–11, 46–55.

Kershner, John. "Leeches, Quicksilver, Megavitamins, and Learning Disabilities." *Journal of Special Education* 12, 1 (1978): 7–15.

Ketchum, James. *Chemical Warfare: Secrets Almost Forgotten: A Personal Story of Medical Testing of Army Volunteers with Incapacitating Chemical Agents during the Cold War (1955–1975)*. Santa Rosa, CA: Chembooks, 2006.

Kety, Seymour. "The Academic Lecture: The Heuristic Aspect of Psychiatry." *American Journal of Psychiatry* 118 (1961): 385–97.

———. "The Biochemical Theories of Schizophrenia. Part I." *Science*, June 5, 1959, 1528–32.

———. "The Biochemical Theories of Schizophrenia. Part II." *Science*, June 12, 1959, 1590–96.

———. "The Biological Roots of Schizophrenia." In *The Sociology of Mental Illness*, edited by Oscar Grusky and Melvin Pollner, 34–45. New York: Holt, Rinehart and Winston, 1981.

———. "Recent Biochemical Theories of Schizophrenia." In *The Etiology of Schizophrenia, edited by Don Jackson, 120–45. New York: Basic Books, 1960.*

Khatchadourian, Raffi. "Operation Delirium." *New Yorker*, December 2012, n. pag. http://www.newyorker.com/magazine/2012/12/17/operation-delirium.

Kirk, Stuart, and Herb Kutchins. *The Selling of DSM: The Rhetoric of Science in Psychiatry*. New York: A. de Gruyter, 1992.

Klein, Donald. "Comment." *American Journal of Psychiatry* 131, 11 (1974): 1263–65.

Kline, N. S., et al. "Controlled Evaluation of Nicotinamide Adenine Dinucleotide in the Treatment of Chronic Schizophrenic Patients." *British Journal of Psychiatry* 113 (1967): 731–42.

Koyré, Alexandre. "Commentary on H. Gueclac's 'Some Historical Assumptions of the History of Science." In *Scientific Change: Historical Studies in the Intellectual, Social, and Technical Conditions for Scientific Discovery and Technical Invention, from Antiquity to the Present*, edited by A. C. Crombie, 847–57. London: Heinemann, 1963.

Krebs, Teri, and Pal-Ørjan Johansen. "Review: Lysergic Acid Diethylamide (LSD) for Alcoholism: Meta-Analysis of Randomized Controlled Trials." *Journal of Psychopharmacology* (March 2012): 1–9. http://jop.sagepub.com/content/early/2012/03/08/0269881112439253.

Kroker, Kenton. "Book Review of Psychedelic Psychiatry: LSD from Clinic to Campus by Erika Dyck." *Canadian Bulletin of Medical History* 27, 1 (2010): 225–27.

Krupitsky, E. M., and A. Y. Grinenko. "Ketamine Psychedelic Therapy (KPT): A Review of the Results of Ten Years of Research." *Journal of Psychoactive Drugs* 29, 2 (1997): 165–83.

Kuhn, Thomas S. *The Structure of Scientific Revolutions*. 2nd ed. Chicago: University of Chicago Press, 1970.

Kurland, A., C. Savage, W. N. Pahnke, S. Grof, and J. E. Olsen. "LSD in the Treatment of Alcoholics." *Pharmakopsychiatry 4* (1971): 83–94.

Kurland, Albert, Sanford Unger, John Shaffer, and Charles Savage. "Psychedelic Therapy in the Treatment of the Alcoholic Patient: A Preliminary Report." *American Journal of Psychiatry 23*, 10 (1967): 1202–09.

Kurtz, Ernest. "Alcoholics Anonymous and the Disease Concept of Alcoholism." *Alcoholism Treatment Quarterly 20*, 3 (2002): 5–39.

———. *Not God: A History of Alcoholics Anonymous*. Center City, MN: Hazelden Pittman Archives Press, 1979.

LaBrosse, E. H., J. D. Mann, and S. S. Kety. "The Physiological and Psychological Effects of Intravenously Administered Epinephrine, and Its Metabolism, in Normal and Schizophrenic Men—III. Metabolism of 7-H3-Epinephrine as Determined in Studies on Blood and Urine." *Journal of Psychiatric Research 1*, 1 (1961): 68–75.

Lame Deer, John (Fire), and Richard Erdoes. *Lame Deer: Seeker of Visions*. New York: Simon and Schuster, 1972.

Langlitz, Nicolas. *Neuropsychedelia*. Berkeley: University of California Press, 2012.

———. "The Persistence of the Subjective in Neuropsychopharmacology: Observations of Contemporary Hallucinogenic Research." *History of the Human Sciences 23*, 1 (2010): 37–57.

Latour, Bruno, and Steve Woolgar. *Laboratory Life: The Construction of Scientific Facts*. Princeton, NJ: Princeton University Press, 1986.

Leary, Timothy. *The Politics of Ecstasy*. New York: G. P. Putnam and Sons, 1965.

Lee, Martin, and Bruce Shlain. *Acid Dreams—The Complete Social History of LSD: The CIA, the Sixties, and Beyond*. New York: Grove Press, 1992.

Lehmann, Heinz. "Case Report: Post-Traumatic Confusional State Treated with Massive Doses of Nicotinic Acid." *Canadian Medical Association Journal 51* (1944): 558–60.

———. "Chlorpromazine in Psychiatric Conditions." *Canadian Medical Association Journal 72* (1955): 91–99.

Lehmann, Heinz, and T. A. Ban. "Author's Reply: Megavitamin Treatment of Schizophrenia." *Canadian Psychiatric Association Journal 20* (1975): 494–95.

"Letters to the Editor." *Canadian Family Physician 21*, 6 (1975): 47–50.

Levine, Harry Gene. "The Alcohol Problem in America: From Temperance to Alcoholism." *British Journal of Addiction 79* (1984): 109–19.

———. "The Discovery of Addiction: Changing Conceptions of Habitual Drunkenness in America." *Journal of Studies on Alcohol 39*, 1 (1978): 143–74.

Levine, Jerome, Arnold Ludwig, and William Lyle. "The Controlled Psychedelic State." *American Journal of Clinical Hypnosis 6*, 2 (1963): 163–64.

Lieberman, Jeffrey, and Daniel Shalev. "Back to the Future: Research Renewed on the Clinical Utility of Psychedelic Drugs." *Journal of Psychopharmacology 30*, 12 (2016): 1198–1200.

Liu, Chih-Min, et al. "Absent Response to Niacin Skin Patch Is Specific to Schizophrenia and Independent of Smoking." *Psychiatry Research 152*, 2–3 (2007): 181–87.

Locke, Jeannine. "The Drug that Brings the Alcoholic Face to Face with Himself." *Star Weekly,* 16 February 1963, n. pag. From URA, Blewett Papers, 91-87, Box 3, File 19.

Ludwig, Arnold. "Letters to the Editor: Dr. Ludwig Replies." *American Journal of Psychiatry 126,* 12 (1970): 1798–99.

———. "Studies in Alcoholism and LSD (1): Influence of Therapist Attitudes on Treatment Outcome." *American Journal of Orthopsychiatry 38,* 4 (1968): 733–37.

Ludwig, Arnold, and Jerome Levine. "A Controlled Comparison of Five Brief Treatment Techniques Employing LSD, Hypnosis, and Psychotherapy." *American Journal of Psychotherapy 19* (1965): 417–65.

Ludwig, Arnold, Jerome Levine, Louis Stark, and Robert Lazar. "A Clinical Study of LSD Treatment in Alcoholism." *American Journal of Psychiatry 126,* 1 (1969): 59–69.

MacDonald, Jake. "Peaking on the Prairies." *Walrus,* June 2007, 77–84.

Mackenzie, Donald, and Barry Barnes. "Scientific Judgement: The Biometry-Mendelism Controversy." In *Natural Order: Historical Studies of Scientific Culture,* edited by Steven Shapin and Barry Barnes, 191–208. Beverly Hills: Sage Publications, 1979.

MacLean, J. R., D. C. MacDonald, U. P. Byrne, and A. M. Hubbard. "The Use of LSD-25 in the Treatment of Alcoholism and Other Psychiatric Problems." *Quarterly Journal of Studies on Alcoholism 22* (1961): 34–45.

Marchildon, Gregory. "A House Divided: Deinstitutionalization, Medicare, and the Canadian Mental Health Association in Saskatchewan, 1944–1964." *Social History 44,* 88 (2011): 305–29.

Marks, John. *The Search for the "Manchurian Candidate."* New York: W. W. Norton and Company, 1991.

Marsh, Richard. "Meaning and the Mind Drugs." *ECT: A Review of General Semantics 22,* 4 (1965): 408–30.

Martin, Sandra. "Obituaries: A Psychiatric Contrarian." *Globe and Mail,* June 20, 2009.

Marwick, Arthur. *The Sixties: Cultural Revolution in Britain, France, Italy, and the United States c. 1958–c. 1974.* Oxford: Oxford University Press, 1998.

Maslow, Abraham. *The Psychology of Science: A Reconnaissance.* South Bend, IN: Gateway Editions, 1966.

———. *Religions, Values, and Peak Experiences.* Columbus: Ohio State University Press, 1964.

Masters, R. E. L., and Jean Houston. *The Varieties of Psychedelic Experience.* New York: Holt, Rinehart and Winston, 1966.

Maté, Gabor. "Postscript—Psychedelics in Unlocking the Unconscious: From Cancer to Addiction." In *The Therapeutic Use of Ayahuasca,* edited by Beatriz Caiuby Labate and Clancy Cavnar, 217–24. Heidelberg: Spinger, 2014.

Matusow, Allen. *The Unraveling of America: A History of Liberalism in the 1960s*. Athens: University of Georgia Press, 2009.

Mayhew, Christopher. *Time to Explain: An Autobiography*. London: Hutchison, 1987.

McFarlane, A. H. "And Another Reader Comments." *Canadian Family Physician 21*, 6 (1975): 50–51.

McGrath, S. D., P. F. O'Brien, P. J. Power, and J. R. Shea. "Short Report: Nicotinamide Treatment of Schizophrenia." *Schizophrenia Bulletin 1*, 5 (1972): 74–76.

McKellar, Peter. "Scientific Theory and Psychosis: The 'Model Psychosis' Experiment and Its Significance." *International Journal of Social Psychiatry 3*, 3 (1957): 170–82.

Merton, Robert K. *The Sociology of Science: Theoretical and Empirical Investigations*. Edited with an Introduction by Norman W. Storer. Chicago: University of Chicago Press, 1973.

Messamore, Erik. "Niacin Subsensitivity Is Associated with Functional Impairment in Schizophrenia." *Schizophrenia Research 137*, 1–3 (2012): 180–84.

Metzl, Jonathon. *Prozac on the Couch: Prescribing Gender in the Era of Wonder Drugs*. Durham, NC: Duke University Press, 2003.

———. "Psychoanalysis and the Miltown Revolution." *Gender and History 15*, 2 (2003): 240–67.

Miller, Christine, and Jeannette Dulay. "The High-Affinity Niacin Receptor HM74A Is Decreased in the Anterior Cingulate Cortex of Individuals with Schizophrenia." *Brain Research Bulletin 77*, 1 (2008): 33–41. doi:10.1016/j.brainresbull.2008.03.015.

Mills, John. "Hallucinogens as Hard Science: The Adrenochrome Hypothesis for the Biogenesis of Schizophrenia." *History of Psychology 13*, 2 (2010): 178–95.

———. "Lessons from the Periphery: Psychiatry in Saskatchewan, Canada, 1944–68." *History of Psychiatry 18*, 2 (2007): 179–201.

Mills, John, and Erika Dyck. "Trust Amply Recompensed: Psychological Research at Weyburn, Saskatchewan, 1957–61." *Journal of the History of Behavioural Sciences 44*, 3 (2008): 199–218.

"Molecules and Mental Health." *British Medical Journal 8* (1975): 296.

Montcrieff, Joanna. "A Critique of the Dopamine Hypothesis of Schizophrenia and Psychosis." *Harvard Review of Psychiatry 17*, 3 (2009): 214–25.

———. "Why Is It So Difficult to Stop Psychiatric Drug Treatment? It May Be Nothing to Do with the Original Problem." *Medical Hypotheses 67*, 3 (2006): 517–23.

Montgomery, Adam. "Book Reviews, Neuropsychedelia: The Revival of Hallucinogen Research since the Decade of the Brain, by Nicolas Langlitz." *Generus 70* (2013): 364–65.

Mosher, Loren. "Nicotinic Acid Side Effects: A Review." *American Journal of Psychiatry 126*, 9 (1970): 1290–96.

Nutt, David, Leslie King, and David Nichols. "Effects of Schedule I Drug Laws on Neuroscience Research and Treatment Innovation." *Nature Reviews Neuroscience 14* (2013): 577–85.

Oken, Donald. "Vitamin Therapy for the Mentally Ill." *Science*, June 14, 1968, 1181–82.

O'Neill, William. *Coming Apart: An Informal History of America in the 1960's*. Chicago: Ivan R. Dee, 2005.

Oram, Matthew. "Efficacy and Enlightenment: LSD Psychotherapy and the Drug Amendments of 1962." *Journal of the History of Medicine and Allied Sciences 69*, 2 (2014): 221–50.

———. "Prohibited or Regulated? LSD Psychotherapy and the United States Food and Drug Administration." *History of Psychiatry 27*, 3 (2016): 1–17.

O'Reilly, P. O., and Genevieve Reich. "Lysergide Acid and the Alcoholic." *Diseases of the Nervous System 23*, 6 (1962): 331–34.

Osmond, Humphry. "The Background to Niacin Treatment." In *Orthomolecular Psychiatry: Treatment of Schizophrenia*, edited by David Hawkins and Linus Pauling, 194–201. San Francisco: W. H. Freeman, 1973.

———. "A Comment on Some Uses of Psychotomimetics in Psychiatry." In *The Uses of LSD in Psychotherapy and Alcoholism*, edited by Harold Abramson, 430–33. New York: Bobbs-Merrill, 1967.

———. "Function as the Basis of Psychiatric Ward Design." *Mental Hospitals 8*, 4 (1957): 23–27.

———. "Inspiration and Method in Schizophrenia Research." *Diseases of the Nervous System 16*, 4 (1955): 2–12.

———. "Models of Madness." *New Scientist 12* (1961): 777–80.

———. "Ololiuqui: The Ancient Aztec Narcotic." *Journal of Mental Science 101* (1955): 526–37.

———. "On Being Mad." In *Psychedelics*, edited by B. Aaronson and H. Osmond, 21–28. New York: Doubleday and Company, 1970.

———. "Peyote Night." In *Psychedelics*, edited by B. Aaronson and H. Osmond, 67–86. New York: Doubleday and Company, 1970.

———. "A Review of the Clinical Effects of Psychotomimetic Agents." *Annals of the New York Academy of Sciences 66*, Article 3 (1957): 418–34.

Osmond, Humphry, and Abram Hoffer. "The Case of Mr. Kovish." *Journal of Mental Science 104* (1958): 302–25.

———. "Massive Niacin Treatment in Schizophrenia: Review of a Nine-Year Study." *Lancet 1* (1962): 316–20.

———. "Schizophrenia: A New Approach (Continued)." *Journal of Mental Science 105* (1959): 653–73.

Osmond, Humphry, and John Smythies. "Schizophrenia: A New Approach." *Journal of Mental Science 98* (1952): 1–8. From a reprint.

Page, Penny Booth. "E.M. Jellinek and the Evolution of Alcohol Studies: A Critical Essay." *Addiction 92*, 12 (1997): 1619–37.

———. "The Origins of Alcohol Studies: E.M. Jellinek and the Documentation of the Alcohol Research Literature." *British Journal of Addiction 83* (1988): 1095–1103.

Parsons, Talcott. "The Institutionalization of Scientific Investigations." In *The Sociology of Science*, edited by B. Barber and W. Hirsch, 7–15. New York: Free Press of Glencoe, 1962.

Pauling, Linus. "On the Orthomolecular Environment of the Mind: Orthomolecular Theory." *American Journal of Psychiatry 131*, 11 (1974): 1251–57.

———. "Orthomolecular Psychiatry." *Science 160* (1968): 265–71.

———. "Preface." In *Orthomolecular Psychiatry: Treatment of Schizophrenia*, edited by David Hawkins and Linus Pauling, v–ix. San Francisco: W. H. Freeman and Company, 1973.

Peck, Abe. *Uncovering the Sixties: The Life and Times of the Underground Press*. New York: Pantheon Books, 1985.

Peters, Charles. "Concepts of Schizophrenia after Kraepelin and Bleuler." In *The Concept of Schizophrenia: Historical Perspectives*, edited by John G. Howells, 93–107. Washington, DC: American Psychiatric Press, 1991.

Pickersgill, Martyn. "From Psyche to Soma? Changing Accounts of Antisocial Personality Disorders in the American Journal of Psychiatry." *History of Psychiatry 21*, 3 (2010): 294–311.

Pilecki, B. C., J. W. Clegg, and D. McKay. "The Influence of Corporate and Political Interests on Models of Illness in the Evolution of the DSM." *European Psychiatry 26* (2011): 194–200.

Pitsula, James. *New World Dawning: The Sixties at Regina Campus*. Regina: Canadian Plains Research Center, 2008.

Polanyi, Michael. "The Autonomy of Science." In *Scientific Thought and Social Reality: Essays by Michael Polanyi*, edited by Fred Schwartz, 15–33. New York: International Universities Press, 1974.

———. "From Copernicus to Einstein." In *Scientific Thought and Social Reality: Essays by Michael Polanyi*, edited by Fred Schwartz, 98–115. New York: International Universities Press, 1974.

———. "The Nature of Scientific Convictions." In *Scientific Thought and Social Reality: Essays by Michael Polanyi*, edited by Fred Schwartz, 49–66. New York: International Universities Press, 1974.

———. *Personal Knowledge: Towards a Post-Critical Philosophy*. Chicago: University of Chicago Press, 1958.

Popper, Karl. "Science: Conjectures and Refutations." In *The Growth of Scientific Knowledge*, by Karl Popper, 2nd ed., 33–65. New York: Basic Books, 1965.

Ramsay, R. A., T. A. Ban, H. E. Lehmann, B. M. Saxena, and Jean Bennett. "Nicotinic Acid as Adjuvant Therapy in Newly Admitted Schizophrenic Patients." *Canadian Medical Association Journal 102* (1970): 939–42.

Rands, Stan. "Community Psychiatric Services in a Rural Area." *Canadian Journal of Public Health 51*, 10 (1960): 404–10.

Rasmussen, Nicolas. "Making the First Anti-Depressant: Amphetamine in American Medicine, 1929–1950." *Journal of the History of Medicine and Allied Sciences 61*, 3 (2006): 288–323.

"Reviving Research into Psychedelic Drugs." *Lancet 367* (2006): 1214.

Rhead, John C. "Psychedelic Medicine: New Evidence for Hallucinogenic Substances as Treatments." *Journal of Psychoactive Drugs 46*, 1 (2014): 78–83.

Richert, Lucas. "Therapy Means Political Change, Not Peanut Butter: American Radical Psychiatry, 1967–1975." *Social History of Medicine 27*, 1 (2014): 104–21.

Richert, Lucas, and Frances Reilly. "Book Reviews: American Psychiatry Scholarship: The Pendulum Maintains Its Momentum." *Medical History 58*, 4 (2014): 614–18.

Riesman, David. *The Lonely Crowd: A Study of the Changing American Character*. New Haven, CT: Yale University Press, 1950.

Rogers, Carl. "Towards a Science of the Person." In *Behaviorism and Phenomenology: Contrasting Bases for Modern Psychology*, edited by T. W. Wann, 109–33. Chicago: University of Chicago Press, 1964.

Roszak, Theodore. *The Making of a Counter Culture: Reflections on the Technocratic Society and Its Youthful Opposition*. Garden City, NY: Anchor Books, 1969.

Roueché, Berton. "Placebo." *New Yorker*, October 15, 1960, 85–103.

Rouse, Joseph. "Kuhn's Philosophy of Scientific Practice." In *Thomas Kuhn*, edited by Thomas Nickles, 101–21. Cambridge, UK: Cambridge University Press, 2003.

Rudnick, Abraham. "The Molecular Turn in Psychiatry: A Philosophical Analysis." *Journal of Medicine and Philosophy 27*, 3 (2002): 287–96.

Sagi, Doug. "White Men Witness Indian Peyote Rites." *Saskatoon Star-Phoenix*, 18 October 1956.

Salzman, Carl. "Controlled Therapy Research with Psychedelic Drugs: A Critique." In *Psychedelic Drugs*, edited by Richard E. Hicks and Paul J. Fink, 23–32. New York: Grune and Stratton, 1969.

Sandison, Ronald. "Psychological Aspects of the LSD Treatment of the Neuroses." *Journal of Mental Science 100* (1954): 508–15.

Sankar, D.V. *LSD—A Total Study*. Westbury, New York: PJD Publications Ltd., 1975.

"The Saskatchewan Plan." *Mental Hospitals*, March 1957: diagram.

Savage, Charles, Oliver McCabe, James Olsson, Sanford Unger, and Albert Kurland. "Research with Psychedelic Drugs." In *Psychedelic Drugs*, edited by Richard E. Hicks and Paul J. Fink, 15–22. New York: Grune and Stratton, 1969.

Scheffler, Israel. *Science and Subjectivity*. Indianapolis: Bobbs-Merrill, 1967.

Schneider, Irving. "Book Review: The Use of LSD in Psychotherapy and Alcoholism." *Psychiatry 31*, 2 (1968): 199–201.

Schor, Larry, and Mike Arons. "Tribute to Duncan Blewett." *Journal of Humanistic Psychology 47*, 4 (2007): 541–54.

Schwarz, B. E., et al. "Effects of Mescaline, LSD-25, and Adrenochrome on Depth Electrograms in Man." *American Medical Association Archives of Neurology and Psychiatry 75* (1956): 579–87.

Scull, Andrew. "The Decarceration of the Mentally Ill: A Critical View." In *The Sociology of Mental Illness*, edited by Oscar Grusky and Melvin Pollner, 417–37. New York: Holt, Rinehart and Winston, 1981.

————. "The Mental Health Sector and the Social Sciences in Post–World War II USA. Part 1: Total War and Its Aftermath." *History of Psychiatry* 22, 1 (2010): 3–19.

————. "The Mental Health Sector and the Social Sciences in Post–World War II USA. Part 2: The Impact of Federal Research Funding and the Drugs Revolution." *History of Psychiatry* 22, 3 (2011): 268–84.

————. "Somatic Treatments and the Historiography of Psychiatry." *History of Psychiatry* 5 (1994): 1–12.

Segal, David, Joel Yager, and John Sullivan. *Foundations of Biochemical Psychiatry*. Boston: Butterworths, 1976.

Sehdev, Haracharan, and Jim Olson. "Nicotinic Acid Therapy in Chronic Schizophrenia." *Comprehensive Psychiatry* 15, 6 (1974): 511–17.

Sessa, Ben. "Is There a Case for MDMA-Assisted Psychotherapy in the UK?" *Journal of Psychopharmacology*, October 18, 2006, doi: 10.1177/0269881106069029.

————. "Why Psychiatry Needs Psychedelics and Psychedelics Need Psychiatry." *Journal of Psychoactive Drugs* 46, 1 (2014): 57–62.

Sewell, Andrew, John H. Halpern, and Harrison G. Pope Jr. "Response of Cluster Headache to Psilocybin and LSD." *Neurology* 66 (2006): 1920–22.

Shapin, Steven. "Cordelia's Love: Credibility and the Social Studies of Science." *Perspectives on Science* 3, 3 (1995): 255–75.

————. "How to Be Antiscientific." In *The One Culture? A Conversation about Science*, edited by Jay Labinger and Harry Collins, 99–115. Chicago: University of Chicago Press, 2001.

————. *The Scientific Revolution*. Chicago: University of Chicago Press, 1996.

————. "Why the Public Ought to Understand Science-in-the-Making." *Public Understanding of Science* 1 (1992): 27–30.

Shepherd, Michael. "The Evaluation of Treatment in Psychiatry." In *Methods of Psychiatric Research*, 2nd ed., edited by Peter Sainsbury and Norman Kreitman, 88–100. London: Oxford University Press, 1975.

Shorter, Edward. *A History of Psychiatry: From the Era of the Asylum to the Age of Prozac*. New York: John Wiley, 1997.

Singh, Ratan. "Correspondence: The Adrenochrome Hypothesis." *Journal of Orthomolecular Medicine* 27, 4 (2012): 198–99.

Slotkin, W. F. *The Peyote Religion: A Study in Indian-White Relations*. Glencoe, IL: Free Press, 1956.

Smart, Reginald G., and Thomas Storm. "The Efficacy of LSD in the Treatment of Alcoholism." *Quarterly Journal on Studies of Alcohol* 25 (1964): 333–38.

Smart, R. G., Thomas Storm, Earl Baker, and Lionel Solursh. "A Controlled Study of Lysergide in the Treatment of Alcoholism." *Quarterly Journal of Studies on Alcohol* 27, 3 (1966): 469–82.

Smith, Colin. "Exploratory and Controlled Studies of Lysergide in the Treatment of Alcoholism." *Quarterly Journal of Studies of Alcohol* 25, 4 (1964): 742–47.

————. "A New Adjunct to the Treatment of Alcoholism: The Hallucinogenic Drugs." *Quarterly Journal of Studies on Alcohol* 19, 3 (1958): 406–17.

——. "Some Reflections on the Possible Therapeutic Effects of the Hallucinogens—With Special Reference to Alcoholism." *Quarterly Journal of Studies on Alcohol 20*, 2 (1959): 292–301.

——. "The Unscientific Bases of Psychiatry." *Journal of Neuropsychiatry 3*, 2 (1961): 67–74.

Smith, David, Glenn E. Raswyck, and Leigh Dickerson Davidson. "From Hofmann to Haight Ashbury, and into the Future: The Past and Potential of LSD." *Journal of Psychoactive Drugs 46*, 1 (2014): 3–10.

Smythies, John. "The Adrenochrome Hypothesis of Schizophrenia Revisited." *Neurotoxicity Research 4*, 2 (2002): 147–50.

——. "The Biochemistry of Schizophrenia." *Postgraduate Medical Journal 39* (1963): 26–33.

——. "Oxidative Reactions and Schizophrenia: A Review—Discussion." *Schizophrenia Research 24* (1997): 357–64.

Snyder, Solomon. "Julius Axelrod (1912–2004): A Biographical Memoir." *Biographical Memoirs 87* (2005): 1–20.

"Sociopetal Building Arouses Controversy." *Mental Hospitals 8*, 5 (1957): 25–32.

Sommer, Robert. "In Memoriam: Humphry Osmond." *Journal of Environmental Psychology 24*, 2 (2004): 257–58.

——. "The Mental Hospital in the Small Community." *Mental Hygiene 42* (1958): 489–96.

——. "Personal Space." *American Institute of Architects Journal* (1962): 81–86.

——. "Psychology in the Wilderness." *Canadian Psychologist 2*, 1 (1961): 26–29.

Sommer, Robert, and Humphry Osmond. "Autobiographies of Former Mental Patients." *Journal of Mental Science 106* (1960): 648–62.

Sourkes, Theodore. "Review Article: The Biochemistry of Mental Disease." *Canadian Medical Association Journal 85* (1961): 487–90.

Stafford, Peter. *Psychedelics Encyclopedia*. 3rd ed. Berkeley: Ronin Publishing, 1992.

Stafford, P. G., and B. H. Golightly. *LSD: The Problem-Solving Psychedelic*. New York: Award Books, 1967.

Stevens, Jay. *Storming Heaven: LSD and the American Dream*. New York: Grove Press, 1987.

Sydenstricker, V. P., and H. M. Cleckley. "Effect of Nicotinic Acid in Stupor, Lethargy, and Various Other Psychotic Disorders." *American Journal of Psychiatry 98* (1941): 83–92.

Szasz, Thomas. *Cruel Compassion: Psychiatric Control of Society's Unwanted*. Syracuse, NY: Syracuse University Press, 1998.

——. *The Manufacture of Madness: A Comparative Study of the Inquisition and the Mental Health Movement*. New York: Dell Publishing Company, 1970.

Tart, Charles. "Science, States of Consciousness, and Spiritual Experiences: The Need for State-Specific Sciences." In *Transpersonal Psychologies*, edited by Charles Tart, 9–58. New York: Harper and Row, 1975.

Taubmann, G., and H. Jantz. "Untersuchung über die dem Adrenochrom Zugeschnie-benen Psychotoxinschen Winkungen." *Der Nervenarzt 20* (1957): 485–88.

Tiebout, Henry. "The Ego Factors in Surrender in Alcoholism." http://silkwoth.net/tiebout_egofactors.html.

Tullock, Gordon. *The Organization of Inquiry*. Rev. ed., with an introduction by Charles Rowley. Indianapolis: Liberty Fund, 2005.

Tupper, Kenneth, Evan Wood, Richard Yensen, and Matthew Johnson. "Psychedelic Medicine: A Re-Emerging Therapeutic Paradigm." *Canadian Medical Association Journal* (September 8, 2015): 1–6. http://www.cmaj.ca/content/early/2015/09/08/cmaj.141124.full.pdf+html.

Tuteur, Werner. "The 'Double Blind' Method: Its Pitfalls and Fallacies." *American Journal of Psychiatry 114* (1958): 921–22.

"The UCLA Interdepartmental Conference—Schizophrenia." *Annals of Internal Medicine 70*, 1 (1969): 107–25.

Unger, Sanford. "Mescaline, LSD, Psilocybin, and Personality Change." *Psychiatry 26* (1963): 111–25.

Valenstein, Elliot. *Blaming the Brain: The Truth about Drugs and Mental Health*. New York: Free Press, 1998.

Ward, P. E., et al. "Niacin Skin Flush in Schizophrenia: A Preliminary Report." *Schizophrenia Research 29*, 3 (1998): 169–74.

Watts, Allan. "Psychedelics and Religious Experience." *California Law Review 56* (1968): 74–85.

———. "Psychedelics and Religious Experience." In *Psychedelics: The Uses and Implications of Hallucinogenic Drugs*, edited by Bernard Aaronson and Humphry Osmond, 131–45. Garden City, NY: Anchor Books, 1970.

Weckowicz, T. E. "Depersonalization." In *Symptoms of Psychopathology: A Handbook*, edited by C. G. Costello, 151–65. New York: Wiley, 1970.

———. "Peyotl Ceremony and Jungian Archetypes." In "The Road to Manitou," by Fannie Kahan, 212–31. Unpublished manuscript, 1958.

Weckowicz, T. E., and R. Sommer. "Body Image and Self-Concept in Schizophrenia: An Experimental Study." *Journal of Mental Science 106* (1960): 17–39.

Weil-Malherbe, Hans, and Stephen Szara. *The Biochemistry of Functional and Experimental Psychoses*. Springfield, IL: Charles C. Thomas, 1971.

Weller, Sheila. "Suddenly That Summer." *Vanity Fair*, July 2012. http://www.vanityfair.com/culture/2012/07/lsd-drugs-summer-of-love-sixties.

Whitaker, Robert. *Anatomy of an Epidemic: Magic Bullets, Psychiatric Drugs, and the Astonishing Rise of Mental Illness in America*. New York: Broadway Paperbacks, 2010.

White, William. "A Disease Concept for the 21st Century." http://www.bbsgsonj.com/documents/White%20-%20A%20Disease%20Concept%20for%20the%20 21st%20Century.pdf.

———. "The Rebirth of the Disease Concept of Alcoholism in the 20th Century." *Counselor 1* (2000): 1–5. https://www.justloveaudio.com/resources/12_Steps_Recovery/Step_1/Disease_Concept.pdf.

Whyte, William H. *Organization Man*. New York: Simon and Schuster, 1956.

Wilson, Mitchell. "The Transformation of American Psychiatry: A History." *American Journal of Psychiatry* 150, 3 (1993): 399–410.

Wittenborn, J. R. "Niacin in the Long-Term Treatment of Schizophrenia." *Archives of General Psychiatry* 28, 3 (1973): 308–15.

———. "A Search for Responders to Niacin Supplementation." *Archives of General Psychiatry* 34 (1974): 547–52.

Xu, X. J., and G. S. Jiang. "Niacin-Respondent Subset of Schizophrenia: A Therapeutic Review." *European Review for Medical and Pharmacological Sciences* 19 (2015): 988–97.

Yao, J. K., et al. "Prevalence and Specificity of the Abnormal Niacin Response: A Potential Endophenotype Marker in Schizophrenia." *Schizophrenia Bulletin* 42, 2 (2016): 369–76.

INDEX

ZED

Zed is a platform for marginalised voices across the globe.

It is the world's largest publishing collective and a world leading example of alternative, non-hierarchical business practice.

It has no CEO, no MD and no bosses and is owned and managed by its workers who are all on equal pay.

It makes its content available in as many languages as possible.

It publishes content critical of oppressive power structures and regimes.

It publishes content that changes its readers' thinking.

It publishes content that other publishers won't and that the establishment finds threatening.

It has been subject to repeated acts of censorship by states and corporations.

It fights all forms of censorship.

It is financially and ideologically independent of any party, corporation, state or individual.

Its books are shared all over the world.

www.zedbooks.net
@ZedBooks

Milton Keynes UK
Ingram Content Group UK Ltd.
UKHW020906161024
449580UK00010B/271